COMMON
LISP

A Gentle Introduction
to Symbolic Computation

COMMON
LISP

A Gentle Introduction
to Symbolic Computation

David S. Touretzky

DOVER PUBLICATIONS, INC.
Mineola, New York

Bibliographical Note

This Dover edition, first published in 2013, is a revised republication of the work originally published by The Benjamin/Cummings Publishing Company, Inc., in 1990.

International Standard Book Number

ISBN-13: 978-0-486-49820-1
ISBN-10: 0-486-49820-4

Manufactured in the United States by LSC Communications
4500051845
www.doverpublications.com

To Phil and Anne

Preface

This book is about learning to program in Lisp. Although widely known as the principal language of artificial intelligence research—one of the most advanced areas of computer science—Lisp is an excellent language for beginners. It is increasingly the language of choice in introductory programming courses due to its friendly, interactive environment, rich data structures, and powerful software tools that even a novice can master in short order.

When I wrote the book I had three types of reader in mind. I would like to address each in turn.

- Students taking their first programming course. The student could be from any discipline, from computer science to the humanities. For you, let me stress the word *gentle* in the title. I assume no prior mathematical background beyond arithmetic. Even if you don't like math, you may find you enjoy computer programming. I've avoided technical jargon, and there are lots of examples. Also you will find plenty of exercises interspersed with the text, and the answers to all of them are included in Appendix C.

- Psychologists, linguists, computer scientists, and other persons interested in Artificial Intelligence. As you begin your inquiry into AI, you will see that almost all research in this field is carried out in Lisp. Most Lisp texts are written exclusively for computer science majors, but I have gone to great effort to make *this* book accessible to everyone. It can be your doorway to the technical literature of AI, as well as a quick introduction to its central tool.

- Computer hobbyists. Prior to about 1984, the Lisps available on personal computers weren't very good due to the small memories of the early machines. Today's personal computers often come with several megabytes of RAM and a hard disk as standard equipment. They run full implementations of the Common Lisp standard, and provide the same high-quality tools as the Lisps in university and industrial research labs. The "Lisp Toolkit" sections of this book will introduce you to the advanced features of the Common Lisp programming environment that have made the language such a productive tool for rapid prototyping and AI programming.

This current volume of the "gentle introduction" uses Common Lisp throughout. Lisp has been changing continuously since its invention 30 years ago. In the past, not only were the Lisp dialects on different machines incompatible, but programs written in one dialect would often no longer run in that same dialect a few years later, because the language had evolved out from under them. Rapid, unconstrained evolution was beneficial in the early days, but demand for a standard eventually grew, so Common Lisp was created. At present, Common Lisp is the *de facto* standard supported by all major computer manufacturers. It is currently undergoing refinement into an official standard. But Lisp will continue to evolve nonetheless, and the standard will be updated periodically to reflect new contributions people have made to the language. Perhaps one of those contributors will be you.

DAVID S. TOURETZKY
PITTSBURGH, PENNSYLVANIA
1989

DOVER EDITION ADDENDUM

This 2013 edition from Dover Publications includes roughly two dozen corrections to the original manuscript, and a few additions to the Further Readings section. With the arrival of ANSI Common Lisp as the official standard, and the availability of several good open source implementations, Lisp will remain an important language for years to come.

Note to Instructors

Much has been learned in the last few years about how to teach Lisp effectively to beginners: where they stumble and what we can do about it. In addition, the switch to Common Lisp has necessitated changes in the way certain topics are taught, especially variables, scoping, and assignment. This version of the "gentle introduction" has been completely revised for Common Lisp, and includes several new teaching tools that I believe you will find invaluable in the classroom. Let me share with you some of the thinking behind this book's novel approach to Lisp.

GRAPHICAL NOTATION

The first two chapters use a graphical box-and-arrow notation for describing primitive functions and function composition. This notation allows students to get comfortable with the basic idea of computation and the three fundamental data structures—numbers, symbols, and lists—before grappling with side issues such as the syntax of a function call or when to use quotes. Although sophisticated Lispers profit from the realization that programs are data, to the beginner this is a major source of confusion. The box-and-arrow notation makes programs and data visually distinct, and thereby eliminates most syntax errors. Another advantage of this notation is its lack of explicit variables; the inputs to a function are simply arrows that enter the function definition from outside. Since there is no computer implementation of function box notation, the first two chapters are designed to be covered rapidly using just pencil and paper. This also shelters the student temporarily from another source of frustration—learning the mechanics of using an actual machine, editing expressions, and coping with the debugger.

Readers who are familiar with other programming languages can flip through Chapter 1 in a minute or so, read the summary at the end, and then skim Chapter 2 to pick up the basic list manipulation primitives.

In Chapter 3 the student is introduced to standard EVAL notation; the concepts of quoting and named variables follow fairly naturally. Now he or she is ready to discard paper and pencil for a real computer (and is probably eager to do so), whereas at the start of the course this might have been viewed with trepidation.

OTHER FEATURES

Three other unique features of the book first appear in Chapter 3: evaltrace notation, Lisp Toolkit sections, and a comprehensive graphical representation for Lisp data structures, including function objects and the internal structure of symbols.

Evaltrace notation shows step-by-step how Lisp expressions are evaluated, how functions are applied to arguments, and how variables are created and bound. The different roles of EVAL and APPLY, the scoping of variables, and the nesting of lexical contours can all be explained graphically using this notation. It makes the process of evaluation transparent to the student by describing it in a visual language which he or she can remember and use.

The Lisp Toolkit sections introduce the various programming aids that Common Lisp provides, such as DESCRIBE, INSPECT, TRACE, STEP, and the debugger. There are also two tools unique to this book; their source code appears in Appendices A and B, and is available on diskette from the publisher. The first tool, SDRAW, draws cons cell diagrams. It is part of a read-eval-draw loop that has proven invaluable for teaching beginners to reason about cons cell structures, particularly the differences among CONS, LIST, and APPEND. The second tool, DTRACE, is a tracing package that generates more detailed output than most implementations of TRACE, and is therefore more useful for teaching beginners.

Finally, the graphical representation of Lisp data structures—particularly the internal structure of symbols with their name, function, value, plist, and package cells—helps students understand the true nature of Lisp interpreters and highlights the distinctions between symbols, functions, variables, and print names.

ORGANIZATION OF LATER CHAPTERS

Applicative operators are introduced in Chapter 7, where the student also learns about lexical closures. In Chapter 8, the dragon stories that were a popular feature of the previous version have been retained, but they are now backed up with a new device—recursion templates—that helps beginners analyze recursive functions to extract the essence of the recursive style. Since some instructors prefer to teach recursion before applicatives, these two chapters have been written so that they may be covered in either order.

The book promotes a clean, side-effect-free style of programming for the first eight chapters. Chapter 9 discusses i/o. Chapter 10 provides a unified picture of assignment that includes ordinary variables, generalized variables, and destructive sequence operations. Chapter 11 covers iteration, and shows how DO and DO* can be used to construct substantial iterative expressions with no explicit assignments. Chapter 12 introduces structures, and Chapter 13 covers arrays, hash tables, and property lists. The final chapter, Chapter 14, is devoted to macros and compilation. It also explains the difference between lexical and dynamic scoping. Evaltrace diagrams clarify the semantics of macros and special variables.

EMPHASIS ON SIMPLICITY

Because Common Lisp is such a complex language, there are a few places where I have chosen to simplify things to better meet the needs of beginners. For example, the 1+ and 1- functions are banished from this book because their names are very confusing. Also, the book relies almost exclusively on EQUAL because this is the most useful equality predicate. EQ, EQL, EQUALP, and = are mentioned in advanced topics sections, but not used very much. In a few places I have chosen to write a function slightly less concisely rather than introduce one of the more obscure primitives like PUSHNEW. And I make no attempt to cover the most advanced features, such as multiple values or the package system.

Some people prefer to teach Scheme in introductory courses because it is so much smaller than Common Lisp. But one can easily teach the subset of Common Lisp that is equivalent to Scheme, so language size isn't really an issue for beginners. A more compelling argument is that there is a certain style of applicative programming, making heavy use of lexical closures, that can be expressed more elegantly in Scheme syntax. But there are also areas where Common Lisp is superior to Scheme, such as its support for user-defined macros, its elegant unification of lists and vectors into a sequence datatype, and its use of keyword arguments to greatly extend the utility of the sequence functions. The combination of tremendous power, extensive manufacturer support, and a built-in object-oriented programming facility make Common Lisp the only ''industrial strength'' Lisp. Although this book does emphasize a side-effect-free, applicative approach to programming with which Scheme afficionados will feel quite at home, it does so in purely Common Lisp style.

This book has been carefully designed to meet the needs of beginning programmers and non-computer science students, but the optional advanced topics sections at the end of each chapter provide enough enrichment material to hold the interest of junior and senior computer science majors.   For advanced undergraduates, Guy L. Steele Jr.'s *Common Lisp: The Language*, 2nd edition (published by Digital Press) or Paul Graham's *ANSI Common Lisp* would be useful companions to the introduction provided here.

Acknowledgements

This book began in 1981 as a set of notes for a programming course for humanities students at Carnegie Mellon University. I am greatly indebted to Phil Miller for the administrative support that made the course possible. John McDermott and Scott Fahlman also helped with administrative matters.

My second major debt is to Anne Rogers, who took it upon herself to edit early drafts of the manuscript. Anne was an irrepressible source of encouragement; her enthusiasm kept the book alive through difficult times.

Loretta Ferro, Maria Wadlow, and Sandy Esch kindly served as test subjects in my first pedagogical experiments. I also thank my students in the first actual Lisp course for the time and energy they put into it. Gail Kaiser, Mark Boggs, Aaron Wohl, and Lynn Baumeister all taught the new Lisp course using my notes. Their feedback helped improve succeeding drafts.

Richard Pattis, author of another fine programming text, was an able publicity agent and ultimately helped me find my first publisher, Harper & Row. Abby Gelles also helped publicize the book. At Harper & Row, John Willig taught me about academic publishing and Mexican food, and remains a good friend.

Throughout the preparation of the previous version I was most fortunate to be supported by a graduate fellowship from the Fannie and John Hertz Foundation.

In 1987, Harper & Row left the computer science publishing business. John Wiley & Sons took over distribution of the previous version while I found a publisher for this volume. The book found a new home at Benjamin/ Cummings thanks to the patience and diligence of executive editor Alan Apt.

I thank Mark Fox, at the time acting president of Carnegie Group, Inc., for permission to include some software in the current volume that I originally developed for his company. I also thank the reviewers who contributed the most valuable advice on improving the current volume: Skona Brittain, Mike

Clancy, Rich Pattis, and Douglas Dankel. Other useful comments were received from Rick Wilson, Sharon Salveter, Terrance Boult, Dick Gabriel, Jos Schreinemakers, and Andre van Meulebrouck.

Cindy Wood helped with the figures. Jos Schreinemakers did the post-copyedit proofreading, and assisted with page makeup. Nahid Capell checked the answers to all the exercises. Brian Harrison nursed the Linotronic. Gillette Elvgren III ported the software to various Lisp implementations. Special technical services were provided by Ignatz G. Bird. I thank everyone for their assistance.

The School of Computer Science at Carnegie Mellon provided the superb computer facilities and stimulating intellectual environment that made this work possible. After eleven years here as a graduate student and faculty member, I can think of no place I'd rather be.

DOVER EDITION ADDENDUM

Thanks to Stephen Horner, Dat Nguyen, Roly Sussex, and Jeff Woodall for helping me correct some errors in the original version of this book. And special thanks to Rochelle Kronzek, science and math editor at Dover, for suggesting this reprint and making it happen.

Contents

Preface   vii

Note to Instructors   ix

Acknowledgements   xiii

1. Functions and Data   1
    1.1. Introduction   1
    1.2. Functions On Numbers   2
    1.3. Three Kinds of Numbers   3
    1.4. Order Of Inputs Is Important   4
    1.5. Symbols   6
    1.6. The Special Symbols T and NIL   7
    1.7. Some Simple Predicates   8
    1.8. The EQUAL Predicate   10
    1.9. Putting Functions Together   12
        1.9.1. Defining ADD1   12
        1.9.2. Defining ADD2   13
        1.9.3. Defining TWOP   15
        1.9.4. Defining ONEMOREP   16
    1.10. The NOT Predicate   18
    1.11. Negating A Predicate   20
    1.12. Number of Inputs to a Function   22
    1.13. Errors   24
  Advanced Topics   27
    1.14. The History of Lisp   27

2. Lists   31
    2.1. Lists Are The Most Versatile Data Type   31
    2.2. What Do Lists Look Like?   31
    2.3. Lists of One Element   33

2.4. Nested Lists	33
2.5. Length of Lists	35
2.6. NIL: The Empty List	37
2.7. Equality of Lists	38
2.8. FIRST, SECOND, THIRD, and REST	39
2.9. Functions Operate On Pointers	41
2.10. CAR and CDR	42
2.10.1. The CDR of a Single-Element List	44
2.10.2. Combinations of CAR and CDR	45
2.10.3. CAR and CDR of Nested Lists	47
2.10.4. CAR and CDR of NIL	50
2.11. CONS	52
2.11.1. CONS and the Empty List	55
2.11.2. Building Nested Lists With CONS	56
2.11.3. CONS Can Build Lists From Scratch	56
2.12. Symmetry of CONS and CAR/CDR	57
2.13. LIST	58
2.14. Replacing the First Element of a List	63
2.15. List Predicates	66
Advanced Topics	70
2.16. Unary Arithmetic with Lists	70
2.17. Nonlist Cons Structures	72
2.18. Circular Lists	74
2.19. Length of Nonlist Cons Structures	75
3. EVAL Notation	**77**
3.1. Introduction	77
3.2. The EVAL Function	78
3.3. EVAL Notation Can Do Anything Box Notation Can Do	79
3.4. Evaluation Rules Define the Behavior of EVAL	80
3.5. Defining Functions in EVAL Notation	82
3.6. Variables	84
3.7. Evaluating Symbols	86
3.8. Using Symbols and Lists as Data	87
3.9. The Problem of Misquoting	88
3.10. Three Ways to Make Lists	89
3.11. Four Ways to Misdefine a Function	91
3.12. More About Variables	92

Lisp on the Computer   96
   3.13. Running Lisp   97
   3.14. The Read-Eval-Print Loop   98
   3.15. Recovering From Errors   98
Lisp Toolkit: ED   100
Keyboard Exercise   101
Advanced Topics   103
   3.16. Functions of No Arguments   103
   3.17. The QUOTE Special Function   104
   3.18. Internal Structure of Symbols   105
   3.19. Lambda Notation   106
   3.20. Scope of Variables   109
   3.21. EVAL and APPLY   110

4. Conditionals   113
   4.1. Introduction   113
   4.2. The IF Special Function   113
   4.3. The COND Macro   116
   4.4. Using T as a Test   117
   4.5. Two More Examples of COND   118
   4.6. COND and Parenthesis Errors   119
   4.7. The AND and OR Macros   122
   4.8. Evaluating AND and OR   122
   4.9. Building Complex Predicates   123
   4.10. Why AND and OR are Conditionals   125
   4.11. Conditionals are Interchangeable   126
Lisp Toolkit: STEP   130
Advanced Topics   132
   4.12. Boolean Functions   132
   4.13. Truth Tables   133
   4.14. DeMorgan's Theorem   134

5. Variables and Side Effects   137
   5.1. Introduction   137
   5.2. Local and Global Variables   137
   5.3. SETF Assigns a Value to a Variable   138
   5.4. Side Effects   140
   5.5. The LET Special Function   141
   5.6. The LET* Special Function   144

5.7. Side Effects Can Cause Bugs 147
Lisp Toolkit: DOCUMENTATION and APROPOS 149
Keyboard Exercise 151
Advanced Topics 153
   5.8. Symbols and Value Cells 153
   5.9. Distinguishing Local from Global Variables 155
   5.10. Binding, Scoping, and Assignment 157

6. List Data Structures 159
   6.1. Introduction 159
   6.2. Parenthesis Notation vs. Cons Cell Notation 160
   6.3. The APPEND Function 161
   6.4. Comparing CONS, LIST, and APPEND 164
   6.5. More Functions on Lists 165
      6.5.1. REVERSE 165
      6.5.2. NTH and NTHCDR 166
      6.5.3. LAST 168
      6.5.4. REMOVE 168
   6.6. Lists as Sets 170
      6.6.1. MEMBER 170
      6.6.2. INTERSECTION 172
      6.6.3. UNION 173
      6.6.4. SET-DIFFERENCE 173
      6.6.5. SUBSETP 174
   6.7. Programming With Sets 175
   6.8. Lists As Tables 179
      6.8.1. ASSOC 179
      6.8.2. RASSOC 180
   6.9. Programming With Tables 181
Lisp Toolkit: SDRAW 186
Keyboard Exercise 188
Advanced Topics 192
   6.10. Trees 192
      6.10.1. SUBST 192
      6.10.2. SUBLIS 193
   6.11. Efficiency of List Operations 193
   6.12. Shared Structure 194
   6.13. Equality of Objects 195

6.14. Keyword Arguments   198

7. Applicative Programming   201
7.1. Introduction   201
7.2. FUNCALL   201
7.3. The MAPCAR Operator   202
7.4. Manipulating Tables With MAPCAR   203
7.5. Lambda Expressions   205
7.6. The FIND-IF Operator   207
7.7. Writing ASSOC With FIND-IF   207
7.8. REMOVE-IF and REMOVE-IF-NOT   210
7.9. The REDUCE Operator   213
7.10. EVERY   214
Lisp Toolkit: TRACE and DTRACE   216
Keyboard Exercise   219
Advanced Topics   224
7.11. Operating on Multiple Lists   224
7.12. The FUNCTION Special Function   225
7.13. Keyword Arguments to Applicative Operators   226
7.14. Scoping and Lexical Closures   226
7.15. Writing An Applicative Operator   229
7.16. Functions That Make Functions   230

8. Recursion   231
8.1. Introduction   231
8.2. Martin and the Dragon   232
8.3. A Function to Search for Odd Numbers   234
8.4. Martin Visits The Dragon Again   236
8.5. A Lisp Version of the Factorial Function   237
8.6. The Dragon's Dream   238
8.7. A Recursive Function for Counting Slices of Bread   240
8.8. The Three Rules of Recursion   241
8.9. Martin Discovers Infinite Recursion   244
8.10. Infinite Recursion in Lisp   246
8.11. Recursion Templates   248
8.11.1. Double-Test Tail Recursion   248
8.11.2. Single-Test Tail Recursion   250
8.11.3. Augmenting Recursion   252
8.12. Variations on the Basic Templates   254

8.12.1. List-Consing Recursion   254
8.12.2. Simultaneous Recursion on Several Variables   256
8.12.3. Conditional Augmentation   258
8.12.4. Multiple Recursion   260
8.13. Trees and CAR/CDR Recursion   262
8.14. Using Helping Functions   266
8.15. Recursion in Art and Literature   268
Lisp Toolkit: The Debugger   272
Keyboard Exercise   275
Advanced Topics   279
8.16. Advantages of Tail Recursion   279
8.17. Writing New Applicative Operators   282
8.18. The LABELS Special Function   282
8.19. Recursive Data Structures   283

9. Input/Output   287
9.1. Introduction   287
9.2. Character Strings   288
9.3. The FORMAT Function   288
9.4. The READ Function   292
9.5. The YES-OR-NO-P Function   293
9.6. Reading Files with WITH-OPEN-FILE   294
9.7. Writing Files with WITH-OPEN-FILE   295
Keyboard Exercise   296
Lisp Toolkit: DRIBBLE   298
Advanced Topics   299
9.8. Parameters to Format Directives   299
9.9. Additional Format Directives   300
9.10. The Lisp 1.5 Output Primitives   301
9.11. Handling End-of-File Conditions   302
9.12. Printing in Dot Notation   303
9.13. Hybrid Notation   304

10. Assignment   307
10.1. Introduction   307
10.2. Updating a Global Variable   308
10.3. Stereotypical Updating Methods   309
10.3.1. The INCF and DECF Macros   309
10.3.2. The PUSH and POP Macros   310

10.3.3. Updating Local Variables   312
10.4. WHEN and UNLESS   313
10.5. Generalized Variables   314
10.6. Case Study: A Tic-Tac-Toe Player   315
Lisp Toolkit: BREAK and ERROR   325
Keyboard Exercise   328
Advanced Topics   332
10.7. Do-It-Yourself List Surgery   332
10.8. Destructive Operations on Lists   334
10.8.1. NCONC   334
10.8.2. NSUBST   336
10.8.3. Other Destructive Functions   336
10.9. Programming With Destructive Operations   337
10.10. SETQ and SET   338

11. Iteration and Block Structure   341
11.1. Introduction   341
11.2. DOTIMES and DOLIST   341
11.3. Exiting the Body of a Loop   342
11.4. Comparing Recursive and Iterative Search   344
11.5. Building Up Results With Assignment   345
11.6. Comparing DOLIST with MAPCAR and Recursion   346
11.7. The DO Macro   347
11.8. Advantages of Implicit Assignment   349
11.9. The DO* Macro   351
11.10. Infinite Loops with DO   352
11.11. Implicit Blocks   353
Keyboard Exercise   355
Lisp Toolkit: TIME   358
Advanced Topics   359
11.12. PROG1, PROG2, and PROGN   359
11.13. Optional Arguments   360
11.14. Rest Arguments   361
11.15. Keyword Arguments   363
11.16. Auxiliary Variables   364

12. Structures and The Type System   365
12.1. Introduction   365
12.2. TYPEP and TYPE-OF   366

12.3. Defining Structures   367
12.4. Type Predicates for Structures   368
12.5. Accessing and Modifying Structures   369
12.6. Keyword Arguments to Constructor Functions   370
12.7. Changing Structure Definitions   371
Lisp Toolkit: DESCRIBE and INSPECT   372
Keyboard Exercise   374
Advanced Topics   377
12.8. Print Functions for Structures   377
12.9. Equality of Structures   379
12.10. Inheritance from Other Structures   380

13. Arrays, Hash Tables, And Property Lists   383
13.1. Introduction   383
13.2. Creating an Array   383
13.3. Printing Arrays   385
13.4. Accessing and Modifying Array Elements   385
13.5. Creating Arrays With MAKE-ARRAY   386
13.6. Strings as Vectors   387
13.7. Hash Tables   388
13.8. Property Lists   389
13.9. Programming With Property Lists   391
Array Keyboard Exercise   393
Hash Table Keyboard Exercise   395
Lisp Toolkit: ROOM   400
Advanced Topics   401
13.10. Property List Cells   401
13.11. More On Sequences   402

14. Macros and Compilation   405
14.1. Introduction   405
14.2. Macros as Shorthand   405
14.3. Macro Expansion   406
14.4. Defining a Macro   408
14.5. Macros as Syntactic Extensions   411
14.6. The Backquote Character   412
14.7. Splicing With Backquote   414
14.8. The Compiler   415
14.9. Compilation and Macro Expansion   417

14.10. Compiling Entire Programs   417
14.11. Case Study: Finite State Machines   418
Lisp Toolkit: PPMX   426
Keyboard Exercise   427
Advanced Topics   429
14.12. The &BODY Lambda-List Keyword   429
14.13. Destructuring Lambda Lists   430
14.14. Macros and Lexical Scoping   433
14.15. Historical Significance of Macros   435
14.16. Dynamic Scoping   435
14.17. DEFVAR, DEFPARAMETER, DEFCONSTANT   439
14.18. Rebinding Special Variables   440

Appendix A. The SDRAW Tool   **A-1**

Appendix B. The DTRACE Tool   **B-1**

Appendix C. Answers to Exercises   **C-1**

Glossary   **G-1**

Further Reading   **FR-1**

Index   **I-1**

1

Functions and Data

1.1 INTRODUCTION

This chapter begins with an overview of the notions of function and data, followed by examples of several built-in Lisp functions. If you already have some experience programming in other languages, you can flip through this chapter in just a few minutes. You'll see arithmetic functions, followed by an introduction to symbols, one of the key datatypes of Lisp, and predicates, which answer yes-or-no questions. When you think you've grasped this material, read the summary section on page 26 to test your understanding.

If you're new to programming, this chapter is designed specifically for you. We'll start by explaining what **functions** and **data** are.* The term data means *information*, such as numbers, words, or lists of things. You can think of a function as a box through which data flows. The function operates on the data in some way, and the **result** is what flows out.

After covering some of the built-in functions provided by Lisp, we will learn how to put existing functions together to make new ones—the essence of computer programming. Several useful techniques for creating new functions will then be presented.

---

*Technical terms like these, which appear in boldface in the text, are defined in the glossary at the back of the book.

1.2 FUNCTIONS ON NUMBERS

Probably the most familiar functions are the simple arithmetic functions of addition, subtraction, multiplication, and division. Here is how we represent the addition of two numbers:

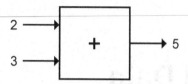

The name of the function is "+." We can describe what's going on in the figure in several ways. From the point of view of the data: The numbers 2 and 3 flow into the function, and the number 5 flows out. From the point of view of the function: The function "+" received the numbers 2 and 3 as inputs, and it produced 5 as its result. From the programmer's point of view: We **called** (or **invoked**) the function "+" on the inputs 2 and 3, and the function **returned** 5. These different ways of talking about functions and data are equivalent; you will encounter all of them in various places in this book.

Here is a table of Lisp functions that do useful things with numbers:

+	Adds two numbers
-	Subtracts the second number from the first
*	Multiplies two numbers
/	Divides the first number by the second
ABS	Absolute value of a number
SQRT	Square root of a number

Let's look at another example of how data flows through a function. The output of the absolute value function, ABS, is the same as its input, except that negative numbers are converted to positive ones.

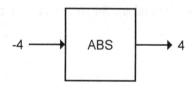

The number −4 enters the ABS function, which computes the absolute value and outputs a result of 4.

1.3 THREE KINDS OF NUMBERS

In this book we will work mostly with **integers**, which are whole numbers. Common Lisp provides many other kinds of numbers. One kind you should know about is **floating point** numbers. A floating point number is always written with a decimal point; for example, the number five would be written 5.0. The SQRT function generally returns a floating point number as its result, even when its input is an integer.

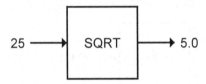

Ratios are yet another kind of number. On a pocket calculator, one-half must be written in floating point notation, as 0.5, but in Common Lisp we can also write one-half as the ratio 1/2. Common Lisp automatically simplifies ratios to use the smallest possible denominator; for example, the ratios 4/6, 6/9, and 10/15 would all be simplified to 2/3.

When we call an arithmetic function with integer inputs, Common Lisp will usually produce an integer or ratio result. If we use a mixture of integers and floating point numbers, the result will be a floating point number:

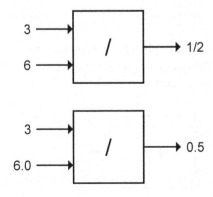

1.4 ORDER OF INPUTS IS IMPORTANT

By convention, when we refer to the ''first'' input to a function, we mean the topmost arrow entering the function box. The ''second'' input is the next highest arrow, and so on. The order in which inputs are supplied to a function is important. For example, dividing 8 by 2 is not the same as dividing 2 by 8:

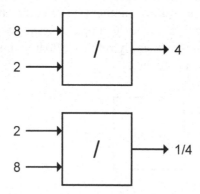

When we divide 8 by 2 we get 4. When we divide 2 by 8 we get the ratio 1/4. By the way, ratios need not always be less than 1. For example:

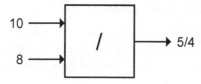

EXERCISE

   1.1. Here are some function boxes with inputs and outputs. In each case one item of information is missing. Use your knowledge of arithmetic to fill in the missing item:

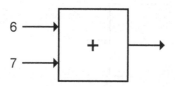

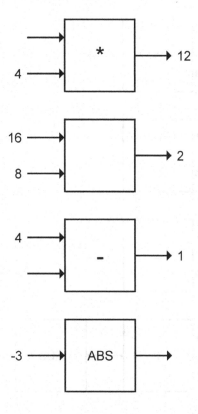

Here are a few more challenging problems. I'll throw in some negative numbers and ratios just to make things interesting.

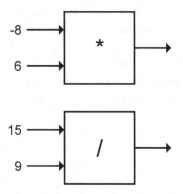

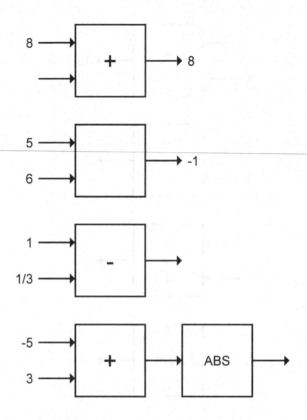

1.5 SYMBOLS

Symbols are another type of data in Lisp. Most people find them more interesting than numbers. Symbols are typically named after English words (such as TUESDAY), or phrases (e.g., BUFFALO-BREATH), or common abbreviations (like SQRT for "square root.") Symbol names may contain practically any combination of letters and numbers, plus some special characters such as hyphens. Here are some examples of Lisp symbols:

```
X                    ZORCH
BANANAS              R2D2
COMPUTER             WINDOW-WASHER
LORETTA              WARP-ENGINES
ABS                  GARBANZO-BEANS
YEAR-TO-DATE         BEEBOP
```

and even

```
ANTIDISESTABLISHMENTARIANISM
```

Notice that symbols may include digits in their names, as in ''R2D2,'' but this does not make them numbers. It is important that you be able to tell the difference between numbers—especially integers—and symbols. These definitions should help:

integer   A sequence of digits ''0'' through ''9,'' optionally preceded by a plus or minus sign.

symbol   Any sequence of letters, digits, and permissible special characters that is not a number.

So FOUR is a symbol, 4 is an integer, +4 is an integer, but + is a symbol. And 7-11 is also a symbol.

EXERCISE

1.2. Next to each of the following, put an ''S'' if it is a symbol, ''I'' if it is an integer, or ''N'' if it is some other kind of number. Remember: English words may sound like integers, but a true Lisp integer contains only the digits 0–9, with an optional sign.

_____	AARDVARK
_____	87
_____	PLUMBING
_____	1-2-3-GO
_____	1492
_____	3.14159265358979
_____	22/7
_____	ZEROP
_____	ZERO
_____	0
_____	-12
_____	SEVENTEEN

1.6 THE SPECIAL SYMBOLS T AND NIL

Two Lisp symbols have special meanings attached to them. They are:

T   Truth, ''yes''

NIL   Falsity, emptiness, ''no''

T and NIL are so basic to Lisp that if you ask a really dedicated Lisp programmer a yes-or-no question, he may answer with T or NIL instead of English. ("Hey, Jack, want to go to dinner?" "NIL. I just ate.") More importantly, certain Lisp *functions* answer questions with T or NIL. Such yes-or-no functions are called **predicates**.

1.7 SOME SIMPLE PREDICATES

A predicate is a question-answering function. Predicates output the symbol T when they mean *yes* and the symbol NIL when they mean *no*. The first predicate we will study is the one that tests whether its input is a number or not. It is called NUMBERP (pronounced "number-pee," as in "number predicate"), and it looks like this:

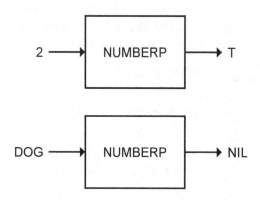

Similarly, the SYMBOLP predicate tests whether its input is a symbol. SYMBOLP returns T when given an input that is a symbol; it returns NIL for inputs that are not symbols.

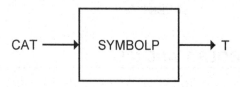

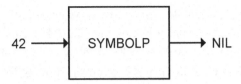

The ZEROP, EVENP, and ODDP predicates work only on numbers. ZEROP returns T if its input is zero.

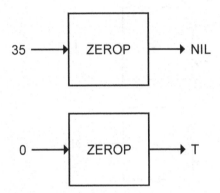

ODDP returns T if its input is odd; otherwise it returns NIL. EVENP does the reverse.

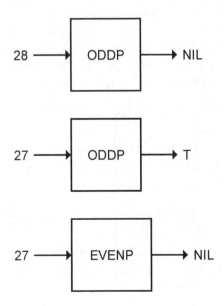

By now you've caught on to the convention of tacking a "P" onto a function name to show that it is a predicate. ("Hey, Jack, HUNGRYP?" "T, I'm starved!") Not all Lisp predicates obey this rule, but most do.

Here are two more predicates: < returns T if its first input is less than its second, while > returns T if its first input is greater than its second. (They are also our first exceptions to the convention that predicate names end with a "P.")

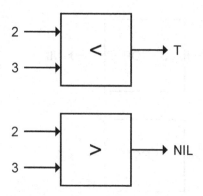

1.8 THE EQUAL PREDICATE

EQUAL is a predicate for comparing two things to see if they are the same. EQUAL returns T if its two inputs are equal; otherwise it returns NIL. Common Lisp also includes predicates named EQ, EQL, and EQUALP whose behavior is slightly different than EQUAL; the differences will not concern us here. For beginners, EQUAL is the right one to use.

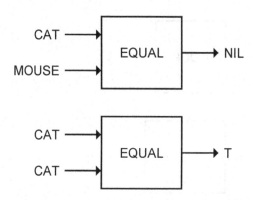

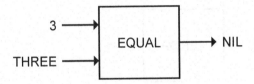

EXERCISE

1.3. Fill in the result of each computation:

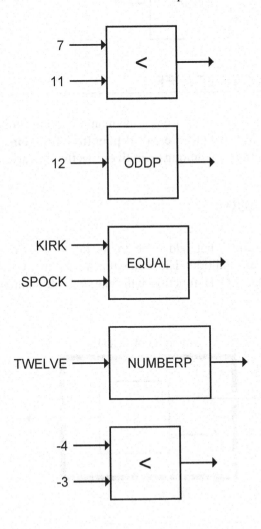

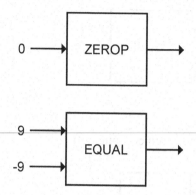

1.9 PUTTING FUNCTIONS TOGETHER

So far we've covered about a dozen of the many functions built into Common Lisp. These built-in functions are called **primitive functions**, or **primitives**. We make new functions by putting primitives together in various ways.

1.9.1 Defining ADD1

Let's define a function that adds one to its input.[**] We already have a primitive function for addition: The + function will add any two numbers it is given as input. Our ADD1 function will take a single number as input, and add one to it.

Definition of ADD1:

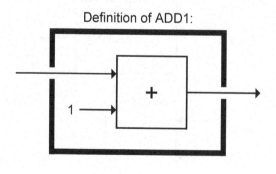

[**]Note to instructors: Common Lisp contains built-in functions `1+` and `1-` that add 1 to or subtract 1 from their input, respectively. But since these unusual names are almost certain to confuse beginning programmers, I will not refer to them in this book.

Now that we've defined ADD1 we can use it to add 1 to any number we like. We just draw a box with the name ADD1 and supply an input, such as 5:

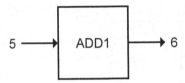

If we look inside the ADD1 box we can see how the function works:

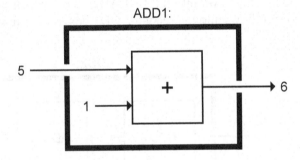

1.9.2 Defining ADD2

Now suppose we want a function that adds 2 to its input. We could define ADD2 the same way we defined ADD1. But in Lisp there is always more than one way to solve a problem; sometimes it is interesting to look at alternative solutions. For example, we could build ADD2 out of two ADD1 boxes:

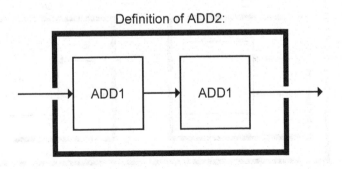
Definition of ADD2:

Once we've defined ADD2, we are free to use it to add 2 to any number. Looking at the ADD2 box from the outside, we have no way of knowing which solution was chosen:

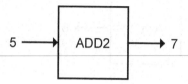

But if we look inside the ADD2 box we can see exactly what's going on. The number 5 flows into the first ADD1 box, which produces 6 as its result. The 6 then flows into the second ADD1 box, and its result is 7.

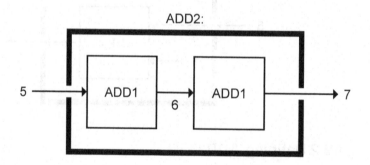

If we want to peer deeper still, we could see the + box inside each ADD1 box, like so:

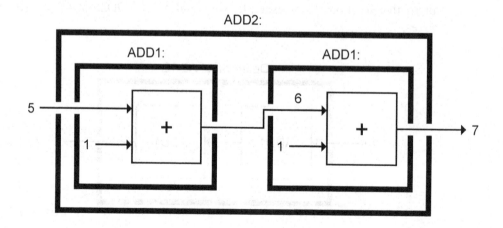

This is as deep as we can go. We can't look inside the + boxes because + is a primitive function.

1.9.3 Defining TWOP

We can use our new knowledge to make our own predicates too, since predicates are just a special type of function. Predicates are functions that return a result of T or NIL. The TWOP predicate defined below returns T if its input is equal to 2.

Definition of TWOP:

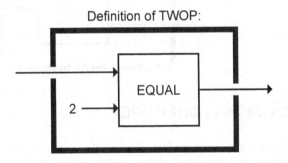

Some examples of the use of TWOP:

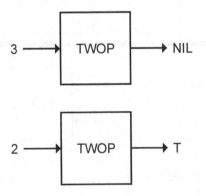

EXERCISES

1.4. Define a SUB2 function that subtracts two from its input.

1.5. Show how to write TWOP in terms of ZEROP and SUB2.

1.6. The HALF function returns a number that is one-half of its input. Show how to define HALF two different ways.

1.7. Write a MULTI-DIGIT-P predicate that returns true if its input is greater than 9.

1.8. What does this function do to a number?

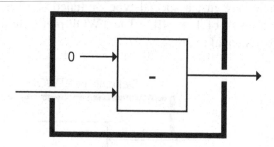

1.9.4 Defining ONEMOREP

Let's try defining a function of two inputs. Here is the ONEMOREP predicate, which tests whether its first input is exactly one greater than its second input.

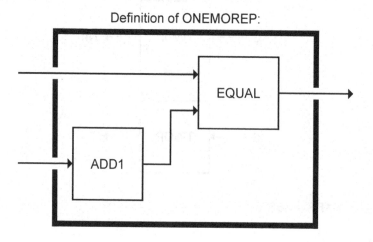

Definition of ONEMOREP:

Do you see how ONEMOREP works? If the first input is one greater than the second input, adding 1 to the second input should make the two equal. In this case, the EQUAL predicate will return T. On the other hand, if the first input to ONEMOREP isn't one greater than the second input, the inputs to EQUAL won't be equal, so it will return NIL. Example:

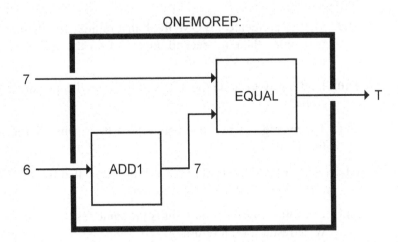

In your mind (or out loud if you prefer), trace the flow of data through ONEMOREP for the preceding example. You should say something like this: "The first input is a 7. The second input, a 6, enters ADD1, which outputs a 7. The two 7's enter the EQUAL function, and since they *are* equal, it outputs a T. T is the result of ONEMOREP." Here is another example to trace:

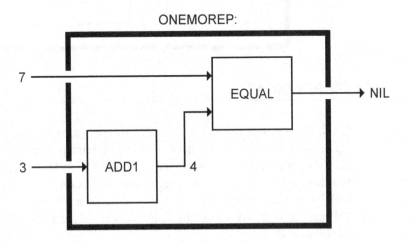

For this second example you should say: "The first input is a 7. The second input, a 3, enters ADD1, which outputs a 4. The 7 and the 4 enter the

EQUAL function, and since they *are not* equal, it outputs a NIL. NIL is the result of ONEMOREP.''

EXERCISES

1.9. Write a predicate TWOMOREP that returns T if its first input is exactly two more than its second input. Use the ADD2 function in your definition of TWOMOREP.

1.10. Find a way to write the TWOMOREP predicate using SUB2 instead of ADD2.

1.11. The average of two numbers is half their sum. Write the AVERAGE function.

1.12. Write a MORE-THAN-HALF-P predicate that returns T if its first input is more than half of its second input.

1.13. The following function returns the same result no matter what its input. What result does it return?

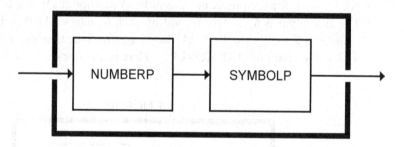

1.10 THE NOT PREDICATE

NOT is the ''opposite'' predicate: It turns *yes* into *no*, and *no* into *yes*. In Lisp terminology, given an input of T, NOT returns NIL. Given an input of NIL, NOT returns T. The neat thing about NOT is that it can be attached to any other predicate to derive its opposite; for example, we can make a ''not equal'' predicate from NOT and EQUAL, or a ''nonzero'' predicate from NOT and ZEROP. We'll see how this is done in the next section. First, some examples of NOT:

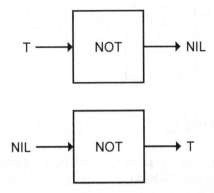

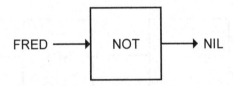

By convention, NIL is the only way to say *no* in Lisp. Everything else is treated as *yes*. So NOT returns NIL for every input except NIL.

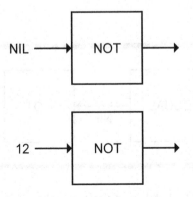

This is not just an arbitrary convention. It turns out to be extremely useful to treat NIL as the only ''false'' object. You'll see why in later chapters.

EXERCISE

1.14. Fill in the results of the following computations:

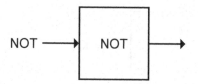

1.11 NEGATING A PREDICATE

Suppose we want to make a predicate that tests whether two things are not equal—the opposite of the EQUAL predicate. We can build it by starting with EQUAL and running its output through NOT to get the opposite result:

Definition of NOT-EQUAL:

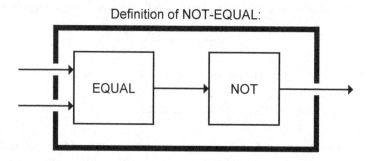

Because of the NOT function, whenever EQUAL would say "T," NOT-EQUAL will say "NIL," and whenever EQUAL would say "NIL," NOT-EQUAL will say "T." Here are some examples of NOT-EQUAL. In the first one, the symbols PINK and GREEN are different, so EQUAL outputs a NIL and NOT changes it to a T.

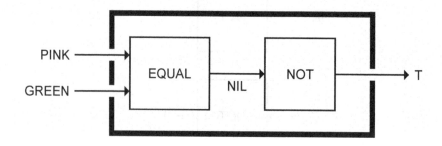

In the second example, PINK and PINK are the same, so EQUAL outputs a T. NOT changes this to NIL.

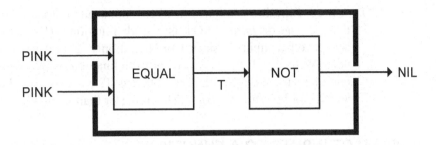

EXERCISES

1.15. Write a predicate NOT-ONEP that returns T if its input is anything other than one.

1.16. Write the predicate NOT-PLUSP that returns T if its input is not greater than zero.

1.17. Some earlier Lisp dialects did not have the EVENP primitive; they only had ODDP. Show how to define EVENP in terms of ODDP.

1.18. Under what condition does this predicate function return T?

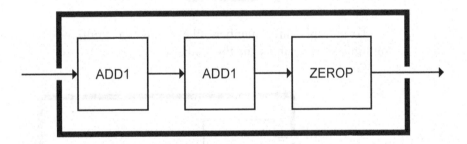

1.19. What result does the function below produce when given the input NIL? What about the input T? Will all data flow through this function unchanged? What result is produced for the input RUTABAGA?

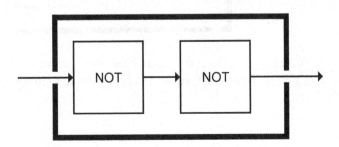

1.20. A **truth function** is a function whose inputs and output are truth values, that is, *true* or *false*. NOT is a truth function. (Even though NOT accepts other inputs besides T or NIL, it only cares if its input is true or not.) Write XOR, the exclusive-or truth function, which returns T when one of its inputs is NIL and the other is T, but returns NIL when both are NIL or both are T. (*Hint:* This is easier than it sounds.)

1.12 NUMBER OF INPUTS TO A FUNCTION

Some functions require a fixed number of inputs, such as ODDP, which accepts exactly one input, and EQUAL, which takes exactly two. But many functions accept a variable number of inputs. For example, the arithmetic functions +, -, *, and / will accept any number of inputs.

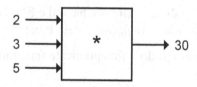

To multiply three numbers, the * function multiplies the first two, then multiplies the result by the third, like so:

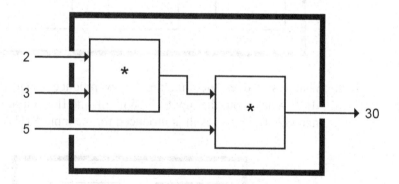

When - or / is given more than two inputs, the result is the first input diminished (or divided, respectively) by the remaining inputs.

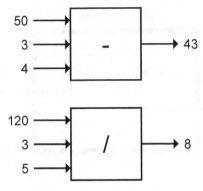

The - and / functions behave differently when given only one input. What - does is negate its input, in other words, it changes the sign from positive to negative or vice versa by subtracting it from zero. When the / function is given a single input, it divides one by that input, which gives the **reciprocal**.

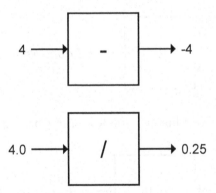

The two-input case is clearly the defining case for the basic arithmetic functions. While they can accept more or fewer than two inputs, they convert those cases to instances of the two-input case. For example, the above computation of the reciprocal of 4.0 is really just a division:

The / function:

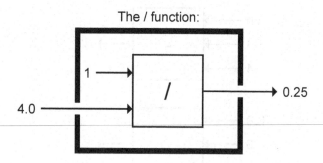

1.13 ERRORS

Even though our system of functions is a very simple one, we can already make several types of errors in it. One error is to give a function the wrong type of data. For example, the + function can add only numbers; it cannot add symbols:

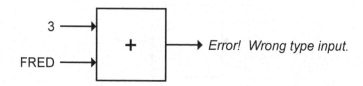

Another error is to give a function too few or too many inputs:

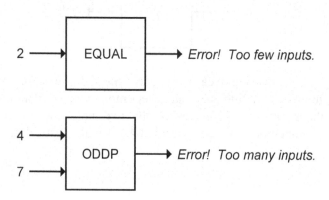

Finally, an error may occur because a function cannot do what is requested of it. This is what happens when we try to divide a number by zero:

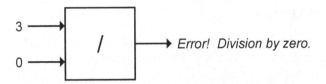

Learning to recognize errors is an important part of programming. You will undoubtedly get lots of practice in this art, since few computer programs are ever written correctly the first time.

EXERCISE

1.21. What is wrong with each of these functions?

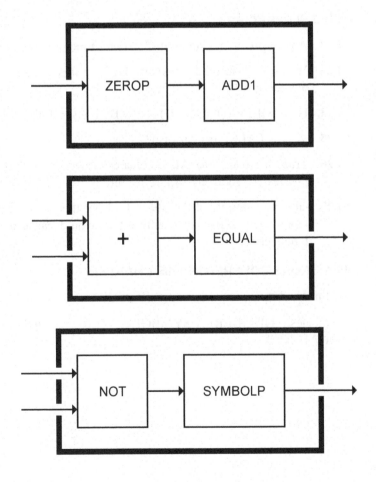

SUMMARY

In this chapter we covered two types of data: numbers and symbols. We also learned several built-in functions that operate on them.

Predicates are a special class of functions that use T and NIL to answer questions about their inputs. The symbol NIL means *false*, and the symbol T means *true*. Actually, anything other than NIL is treated as *true* in Lisp.

A function must have a definition before we can use it. We can make new functions by putting old ones together in various ways. A particularly useful combination, used quite often in programming, is to feed the output of a predicate through the NOT function to derive its opposite, as the NOT-EQUAL predicate was derived from EQUAL.

REVIEW EXERCISES

1.22. Are all predicates functions? Are all functions predicates?

1.23. Which built-in predicates introduced in this chapter have names that do not end in ''P''?

1.24. Is NUMBER a number? Is SYMBOL a symbol?

1.25. Why is FALSE true in Lisp?

1.26. True or false: (a) All predicates accept T or NIL as input; (b) all predicates produce T or NIL as output.

1.27. Give an example of the use of EVENP that would cause a wrong-type-input error. Give an example that would cause a wrong-number-of-inputs error.

FUNCTIONS COVERED IN THIS CHAPTER

Arithmetic functions: +, -, *, /, ABS, SQRT.

Predicates: NUMBERP, SYMBOLP, ZEROP, ODDP, EVENP, <, >, EQUAL, NOT.

1   Advanced Topics

The Advanced Topics sections at the end of each chapter have been added not only to introduce advanced programming material, but also to show computer programming in its broader mathematical and logical perspective.

These sections are entirely optional. Beginning programmers may wish to skip them on their first trip through the book. Some of the later chapters do, in a few places, refer to material introduced in earlier advanced topics sections, but those instances are clearly marked, so it is easy to go back and read the appropriate advanced-topics section before continuing.

1.14   THE HISTORY OF LISP

The origins of Lisp date back to 1956, when a summer research meeting on artificial intelligence was held at Dartmouth College. At the meeting, John McCarthy learned about a technique called ''list processing'' that Allen Newell, J. C. Shaw, and Herbert Simon had developed. Most programming in the 1950s was done in assembly language, a primitive language defined directly by the circuitry of the computer. Newell, Shaw, and Simon had created something more abstract, called IPL (for Information Processing Language), that manipulated symbols and lists, two important datatypes in artificial intelligence programming. But IPL's syntax was similar to (and as akward as) assembly language.

Elsewhere in the 1950s a new language called FORTRAN was being developed. FORTRAN was designed for the sort of numerical calculations that are common in scientific computing. It allowed the programmer to think in terms of algebraic *expressions* such as A=(X+Y)*Z instead of writing assembly language instructions. The idea that programmers should expresss their ideas in familiar mathematical notation, and the computer should be the one to translate these expressions into assembly language, was a radical innovation. It made FORTRAN a powerful numerical computing language. McCarthy wanted to build an equally powerful language for symbolic computing.

One approach he suggested was to build on top of FORTRAN, by creating a set of special subroutines for list manipulation. This idea was pursued by Herbert Gelerntner and Carl Gerberich at IBM, and was called FLPL, for FORTRAN List Processing Language. But McCarthy himself, working first at Dartmouth and later at the Massachusetts Institute of Technology, designed a new language, LISP (for LISt Processor), that drew on ideas from IPL, FORTRAN, and FLPL. The first version, Lisp 1, was developed for the IBM 704 computer.

Lisp 1.5 was the first Lisp dialect to be widely used. The *Lisp 1.5 Programmer's Manual* by McCarthy et al. appeared in 1962. By 1964 Lisp was running on several types of computers, including an IBM 7094 under MIT's Compatible Timesharing System; it was thus one of the first interactive programming languages. Digital Equipment Corporation (DEC) also played a prominent role in Lisp's history. One of the early Lisp implementations ran on its first computer, the PDP-1. The PDP-6 and PDP-10 (later DECSystem-20) computers were specifically designed to implement Lisp efficiently.

After the mid-1960s, Lisp implementations began to diverge. MIT developed MacLisp, while Bolt, Beranek and Newman and the Xerox Corporation jointly developed Interlisp. Stanford Lisp 1.6 was an offshoot of an early version of MacLisp; it eventually gave rise to UCI Lisp. Each of these dialects substantially extended the original Lisp 1.5, but they did so in incompatible ways.

In the 1970s Guy Steele and Gerald Sussman defined a new kind of Lisp, called Scheme, that combined some of the elegant ideas from the Algol family of programming languages with the power of Lisp's syntax and data structures. Extended dialects of Scheme began evolving, paralleling the development of Lisp.

By the early 1980s there were dozens of incompatible Lisp implementations in existence, with about half a dozen major dialects. A project was begun, led by Scott Fahlman, Daniel Weinreb, David Moon, Guy Steele, and Richard Gabriel, to define a Common Lisp that would merge the best features of existing dialects into a coherent whole. The first edition of the Common Lisp standard appeared in 1984; a revised standard appeared in 1990. The ANSI Common Lisp standard was released in 1994 and was revised in 2004. Common Lisp rapidly became the Lisp of choice in both academic and industrial settings. The other dialects have mostly died out, except for Scheme, which continues to enjoy a modest popularity for educational applications.

Many of the more important ideas in programming systems first arose in connection with Lisp. These include mixing of interpreted and compiled functions, garbage collection, recursive function calls, source-level tracing and debugging, and syntax-directed editors. Today Lisp is a leading language for sophisticated research on functional, object-oriented, and parallel programming styles.

For additional information on the history of Lisp, see the articles by McCarthy and Gabriel cited in the Further Readings section at the back of the book, and Paul Graham's Lisp pages at www.paulgraham.com.

2

Lists

2.1 LISTS ARE THE MOST VERSATILE DATA TYPE

The name "Lisp" is an acronym for List Processor. Even though the language has matured in many ways over the years, lists remain its central data type. Lists are important because they can be made to represent practically anything: sets, tables, and graphs, and even English sentences. Functions can also be represented as lists, but we'll save that topic for the next chapter.

2.2 WHAT DO LISTS LOOK LIKE?

Every list has two forms: a printed representation and an internal one. The printed representation is most convenient for people to use, because it's compact and easy to type on a computer keyboard. The internal representation is the way the list actually exists in the computer's memory. We will use a graphical notation when we want to refer to lists in their internal form.

In its printed form, a list is a bunch of items enclosed in parentheses. These items are called the **elements** of the list. Here are some examples of lists written in parenthesis notation:

    (RED GREEN BLUE)

    (AARDVARK)

```
(2 3 5 7 11 13 17)

(3 FRENCH HENS 2 TURTLE DOVES 1 PARTRIDGE
 1 PEAR TREE)
```

The internal representation of lists does not involve parentheses. Inside the computer's memory, lists are organized as chains of **cons cells**, which we'll draw as boxes. The cons cells are linked together by **pointers**, which we'll draw as arrows. Each cons cell has two pointers. One of them always points to an element of the list, such as RED, while the other points to the next cons cell in the chain.[*] When we say "lists may include symbols or numbers as elements," what we are really saying is that cons cells may contain pointers to symbols or numbers, as well as pointers to other cons cells. The computer's internal representation of the list (RED GREEN BLUE) is drawn this way:[**]

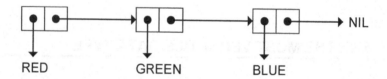

Looking at the rightmost cell, you'll note that the cons cell chain ends in NIL. This is a convention in Lisp. It may be violated in some circumstances, but most of the time lists will end in NIL. When the list is written in parenthesis notation, the NIL at the end of the chain is omitted, again by convention.

EXERCISE

2.1. Show how the list (TO BE OR NOT TO BE) would be represented in computer memory by drawing its cons cell representation.

---

[*]What each cons cell actually is, internally, is a small piece of memory, split in two, big enough to hold two addresses (pointers) to other places in memory where the actual data (like RED, or NIL, or another cons cell) is stored. On most computers pointers are four bytes long, so each cons cells is eight bytes.

[**]Note to instructors: If students are already using the computer, this would be a good time to introduce the SDRAW tool appearing in the appendix.

2.3 LISTS OF ONE ELEMENT

A symbol and a list of one element are not the same. Consider the list (AARDVARK) shown below; it is represented by a cons cell. One of the cons cell's pointers points to the symbol AARDVARK; the other points to NIL. So you see that the list (AARDVARK) and the symbol AARDVARK are different objects. The former is a cons cell that points to the latter.

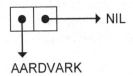

AARDVARK

2.4 NESTED LISTS

A list may contain other lists as elements. Given the three lists

        (BLUE SKY)

        (GREEN GRASS)

        (BROWN EARTH)

we can make a list of them by enclosing them within another pair of parentheses. The result is shown below. Note the importance of having two levels of parentheses: This is a list of *three lists*, not a list of six symbols.

        ((BLUE SKY) (GREEN GRASS) (BROWN EARTH))

We can display the three elements of this list vertically instead of horizontally if we choose. Spacing and indentation don't matter as long as the elements themselves and the parenthesization aren't changed. For example, the list of three lists could have been written like this:

        ((BLUE SKY)
         (GREEN GRASS)
          (BROWN EARTH))

The first element of this list is still (BLUE SKY). In cons cell notation, the list would be written as shown below. Since it has three elements, there are three cons cells in the top-level chain. Since each element is a list of two symbols, each top-level cell points to a lower-level chain of two cons cells.

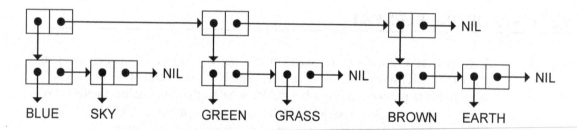

Lists that contain other lists are called **nested lists**. In parenthesis notation, a nested list has one or more sets of parentheses nested within the outermost pair. In cons cell notation, a nested list has at least one level of cons cells below the top-level chain. Lists that are not nested are called **flat lists**. A flat list has only a top-level cons cell chain.

Lists aren't always uniform in shape. Here's a nested list whose elements are a list, a symbol, and a list:

```
((BRAIN SURGEONS) NEVER (SAY OOPS))
```

You can see the pattern of parenthesization reflected in the cons cell diagram below.

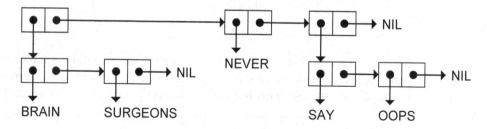

Anything we write in parenthesis notation will have an equivalent description inside the computer as a cons cell structure—if the parentheses balance properly. If they don't balance, as in the malformed expression "(RED (GREEN BLUE," the computer cannot make a cons cell chain corresponding to that expression. The computer will respond with an error message if it reads an expression with unbalanced parentheses.

EXERCISES

2.2. Which of these are well-formed lists? That is, which ones have properly balanced parentheses?

```
(A B (C)
```

((A) (B))

A B) (C D)

(A (B (C))

(A (B (C)))

(((A) (B)) (C))

2.3. Draw the cons cell representation of the list (PLEASE (BE MY) VALENTINE).

2.4. What is the parenthesis notation for this cons cell structure?

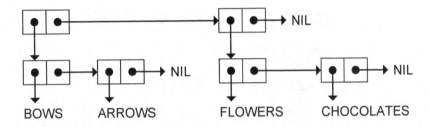

2.5 LENGTH OF LISTS

The length of a list is the number of elements it has, for example, the list (HI MOM) is of length two. But what about lists of lists? When a list is written in parenthesis notation, its elements are the things that appear inside only *one* level of parentheses. For example, the elements of the list (A (B C) D) are A, the list (B C), and D. The symbols B and C are not elements themselves, they are merely components of the element (B C).

Remember that the computer does not use parentheses internally. From the computer's point of view, the list (A (B C) D) contains three elements because its internal representation contains three top-level cons cells, like this:

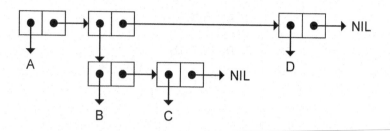

So you see that the length of a list is independent of the complexity of its elements. The following lists all have exactly three elements, even though in some cases the elements are themselves lists. The three elements are underlined.

(RED GREEN BLUE)

((BLUE SKY) (GREEN GRASS) (BROWN EARTH))

(A (B X Y Z) C)

(FOO 937 (GLEEP GLORP))

(ROY (TWO WHITE DUCKS) ((MELTED) (BUTTER)))

The primitive function LENGTH computes the length of a list. It is an error to give LENGTH a symbol or number as input.

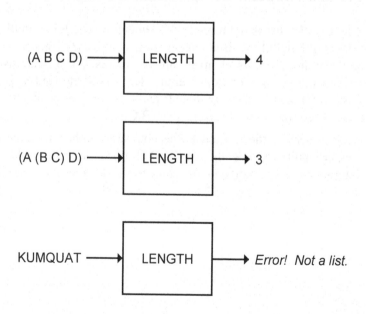

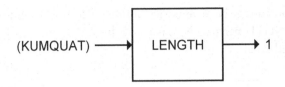

EXERCISE

2.5. How many elements do each of the following lists have?

_____ (OPEN THE POD BAY DOORS HAL)

_____ ((OPEN) (THE POD BAY DOORS) HAL)

_____ ((1 2 3) (4 5 6) (7 8 9) (10 11 12))

_____ ((ONE) FOR ALL (AND (TWO (FOR ME))))

_____ ((Q SPADES)
(7 HEARTS)
(6 CLUBS)
(5 DIAMONDS)
(2 DIAMONDS))

_____ ((PENNSYLVANIA (THE KEYSTONE STATE))
(NEW-JERSEY (THE GARDEN STATE))
(MASSACHUSETTS (THE BAY STATE))
(FLORIDA (THE SUNSHINE STATE))
(NEW-YORK (THE EMPIRE STATE))
(INDIANA (THE HOOSIER STATE)))

2.6 NIL: THE EMPTY LIST

A list of zero elements is called an **empty list**. It has no cons cells. It is written as an empty pair of parentheses:

()

Inside the computer the empty list is represented by the symbol NIL. This is a tricky point: the symbol NIL _is_ the empty list; that's why it is used to mark the end of a cons cell chain. Since NIL and the empty list are identical, we are always free to write NIL instead of (), and vice versa. Thus (A NIL B) can also be written (A () B). It makes no difference which printed form is used; inside the computer the two are the same.

The length of the empty list is zero. Even though NIL is a symbol, it is still a valid input to LENGTH because NIL is also a list. NIL is the only thing that is both a symbol and a list.

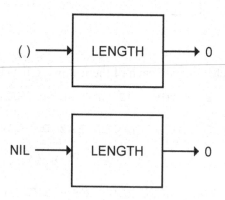

EXERCISE

2.6. Match each list on the left with a corresponding list on the right by substituting NIL for ( ) wherever possible. Pay careful attention to levels of parenthesization.

( )	((NIL))
(( ))	NIL
((( )))	(NIL)
(( ) ( ))	(NIL (NIL))
(( ) (( )))	(NIL NIL)

2.7 EQUALITY OF LISTS

Two lists are considered EQUAL if their corresponding elements are EQUAL. Consider the lists (A (B C) D) and (A B (C D)) shown below.

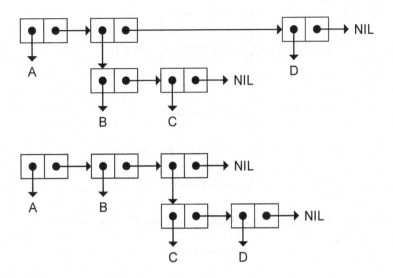

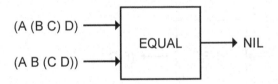

These two lists have the same number of elements (three), but they are not EQUAL. The second element of the former is (B C), while the second element of the latter is B. And neither list is equal to (A B C D), which has four elements. If two lists have different numbers of elements, they are never EQUAL.

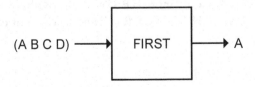

2.8 FIRST, SECOND, THIRD, AND REST

Lisp provides primitive functions for extracting elements from a list. The functions FIRST, SECOND, and THIRD return the first, second, and third element of their input, respectively.

(A B C D) ────→ | FIRST | ──→ A

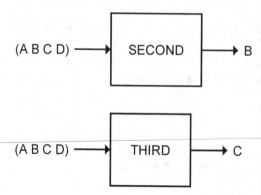

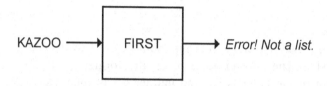

It is an error to give these functions inputs that are not lists.

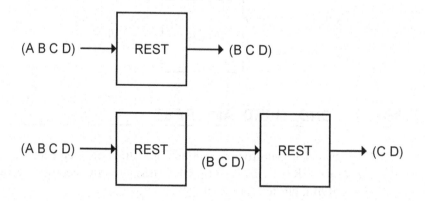

The REST function is the complement of FIRST: It returns a list containing everything *but* the first element.

Using just FIRST and one or more RESTs, it is possible to construct our own versions of SECOND, THIRD, FOURTH, and so on. For example:

Definition of MY-SECOND:

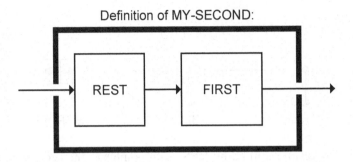

If the input to MY-SECOND is (PENGUINS LOVE ITALIAN ICES), the REST function will output the list (LOVE ITALIAN ICES), and the FIRST element of that is LOVE.

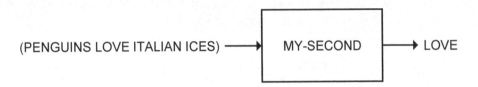

EXERCISES

2.7. What goes on inside the MY-SECOND box when it is given the input (HONK IF YOU LIKE GEESE)?

2.8. Show how to write MY-THIRD using FIRST and two RESTs.

2.9. Show how to write MY-THIRD using SECOND.

2.9 FUNCTIONS OPERATE ON POINTERS

When we say that an object such as a list or symbol is an input to a function, we are speaking informally. Inside the computer, everything is done with pointers, so the real input to the function isn't the object itself, but a pointer to the object. Likewise, the result returned by a function is really a pointer.

Suppose (THE BIG BOPPER) is supplied as input to REST. What REST actually receives is a *pointer* to the first cons cell. This pointer is shown below, drawn as a wavy line. The line is wavy because the pointer's location isn't specified. In other words, it does not live inside any cons cell; it lives elsewhere in the computer. Computer scientists would say that the pointer lives "in a register" or "on the stack," but these details need not concern us.

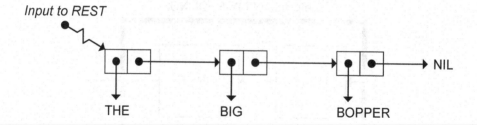

Input to REST

THE   BIG   BOPPER

The result returned by REST is a pointer to the second cons cell, which is the first cell of the list (BIG BOPPER). Where did this pointer come from? What REST did was extract the pointer from the right half of the first cons cell, and return that pointer as its result. So the result of REST is a pointer into the same cons cell chain as the input to REST. (See the figure below.) No new cons cells were created by REST when it returned (BIG BOPPER); all it did was extract and return a pointer.

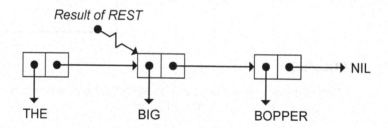

Result of REST

THE   BIG   BOPPER

Note: We show a cons cell pointing to THE in the above figure to emphasize that the result is part of the same chain as the input to REST. But the cons cell that points to THE is not part of the result of REST. There is no way to reach this cell from the pointer returned by REST. (You can't follow pointers backward, only forward.)

2.10 CAR AND CDR

By now you know that each half of a cons cell points to something. The two halves have obscure names. The left half is called the CAR, and the right half is called the CDR (pronounced ''cou-der,'' rhymes with ''good-er''). These names are relics from the early days of computing, when Lisp first ran on a machine called the IBM 704. The 704 was so primitive it didn't even have transistors—it used vacuum tubes. Each of its ''registers'' was divided into several components, two of which were the address portion and the decrement portion. Back then, the name CAR stood for Contents of Address portion of Register, and CDR stood for Contents of Decrement portion of Register. Even though these terms don't apply to modern computer hardware, Common Lisp still uses the acronyms CAR and CDR when referring to cons cells, partly for historical reasons, and partly because these names can be composed to form longer names such as CADR and CDDAR, as you will see shortly.

Besides naming the two halves of a cons cell, CAR and CDR are also the names of built-in Lisp functions that return whatever pointer is in the CAR or CDR half of the cell, respectively. Consider again the list (THE BIG BOPPER). When this list is used as input to a function such as CAR, what the function actually receives is not the list itself, but rather a pointer to the first cons cell:

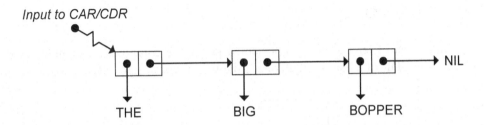

CAR follows this pointer to get to the actual cons cell and extracts the pointer sitting in the CAR half. So CAR returns as its result a pointer to the symbol THE. What does CDR return when given the same list as input?

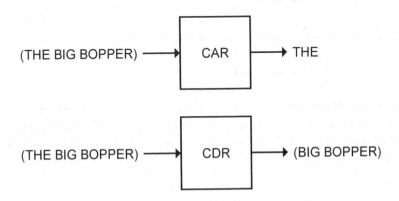

CDR follows the pointer to get to the cons cell, and extracts the pointer sitting in the CDR half, which it returns. So the result of CDR is a pointer to the list (BIG BOPPER). From this example you can see that CAR is the same as FIRST, and CDR is the same as REST. Lisp programmers usually prefer to express it the other way around: FIRST returns the CAR of a list, and REST returns the CDR.

2.10.1 The CDR of a Single-Element List

We saw previously that the list (AARDVARK) is not the same thing as the symbol AARDVARK. The list (AARDVARK) looks like this:

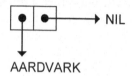

AARDVARK

Since a list of length one is represented inside the computer as a single cons cell, the CDR of a list of length one is the list of length zero, NIL.

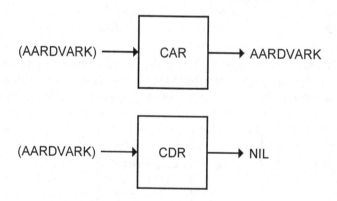

The list ((PHONE HOME)) has only one element. Remember that the elements of a list are the items that appear inside only one level of parentheses, in other words, the items pointed to by top-level cons cells. ((PHONE HOME)) looks like this:

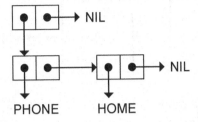

PHONE   HOME

Since the CAR and CDR functions extract their respective pointers from the first cons cell of a list, the CAR of ((PHONE HOME)) is (PHONE HOME), and the CDR is NIL.

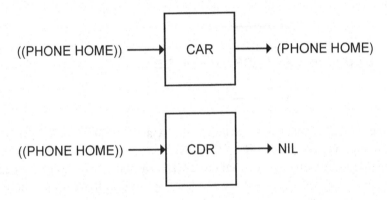

EXERCISES

2.10. Draw the cons cell representation of the list (((PHONE HOME))), which has three levels of parentheses. What is the CAR of this list? What is the CDR?

2.11. Draw the cons cell representation of the list (A (TOLL) ((CALL))).

2.10.2 Combinations of CAR and CDR

Consider the list (FEE FIE FOE FUM), the first element of which is FEE. The second element of this list is the FIRST of the REST, or, in our new terminology, the CAR of the CDR.

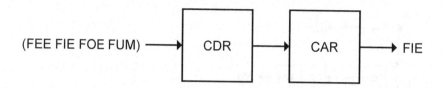

If you read the names of these function boxes from left to right, you'll read "CDR" and then "CAR." But since the input to the CAR function is the output of the CDR function, we say in English that we are computing "the CAR of the CDR" of the list, not the other way around. In Lisp, the CADR function is an abbreviation for "the CAR of the CDR." CADR is pronounced "kae-der."

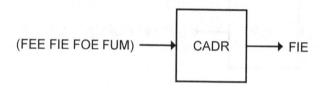

What would happen if we switched the A and the D? The CDAR ("cou-dar") function takes the CDR of the CAR of a list. The CAR of (FEE FIE FOE FUM) is FEE; if we try to take the CDR of that we get an error message. Obviously, CDAR doesn't work on lists of symbols. It works perfectly well on lists of lists, though.

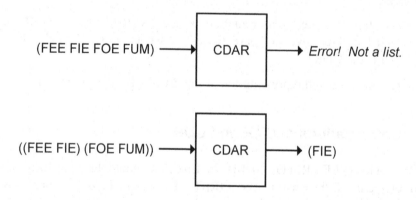

The CADDR ("ka-dih-der") function returns the THIRD element of a list. (If you're having trouble with these strange names, see the pronunciation guide on page 48.) Once again, the name indicates how the function works: It takes the CAR of the CDR of the CDR of the list.

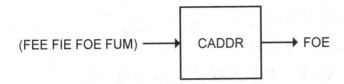

To really understand how CADDR works, you have to read the As and Ds from right to left. Starting with the list (FEE FIE FOE FUM), first take the CDR, yielding (FIE FOE FUM). Then take the CDR of that, which gives (FOE FUM). Finally take the CAR, which produces FOE.

Here's another way to look at CADDR. Start with the CDDR (''cou-dih-der'') function, which takes the CDR of the CDR, or the REST of the REST. The CDDR of (FEE FIE FOE FUM) is (FOE FUM), and the CAR of that is FOE. The CAR of the CDDR is the CADDR!

Common Lisp provides built-in definitions for all combinations of CAR and CDR up to and including four As and Ds in the function name. So CAADDR is built in, but not CAADDAR. Common Lisp also provides built-in definitions for FIRST through TENTH.

EXERCISE

2.12. What C...R name does Lisp use for the function that returns the fourth element of a list? How would you pronounce it?

2.10.3 CAR and CDR of Nested Lists

CAR and CDR can be used to take apart nested lists just as easily as flat ones. Let's see how we can get at the various components of the nested list ((BLUE CUBE) (RED PYRAMID)), which looks like this:

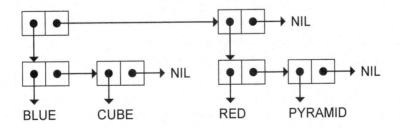

CAR/CDR Pronunciation Guide

Function	Pronunciation	Alternate Name
CAR	*kar*	FIRST
CDR	*cou-der*	REST
CAAR	*ka-ar*	
CADR	*kae-der*	SECOND
CDAR	*cou-dar*	
CDDR	*cou-dih-der*	
CAAAR	*ka-a-ar*	
CAADR	*ka-ae-der*	
CADAR	*ka-dar*	
CADDR	*ka-dih-der*	THIRD
CDAAR	*cou-da-ar*	
CDADR	*cou-dae-der*	
CDDAR	*cou-dih-dar*	
CDDDR	*cou-did-dih-der*	
CADDDR	*ka-dih-dih-der*	FOURTH

and so on

The CAR of this list is (BLUE CUBE). To get to BLUE, we must take the CAR of the CAR. The CAAR function, pronounced ''ka-ar,'' does this.

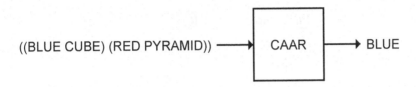

What about getting to the symbol CUBE? Put your finger on the first cons cell of the list. Following the CAR pointer from the first cell takes us to the list (BLUE CUBE). Following the CDR pointer from that cell takes us to the list (CUBE), and following the CAR pointer from there takes us to the symbol CUBE. So CUBE is the CAR of the CDR of the CAR of the list, or, in short, the CADAR (''ka-dar'').

Here's another way to think about it. The first element of the nested list is (BLUE CUBE), so CUBE is the SECOND of the FIRST of the list. This is the CADR of the CAR, which is precisely the CADAR.

Now let's try to get to the symbol RED. RED is the FIRST of the SECOND of the list. You know by now that this is the CAR of the CADR. Putting the two names together yields CAADR, which is pronounced ''ka-ae-der.'' Reading from right to left, put your finger on the first cons cell and follow the CDR pointer, then the CAR pointer, and then the CAR pointer again; you will end up at RED.

Let's build a table of the steps to follow to get to PYRAMID:

Step	Result
start	((BLUE CUBE) (RED PYRAMID))
C...DR	((RED PYRAMID))
C..ADR	(RED PYRAMID)
C.DADR	(PYRAMID)
CADADR	PYRAMID

EXERCISES

2.13. Write down tables similar to the one above to illustrate how to get to each word in the list (((FUN)) (IN THE) (SUN)).

2.14. What would happen if you tried to explain the operation of the CAADR function on the list ((BLUE CUBE) (RED PYRAMID) by reading the

As and Ds from left to right instead of from right to left?

2.15. Using the list ((A B) (C D) (E F)), fill in the missing parts of this table.

Function	Result
CAR	(A B)
CDDR	_____
CADR	_____
CDAR	_____
_____	B
CDDAR	_____
_____	A
CDADDR	_____
_____	F

2.16. What does CAAR do when given the input (FRED NIL)?

2.10.4 CAR and CDR of NIL

Here is another interesting fact about NIL: The CAR and CDR of NIL are defined to be NIL. At this point it's probably not obvious why this should be so. In some earlier Lisp dialects it was actually an error to try to take the CAR or CDR of NIL. But experience shows that defining the CAR and CDR of NIL to be NIL has useful consequences in certain programming situations. You'll see some examples in later chapters.

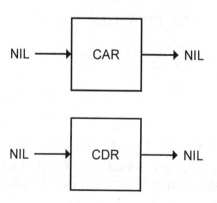

Since FIRST, SECOND, THIRD, and so on are defined in terms of CAR and CDR, you now know what will happen if you try to extract an element of a list that is too short, such as taking the third element of the list (DING ALING). THIRD is CADDR. The CDR of (DING ALING) is (ALING); the CDR of (ALING) is NIL, and the CAR of that is NIL, so:

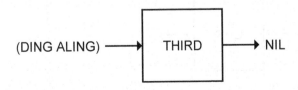

EXERCISE

2.17. Fill in the results of the following computations.

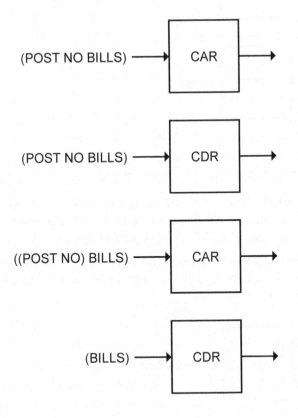

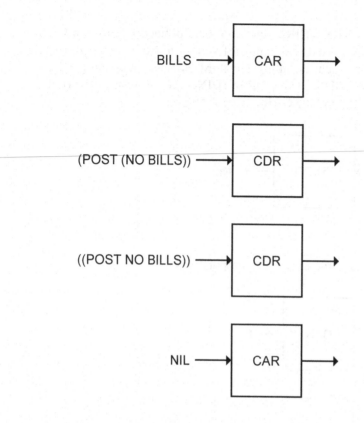

2.11 CONS

The CONS function creates cons cells. It takes two inputs and returns a pointer to a new cons cell whose CAR points to the first input and whose CDR points to the second. The term "CONS" is short for CONStruct.

If we try to explain CONS using parenthesis notation, we might say that CONS adds an element to the front of a list. For example, we can add the symbol A to the front of the list (B C D):

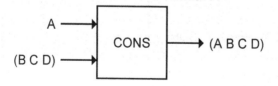

Another example: adding the symbol SINK onto the list (OR SWIM).

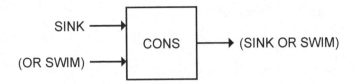

Here is a function GREET that adds the symbol HELLO onto whatever list it is given as input:

Definition of GREET:

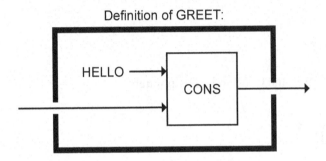

Examples of GREET:

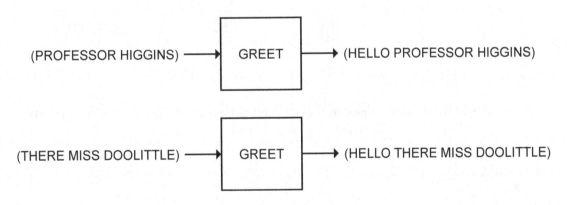

To really understand what CONS does, it is better to think about it using cons cell notation. CONS is a very simple function: It doesn't know anything about the "front of a list." (Remember, inside the computer there are no parentheses.) All CONS does is create one new cons cell. But if the second input to CONS is a cons cell chain of length n, the new cell will form the head of a cons cell chain of length $n+1$. See Figure 2-1. So even though CONS just returns a pointer to the cell it created, in effect it returns a cons cell chain one longer than its second input.

CONS creates a new cons cell:

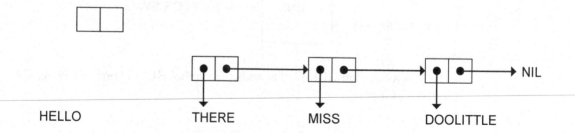

It fills in the CAR and CDR pointers:

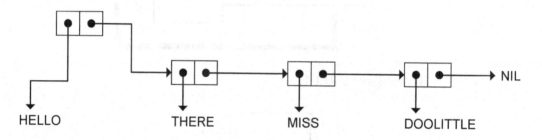

And it returns a pointer to the new cell, which is now the head of a cons cell chain one longer than CONS's second input:

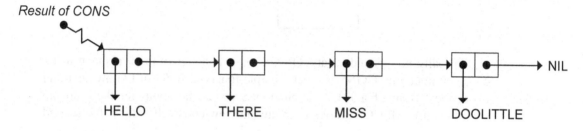

Figure 2-1 Creating a new cons cell with CONS.

2.11.1 CONS and the Empty List

Since NIL is the empty list, if we use CONS to add something onto NIL we get a list of one element.

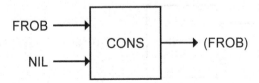

You should be able to confirm this result by looking at the cons cell notation for the list (FROB). The CAR of (FROB) is the symbol FROB and the CDR of (FROB) is NIL, so CONS must have built the list from the inputs FROB and NIL.

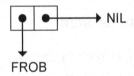

Here's another example that's very similar, except that NIL has been substituted for FROB:

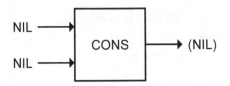

In printed notation, consing something onto NIL is equivalent to throwing an extra pair of parentheses around it.

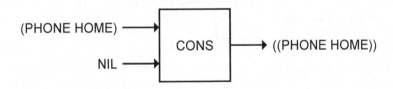

2.11.2 Building Nested Lists With CONS

Any time the first input to CONS is a nonempty list, the result will be a nested list, that is, a list with more than one level of cons cells. Examples:

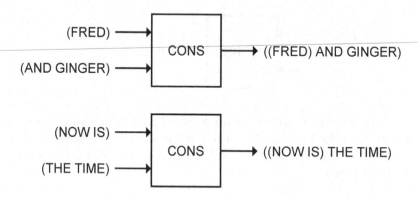

2.11.3 CONS Can Build Lists From Scratch

Suppose we wish to construct the list (FOO BAR BAZ) from scratch. We could start by adding the symbol BAZ onto the empty list. This gives us the list (BAZ).

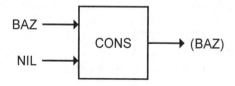

Then we can add BAR onto that:

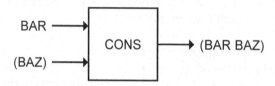

Finally we add the FOO:

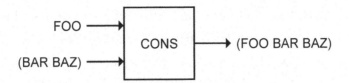

We have cascaded three CONSs together to build the list (FOO BAR BAZ) from scratch. Here is a diagram of the cascade:

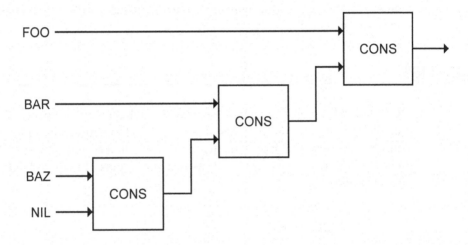

If you turn this diagram sideways you will see that it is almost identical to the cons cell diagram for the list (FOO BAR BAZ). This should give you a clue as to why cons cells and the CONS function share the same name.

EXERCISE

2.18. Write a function that takes any two inputs and makes a list of them using CONS.

2.12 SYMMETRY OF CONS AND CAR/CDR

There is an interesting symmetry between CONS and CAR/CDR. Given some list x, if we know the CAR of x and the CDR of x we can CONS them together to figure out what x is. For example, if the CAR of x is the symbol A and the CDR of x is the list (E I O U), we know that x must be the list (A E I O U).

The symmetry between CONS and CAR/CDR can be expressed formally as:

$$x = \text{CONS of (CAR of } x\text{) and (CDR of } x\text{)}$$

However, this relationship only holds for nonempty lists. When x is NIL, the CAR and CDR of x are also NIL. If we try to reconstruct x by consing together its CAR and CDR portions—that is, CONS of NIL and NIL—we get the list (NIL), not the empty list NIL. This should not be taken to mean that NIL and (NIL) are identical, for we know that they are not. Instead it serves to remind us that although NIL is a list, it's a very unusual one. Certain facts about lists apply only to nonempty ones, in other words, those containing at least one cons cell.

2.13 LIST

Creating a list from a bunch of elements is such a common operation that Lisp has a built-in function to do just that. The LIST function takes any number of inputs and makes a list of them. That is, it makes a new cons cell chain, ending in NIL, with as many cells as there are inputs. Figure 2-2 demonstrates this process.

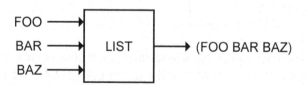

Recall that CONS always makes a single new cons cell; it appears to add its first input onto the list that is its second input. The LIST function, on the other hand, makes an entirely new cons cell chain. In parenthesis notation, it appears to throw a pair of parentheses around its inputs, however many there are. The result of LIST always has one more level of parenthesization than any input had.

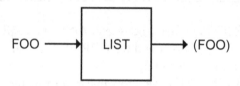

LIST allocates three new cons cells:

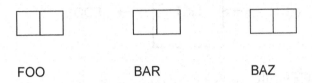

It fills in the CAR pointers:

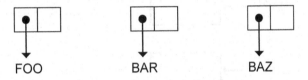

Then it fills in the CDR pointers to form a chain, and returns a pointer to the first cell:

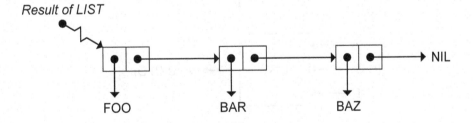

Figure 2-2 How LIST builds a new list.

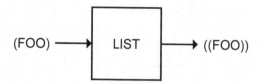

LIST actually works by building a new chain of cons cells. The CAR halves of the cells point to the inputs LIST received. The result of LIST is a pointer to the first cell in the chain. Examples of LIST:

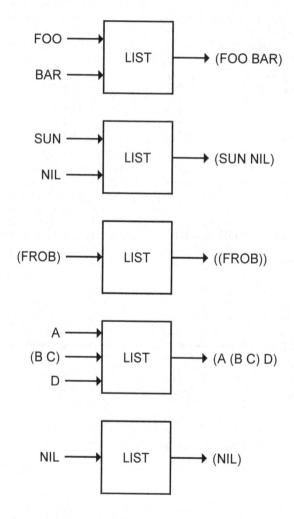

Here is a function called BLURT that takes two inputs and uses them to fill in the blanks in a sentence constructed with LIST.

Definition of BLURT:

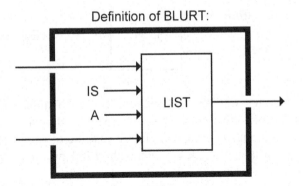

Example of BLURT:

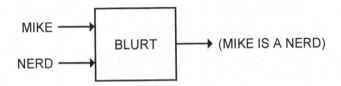

Let's look again at the difference between CONS and LIST. CONS makes a single cons cell. LIST makes a new cons cell chain list out of however many inputs it receives.

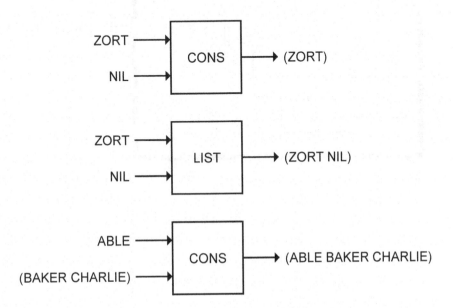

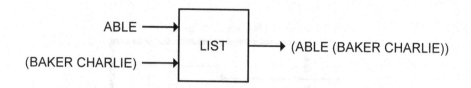

Another way to understand LIST is to think of it as expanding into a cascade of CONS boxes, much the way a call to an arithmetic function like ''+ of 2, 3, 7, and 12'' expands into a cascade of calls to the two-input version of +. So, what really goes on inside the LIST primitive, given an expression like

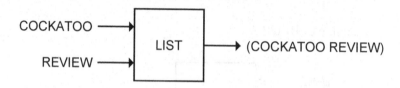

is that several cascaded calls to CONS are made:

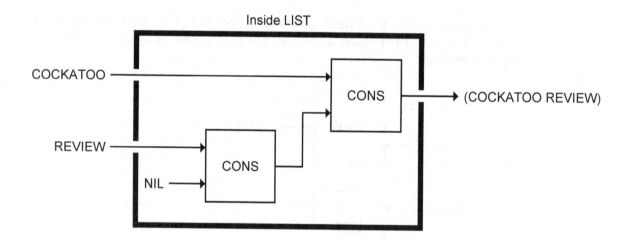

EXERCISE

2.19. Fill in the results of the following computations.

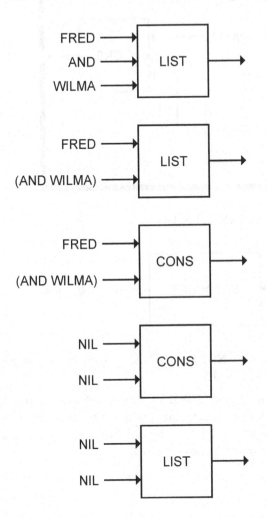

2.14 REPLACING THE FIRST ELEMENT OF A LIST

Suppose we want to replace the first element of a list with the symbol WHAT. The REST function can be used to obtain the sublist beyond the first element; then we can use CONS to add the symbol WHAT to the front of that sublist. We'll call our function SAY-WHAT.

Definition of SAY-WHAT:

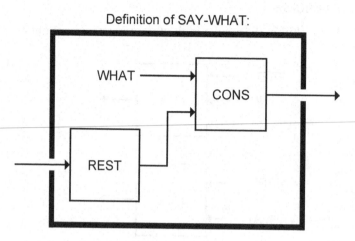

Here's an example of SAY-WHAT:

The REST of (TAKE A NAP) is (A NAP). Consing the symbol WHAT onto that yields (WHAT A NAP).

As you can see now, the SAY-WHAT function doesn't really replace any part of the list. What it does is generate a new list by making a new cons cell whose CDR half points to a portion of the old list. The input to SAY-WHAT and the result it returns are both shown below.

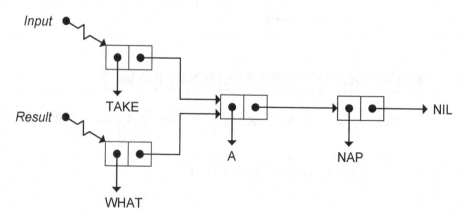

EXERCISES

2.20. What results are returned by the following?

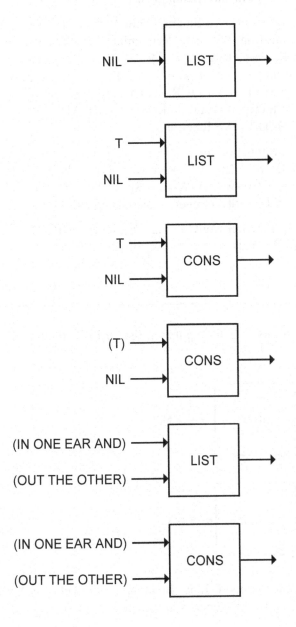

2.21. Write a function that takes four inputs and returns a two-element nested list. The first element should be a list of the first two inputs, and the second element a list of the last two inputs.

2.22. Suppose we wanted to make a function called DUO-CONS that added two elements to the front of a list. Remember that the regular CONS function adds only one element to a list. DUO-CONS would be a function of three inputs. For example, if the inputs were the symbol PATRICK, the symbol SEYMOUR, and the list (MARVIN), DUO-CONS would return the list (PATRICK SEYMOUR MARVIN). Show how to write the DUO-CONS function.

2.23. TWO-DEEPER is a function that surrounds its input with two levels of parentheses. TWO-DEEPER of MOO is ((MOO)). TWO-DEEPER of (BOW WOW) is (((BOW WOW))). Show how to write TWO-DEEPER using LIST. Write another version using CONS.

2.24. What built-in Lisp function would extract the symbol NIGHT from the list (((GOOD)) ((NIGHT)))?

2.15 LIST PREDICATES

The LISTP predicate returns T if its input is a list. LISTP returns NIL for non-lists.

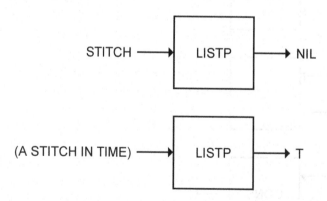

The CONSP predicate returns T if its input is a cons cell. CONSP is almost the same as LISTP; the difference is in their treatment of NIL. NIL is a list, but it is not a cons cell.

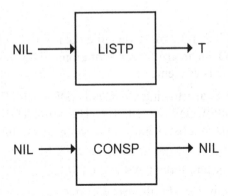

The ATOM predicate returns T if its input is anything other than a cons cell. ATOM and CONSP are opposites; when one returns T, the other always returns NIL.

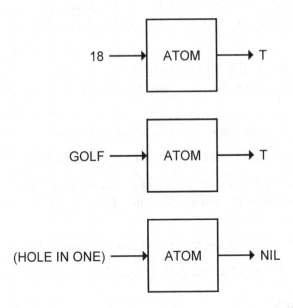

The word "atom" comes from the Greek *atomos*, meaning indivisible. Numbers and symbols are atomic because they cannot be taken apart. Nonempty lists aren't atomic: FIRST and REST take them apart.

The NULL predicate returns T if its input is NIL. Its behavior is the same as the NOT predicate. By convention, Lisp programmers reserve NOT for logical operations: changing *true* to *false* and *false* to *true*. They use NULL when they want to test whether a list is empty.

SUMMARY

This chapter introduced the most versatile data type in Lisp: lists. Lists have both a printed and an internal representation. They may contain numbers, symbols, or other lists as elements.

We can take lists apart using CAR and CDR ("first" and "rest") and put them together with CONS or LIST. The LENGTH function counts the number of elements in a list, which is the same as its number of top-level cons cells.

The important points about CAR and CDR are:

- CAR and CDR accept only lists as input.

- FIRST and REST are the same as CAR and CDR.

- SECOND and THIRD are the same as CADR and CADDR.

- Common Lisp provides built-in C...R functions for all combinations of CAR and CDR up to and including four As and Ds.

The symbol NIL has several interesting properties:

- NIL is a symbol. It is the only way to say "no" or "false" in Lisp.

- NIL is a list. It is the empty list; its LENGTH is zero.

- NIL is the only Lisp object that is both a symbol and a list.

- NIL marks the end of a cons cell chain. When lists are printed in parenthesis notation, the NILs at the end of chains are omitted by convention.

- NIL and () are interchangeable notations for the same object.

- The CAR and CDR of NIL are defined to be NIL.

REVIEW EXERCISES

2.25. Why do cons cells and the CONS function share the same name?

2.26. What do these two functions do when given the input (A B C)?

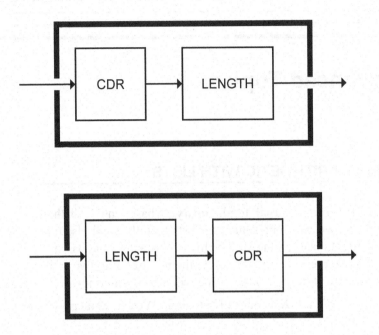

2.27. When does the internal representation of a list involve more cons cells than the list has elements?

2.28. Using just CAR and CDR, is it possible to write a function that returns the *last* element of a list, no matter how long the list is? Explain.

FUNCTIONS COVERED IN THIS CHAPTER

List functions: FIRST, SECOND, THIRD, FOURTH, REST, CAR, CDR, CONS, LIST, LENGTH.

Compositions of CAR and CDR: CADR, CADDR, and so on.

Predicates: LISTP, CONSP, ATOM, NULL.

2   Advanced Topics

2.16   UNARY ARITHMETIC WITH LISTS

Lists can be used to do unary ("base one") arithmetic. In this system, numbers are represented by lists of tally symbols, like the marks a prisoner might make on the wall of his cell to record the passage of time. The number 1 is represented by one tally, the number 2 by two tallies, and so on. We can represent 0 by no tallies. We will not consider negative numbers.

Let's use X as our tally symbol. We can write down unary numbers as lists of Xs:

0   is represented as   NIL

1   is represented as   (X)

2   is represented as   (X X)

3   is represented as   (X X X)

Having defined unary numbers in terms of lists, we may proceed to investigate what effects list-manipulation functions have on them. The REST function subtracts 1 in unary, just as a SUB1 function defined using - would take 1 away from an ordinary integer. Let's subtract 1 from 3:

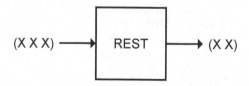

Subtracting 1 from 1 yields 0:

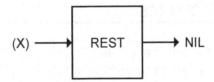

But subtracting 1 from 0 yields 0, not -1. Remember that our unary number scheme was only defined for nonnegative integers.

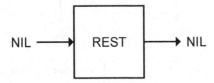

The LENGTH function converts unary numbers to regular integers. Here is an instance of LENGTH operating on the unary number (X X X X):

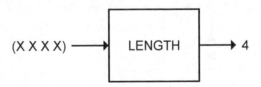

Not all primitive list functions translate into interesting unary arithmetic functions. The CAR function does not, for example. However, it is possible to write our own nonprimitive functions that perform useful unary operations.

EXERCISES

2.29. Write a function UNARY-ADD1 that increases a unary number by one.

2.30. What does the CDDR function do to unary numbers?

2.31. Write a UNARY-ZEROP predicate.

2.32. Write a UNARY-GREATERP predicate, analogous to the > predicate on ordinary numbers.

2.33. CAR can be viewed as a predicate on unary numbers. Instead of returning T or NIL, CAR returns X or NIL. Remember that X or any other non-NIL object is taken as *true* in Lisp. What question about a unary number does CAR answer?

2.17 NONLIST CONS STRUCTURES

A **proper list** is a cons cell chain ending in NIL. The convention is to omit this NIL when writing lists in parenthesis notation, so the structure below is written (A B C).

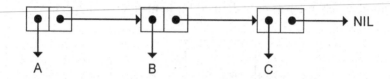

There are other sorts of cons cell structures that are not proper lists, because their chains do not end in NIL. How can the structure below be represented in parenthesis notation?

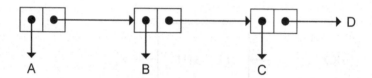

When printing a list in parenthesis notation, Lisp starts by printing a left parenthesis followed by all the elements, separated by spaces. Then, if the list ends in NIL, Lisp prints a right parenthesis. If it does not end in NIL, before printing the right parenthesis Lisp prints a space, a period, another space, and the atom that ends the chain. The list above, which is called a **dotted list** rather than a proper list, is written like this:

(A B C . D)

So far, the only way we have to produce a cons cell structure that doesn't end in NIL is to use CONS.

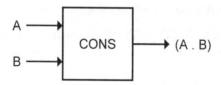

The result of the CONS of A and B is called a **dotted pair**. It is written

(A . B) in parenthesis notation, while in cons cell notation it looks like this:

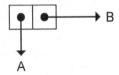

A dotted pair is a single cons cell whose CDR is not NIL. The dotted list (A B . C) contains two cons cells, and is constructed this way:

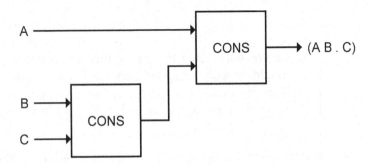

In cons cell notation, (A B . C) looks like this:

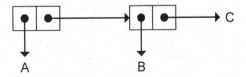

Although LIST is often a more convenient tool than CONS for constructing lists, the LIST function can only build proper lists, since it always constructs a chain ending in NIL. For dotted lists CONS must be used.

EXERCISES

2.34. Write an expression involving cascaded calls to CONS to construct the dotted list (A B C . D).

2.35. Draw the dotted list ((A . B) (C . D)) in cons cell notation. Write an expression to construct this list.

2.18 CIRCULAR LISTS

Dotted lists may look a bit strange, but even stranger structures are possible. For example, here is a **circular list:**

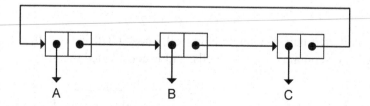

If the computer tried to display this list in printed form, one of several things might happen, depending on the setting of certain printer parameters that will be discussed later. The computer could go into an infinite loop. Or it might try to print part of the list, using ellipsis (three dots), as in:

    (A B C A B C A B ...)

This way of writing the list is incorrect, because it suggests that the list contains more than ten elements, when in fact it contains only three.

Common Lisp does provide a completely correct way to print circular structures, using something called "sharp-equal notation," based on the # (sharp-sign) character. Essentially, to write circular structures we need a way to assign a label to a cons cell so we can refer back to it later. (For example, in the circular list above, the CDR of the third cons cell refers back to the first cell.) We will use integers for labels, and the notation #n= to label an object. We'll write #n# to refer to the object later on in the expression. The list above is therefore written this way:

    #1=(A B C . #1#)

EXERCISE

2.36. Prove by contradiction that this list cannot be constructed using just CONS. *Hint:* Think about the order in which the cells are created.

An even more deviant structure is the one below, in which the CAR of a cons cell points directly back to the cell itself.

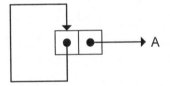

If the computer tried to print this structure, it might end up printing an infinite series of left parentheses. But if the printer is instructed to use sharp-equal notation, the list would print this way:

```
#1=(#1#  .  A)
```

2.19 LENGTH OF NONLIST CONS STRUCTURES

The LENGTH of a list is the number of top-level cons cells in the chain. Therefore the length of (A B C . D) is 3, not 4. It is the same length as the chain (A B C), which can also be written (A B C . NIL).

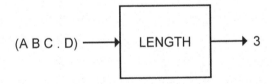

If given a circular list such as #1=(A B C . #1#) as input, LENGTH may not return a value at all. In most implementations it will go into an infinite loop.

3

EVAL Notation

3.1 INTRODUCTION

Before progressing further in our study of Lisp, we must switch to a more flexible notation, called **EVAL notation**. Instead of using boxes to represent functions, we will use lists. Box notation is easy to read, but EVAL notation has several advantages:

- Programming concepts that are too sophisticated to express in box notation can be expressed in EVAL notation.

- EVAL notation is easy to type on a computer keyboard; box notation is not.

- From a mathematical standpoint, representing functions as ordinary lists is an elegant thing to do, because then we can use exactly the same notation for functions as for data.

- In Lisp, functions *are* data, and EVAL notation allows us to write functions that accept other functions as inputs. We'll explore this possibility further in chapter 7.

- When you have mastered EVAL notation, you will know most of what you need to begin conversing in Lisp with a computer.

3.2 THE EVAL FUNCTION

The EVAL function is the heart of Lisp. EVAL's job is to evaluate Lisp **expressions** to compute their result. Most expressions consist of a function followed by a set of inputs. If we give EVAL the expression (+ 2 3), for example, it will invoke the built-in function + on the inputs 2 and 3, and + will return 5. We therefore say the expression (+ 2 3) **evaluates to** 5.

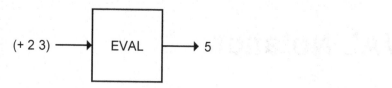

From now on, instead of drawing an EVAL box we'll just use an arrow. The preceding example will be written like this:

(+ 2 3) ⇒ 5

When we want to be slightly more verbose, we'll use a two-headed arrow:

    ⌐→ (+ 2 3)
    ⌐↳ 5

And when we want to show as much detail as possible, we will use a three-headed arrow, like this:

    ⌐→ (+ 2 3)
    ▶ Enter + with inputs 2 and 3
    ↳ Result of + is 5

The meanings of the thin and thick lines will be explained later. Here are some more examples of expressions in EVAL notation:

(+ 1 6) ⇒ 7

(oddp (+ 1 6)) ⇒ t

(* 3 (+ 1 6)) ⇒ 21

(/ (* 2 11) (+ 1 6)) ⇒ 22/7

3.3 EVAL NOTATION CAN DO ANYTHING BOX NOTATION CAN DO

It should be obvious that any expression we write in box notation can also be written in EVAL notation. The expression

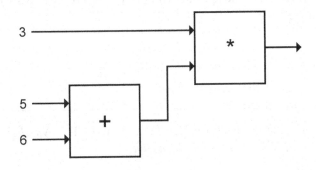

can be represented in EVAL notation as

```
(* 3 (+ 5 6))
```

Similarly, the EVAL notation expression

```
(not (equal 5 6))
```

is represented in box notation as

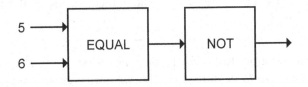

You may notice that EVAL notation appears to read opposite to box notation, in other words, if you read the box notation expression above as "five six, EQUAL, NOT," the corresponding EVAL notation expression reads "NOT EQUAL five six." In the box notation version the computation starts on the left and flows rightward. In EVAL notation the inputs to a function are processed left to right, but since expressions are nested, evaluation actually starts at the innermost expression and flows outward, making the order of function calls in this example right to left.

3.4 EVALUATION RULES DEFINE THE BEHAVIOR OF EVAL

EVAL works by following a set of evaluation rules. One rule is that numbers and certain other objects are "self-evaluating," meaning they evaluate to themselves. The special symbols T and NIL also evaluate to themselves.

```
23   ⇒   23

t  ⇒  t

nil  ⇒  nil
```

> Evaluation Rule for Numbers, T, and NIL: *Numbers, and the symbols T and NIL, evaluate to themselves.*

There is also a rule for evaluating lists. The first element of a list specifies a function to call; the remaining elements are the unevaluated **arguments** to the function. These arguments must be evaluated, in left to right order, to determine the inputs to the function. For example, to evaluate the expression (ODDP (+ 1 6)) the first thing we must do is evaluate ODDP's argument: the list (+ 1 6). To do that, we start by evaluating the arguments to +. 1 evaluates to 1, and 6 evaluates to 6. Now we can call the + function with those inputs and get back the result 7. The 7 then serves as the input to ODDP, which returns T.

> Evaluation Rule for Lists: *The first element of the list specifies a function to be called. The remaining elements specify arguments to the function. The function is called on the evaluated arguments.*

The following diagram, called an **evaltrace diagram**, shows how the evaluation of (ODDP (+ 1 6)) takes place. Notice that evaluation proceeds from the inner nested expression, (+ 1 6), to the outer expression, ODDP. This inner-to-outer quality is reflected in the shape of the evaltrace diagram.

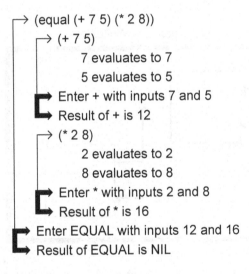

Here's another example of the arguments to a function getting evaluated before the function is called: an evaltrace for the expression (EQUAL (+ 7 5) (* 2 8)):

EXERCISES

3.1. What does (NOT (EQUAL 3 (ABS -3))) evaluate to?

3.2. Write an expression in EVAL notation to add 8 to 12 and divide the result by 2.

3.3. You can square a number by multiplying it by itself. Write an expression in EVAL notation to add the square of 3 and the square of 4.

3.4. Draw an evaltrace diagram for each of the following expressions.

    (- 8 2)

```
(not (oddp 4))

(> (* 2 5) 9)

(not (equal 5 (+ 1 4)))
```

3.5 DEFINING FUNCTIONS IN EVAL NOTATION

In box notation we defined a function by showing what went on inside the box. The inputs to the function were depicted as arrows, In EVAL notation we use lists to define functions, and we refer to the function's arguments by giving them names. We can name the inputs to box notation functions too, by writing the name next to the arrow like this:

Definition of AVERAGE:

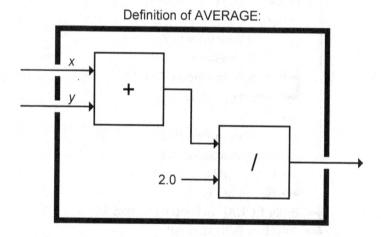

The AVERAGE function is defined in EVAL notation this way:

```
(defun average (x y)
  (/ (+ x y) 2.0))
```

DEFUN is a special kind of function, called a **macro function**, that does not evaluate its arguments. Therefore they do not have to be quoted. DEFUN is used to define other functions. The first input to DEFUN is the name of the function being defined. The second input is the **argument list**: It specifies the names the function will use to refer to its arguments. The remaining inputs to DEFUN define the **body** of the function: what goes on "inside the box." By the way, DEFUN stands for *define function*.

Once you've typed the function definition for AVERAGE into the computer, you can call AVERAGE using EVAL notation. When you type (AVERAGE 6 8), for example, AVERAGE uses 6 as the value for X and 8 as the value for Y. The result, naturally, is 7.0.

Here is another example of function definition with DEFUN:

```
(defun square (n) (* n n))
```

The function's name is SQUARE. Its argument list is (N), meaning it accepts one argument which it refers to as N. The body of the function is the expression (* N N). The right way to read this definition aloud (or in your head) is: "DEFUN SQUARE of N, times N N."

Almost any symbol except T or NIL can serve as the name of an argument. X, Y, and N are commonly used, but BOZO or ARTICHOKE would also work. Functions are more readable when their argument names mean something. A function that computed the total cost of a merchandise order might name its arguments QUANTITY, PRICE, and HANDLING-CHARGE.

```
(defun total-cost (quantity price handling-charge)
    (+ (* quantity price) handling-charge))
```

EXERCISES

3.5. Write definitions for HALF, CUBE, and ONEMOREP using DEFUN. (The CUBE function should take a number n as input and return n^3.)

3.6. Define a function PYTHAG that takes two inputs, x and y, and returns the square root of x^2+y^2. You may recognize this as Pythagoras's formula for computing the length of the hypotenuse of a right triangle given the lengths of the other two sides. (PYTHAG 3 4) should return 5.0.

3.7. Define a function MILES-PER-GALLON that takes three inputs, called INITIAL-ODOMETER-READING, FINAL-ODOMETER-READING, and GALLONS-CONSUMED, and computes the number of miles traveled per gallon of gas.

3.8. How would you define SQUARE in box notation?

3.6 VARIABLES

A **variable** is a place where data is stored.* Let's consider the AVERAGE function again. When we call AVERAGE, Lisp creates two new variables to hold the inputs so that the expression in the body can refer to them by name. The names of the variables are X and Y. It is important to distinguish here between variables and symbols. Variables are not symbols; variables are *named by* symbols. Functions are also named by symbols.

The value of a variable is the data it holds. When we evaluate (AVERAGE 3 7), Lisp creates variables named X and Y and assigns them the values 3 and 7, respectively. In the body of AVERAGE, the symbol X refers to the first variable and the symbol Y refers to the second. These variables can *only* be referenced inside the body; outside of AVERAGE they are inaccessible. Of course the symbols X and Y still exist outside of AVERAGE, but they don't have the same meanings outside as they have inside. The evaltrace diagram below shows how AVERAGE computes its result.

```
┌─→ (average 3 7)
│        3 evaluates to 3
│        7 evaluates to 7
├─▶ Enter AVERAGE with inputs 3 and 7
│      create variable X, with value 3
│      create variable Y, with value 7
│      ┌─→ (/ (+ x y) 2.0)
│      │      ┌─→ (+ x y)
│      │      │      X evaluates to 3
│      │      │      Y evaluates to 7
│      │      └─→ 10
│      │      2.0 evaluates to 2.0
│      └─→ 5.0
└─▶ Result of AVERAGE is 5.0
```

---

*This use of the term "variable" is peculiar to computer programming. In mathematics, a variable is a notation for an unknown quantity, not a physical place in computer memory. But these two meanings are not incompatible, since the inputs to a function are in fact unknown quantities at the time the function is defined.

Now I can explain the meaning of the thick and thin arrows. A thin arrow connects an expression with its value. You see, for example, that the value of the expression (+ X Y) is 10. A thick arrow is used to show entry into the body of a function and exit from that body. Within the scope of the thick arrow we show what goes on inside the body. In the body of AVERAGE, variables are created and expressions are evaluated. We can't see inside the bodies of + or / because they're primitive, so there's not much point in using a thick arrow for those functions, although we could if we wanted to show their entry and exit. For user-defined functions like AVERAGE we start with a thin arrow showing the expression generating the function call, and attach to it a thick arrow showing the entry to and exit from the body. The abstract syntax for this kind of display is:

> (function arg-1 ... arg-n)
>   evaluate the arguments
> Enter FUNCTION with inputs (evaluated arguments)
>   create variables to hold the inputs
>   > body of the function
>   > value of the body
> Result of FUNCTION is (value)

Evaltrace notation is flexible: We can suppress detail when appropriate, such as by not showing function bodies. Another way to simplify an evaltrace is to not display the evaluation of numbers, since they always evaluate to themselves. Sometimes we will also omit the evaluation of symbols. Here is an evaltrace of ONEMOREP using a fairly brief format:

> (onemorep 7 6)
> Enter ONEMOREP with inputs 7 and 6
>   create variable X, with value 7
>   create variable Y, with value 6
>   > (equal x (+ y 1))
>   > (+ y 1)
>   > 7
>   > T
> Result of ONEMOREP is T

3.7 EVALUATING SYMBOLS

The names a function uses for its arguments are independent of the names any other function uses. Two functions such as HALF and SQUARE might both call their argument N, but when N appears in HALF it can only refer to the input of HALF; it has no relation to the use of N in SQUARE.

The rule EVAL uses for evaluating symbols is simple:

Evaluation Rule for Symbols: *A symbol evaluates to the value of the variable it refers to.*

Outside the bodies of HALF and SQUARE, the symbol N refers to the **global variable** named N. A global variable is one that is not associated with any function. PI is an example of a global variable that is built in to Common Lisp.

```
pi   ⇒   3.14159
```

Informally, Lisp programmers sometimes talk of evaluating variables. They might say "variables evaluate to their values." What they really mean is that a *symbol* evaluates to the value of the variable it refers to. Since there can be many variables named N, which one you get depends on where the symbol N appears. If it appears inside the body of SQUARE, you get the variable that holds the input to SQUARE. If it appears outside of any function, you get the global variable named N.

Lisp will complain if you ask it for the value of a variable that has not been assigned a value. We refer to this as an "unassigned variable error."** For example, there is no built-in variable named EGGPLANT in Common Lisp. Evaluating the symbol EGGPLANT causes an unassigned variable error, unless, of course, you evaluate it inside the body of some function that calls one of its inputs EGGPLANT.

```
eggplant   ⇒   Error! EGGPLANT unassigned variable.
```

There is also no built-in variable named N in Common Lisp, so evaluating N outside the body of HALF or SQUARE will cause the same error.

**Most books call this an *unbound* variable error, but this is a historical artifact and is not really appropriate for Common Lisp. Following a suggestion of Robert Wilensky, we use the term "unassigned" instead. This is discussed further in section 5.10.

3.8 USING SYMBOLS AND LISTS AS DATA

Suppose we want to call EQUAL on the symbols KIRK and SPOCK. In box notation this was easy, because symbols and lists were always treated as data. But in EVAL notation symbols are used to name variables, so if we write

```
(equal kirk spock)
```

Lisp will think we are trying to compare the value of the global variable named KIRK with the value of the global variable named SPOCK. Since we haven't given any values to these variables, this will cause an error:

```
(equal kirk spock)   ⇒   Error! KIRK unassigned variable.
```

What we really want to do is compare the symbols themselves. We can tell Lisp to treat KIRK and SPOCK as data rather than as variable references by putting a quote before each one.

```
(equal 'kirk 'spock)   ⇒   nil
```

Because the symbols T and NIL evaluate to themselves, they don't need to be quoted to use them as data. Most other symbols do, though.

```
(list 'james t 'kirk)   ⇒   (james t kirk)
```

Whether symbols are used as data in a function definition, or are passed as inputs when the function is called, they must be quoted to prevent evaluation.

```
(defun riddle (x y)
  (list 'why 'is 'a x 'like 'a y))

(riddle 'raven 'writing-desk)   ⇒
  (why is a raven like a writing-desk)
```

Lists also need to be quoted to use them as data; otherwise Lisp will try to evaluate them, which typically results in an "undefined function" error.

```
(first (we hold these truths))
  ⇒   Error! WE undefined function.

(first '(we hold these truths))   ⇒   we
```

> Evaluation Rule for Quoted Objects: *A quoted object evaluates to the object itself, without the quote.*

Here are some more examples of the difference between quoting and not quoting a list:

```
(third (my aunt mary))   ⇒   Error! MY undefined function.

(third '(my aunt mary))   ⇒   mary


(+ 1 2)   ⇒   3

'(+ 1 2)   ⇒   (+ 1 2)


(oddp (+ 1 2))   ⇒   t

(oddp '(+ 1 2))   ⇒   Error! Wrong type input to ODDP.
```

The error in the last example occurs because ODDP is called with *the list* (+ 1 2) as input. Quoting prevented the list from being evaluated. ODDP can't accept lists as inputs; it can only accept numbers.

Now let's see an evaltrace of an expression involving quotes:

```
→ (length (cons 'fish '(beef chicken)))
    → (cons 'fish '(beef chicken))
        'fish evaluates to FISH
        '(beef chicken) evaluates to (BEEF CHICKEN)
      Enter CONS with inputs FISH and (BEEF CHICKEN)
      Result of CONS is (FISH BEEF CHICKEN)
  Enter LENGTH with input (FISH BEEF CHICKEN)
  Result of LENGTH is 3
```

3.9 THE PROBLEM OF MISQUOTING

It is easy for beginning Lisp programmers to get confused about quoting and either put quotes in the wrong place or leave them out where they are needed. The error messages Lisp gives are a good hint about what went wrong. An unassigned variable or undefined function error usually indicates that a quote was left out:

```
(list 'a 'b c)   ⇒   Error! C unassigned variable.
```

```
(list 'a 'b 'c)  ⇒  (a b c)
```

```
(cons 'a (b c))  ⇒   Error! B undefined function.
```

```
(cons 'a '(b c))  ⇒  (a b c)
```

On the other hand, wrong-type input errors or funny results may be an indication that a quote was put in where it doesn't belong.

```
(+ 10 '(- 5 2))  ⇒   Error! Wrong type input to +.
```

```
(+ 10 (- 5 2))  ⇒  13
```

```
(list 'buy '(* 27 34) 'bagels)
  ⇒  (buy (* 27 34) bagels)
```

```
(list 'buy (* 27 34) 'bagels)
  ⇒  (buy 918 bagels)
```

When we quote a list, the quote must go *outside* the list to prevent the list from being evaluated. If we put the quote inside the list, EVAL will try to evaluate the list and an error will result:

```
('foo 'bar 'baz)  ⇒   Error! 'FOO undefined function.
```

```
'(foo bar baz)  ⇒  (foo bar baz)
```

3.10 THREE WAYS TO MAKE LISTS

We have three ways to make lists using EVAL notation. We can write the list out directly, using a quote to prevent its evaluation, like this:

```
'(foo bar baz)  ⇒  (foo bar baz)
```

Or we can use LIST or CONS to build the list up from individual elements. If we use this method, we must quote each argument to the function:

```
(list 'foo 'bar 'baz)  ⇒  (foo bar baz)
```

```
(cons 'foo '(bar baz))  ⇒  (foo bar baz)
```

One advantage of building the list up from individual elements is that some of the elements can be *computed* rather than specified directly.

```
(list 33 'squared 'is (* 33 33))
  ⇒  (33 squared is 1089)
```

If we quote a list, nothing inside it will get evaluated:

```
'(33 squared is (* 33 33))
  ⇒  (33 squared is (* 33 33))
```

We have seen several ways things can go wrong if quotes are not used properly when building a list:

```
(list foo bar baz)    ⇒     Error! FOO unassigned variable.
```

```
(foo bar baz)    ⇒     Error! FOO undefined function.
```

```
('foo 'bar 'baz)    ⇒     Error! 'FOO undefined function.
```

EXERCISES

3.9. The following expressions evaluate without any errors. Write down the results.

```
(cons 5 (list 6 7))
```

```
(cons 5 '(list 6 7))
```

```
(list 3 'from 9 'gives (- 9 3))
```

```
(+ (length '(1 foo 2 moo))
   (third '(1 foo 2 moo)))
```

```
(rest '(cons is short for construct))
```

3.10. The following expressions all result in errors. Write down the type of error that occurs, explain how the error arose (for example, missing quote, quote in wrong place), and correct the expression by changing *only* the quotes.

```
(third (the quick brown fox))
```

```
(list 2 and 2 is 4)
```

```
(+ 1 '(length (list t t t t)))
```

```
(cons 'patrick (seymour marvin))
```

```
(cons 'patrick (list seymour marvin))
```

3.11. Define a predicate called LONGER-THAN that takes two lists as input and returns T if the first list is longer than the second.

3.12. Write a function ADDLENGTH that takes a list as input and returns a new list with the length of the input added onto the front of it. If the input is (MOO GOO GAI PAN), the output should be (4 MOO GOO GAI PAN). What is the result of (ADDLENGTH (ADDLENGTH '(A B C)))?

3.13. Study this function definition:

```
(defun call-up (caller callee)
    (list 'hello callee 'this 'is
        caller 'calling))
```

How many arguments does this function require? What are the names of the arguments? What is the result of (CALL-UP 'FRED 'WANDA)?

3.14. Here is a variation on the CALL-UP function from the previous problem. What is the result of (CRANK-CALL 'WANDA 'FRED)?

```
(defun crank-call (caller callee)
    '(hello callee this is caller calling))
```

3.11 FOUR WAYS TO MISDEFINE A FUNCTION

Beginning users of EVAL notation sometimes have trouble writing syntactically correct function definitions. Let's take a close look at a proper definition for the function INTRO:

```
(defun intro (x y) (list x 'this 'is y))
```

```
(intro 'stanley 'livingstone)  ⇒
    (stanley this is livingstone)
```

Notice that INTRO's argument list consists of two symbols, X and Y, with neither quotes nor parentheses around them, and the variables X and Y are not quoted or parenthesized in the body, either.

The first way to misdefine a function is to put something other than plain, unadorned symbols in the function's argument list. If we put quotes or extra levels of parentheses in the argument list, the function won't work. Beginners are sometimes tempted to do this when they write a function that is to be called with a list instead of a symbol as input. This is always a mistake.

```
(defun intro ('x 'y)        bad argument list
  (list x 'this 'is y))

(defun intro ((x) (y))      bad argument list
  (list x 'this 'is y))
```

The second way to misdefine a function is to put parentheses around variables where they appear in the body. Only function calls should have parentheses around them. Putting parentheses around a variable will cause an undefined function error:

```
(defun intro (x y) (list (x) 'this 'is (y)))

(intro 'stanley 'livingstone)
  ⇒    Error! X undefined function.
```

The third way to misdefine a function is to quote a variable. Symbols *must* be left unquoted when they refer to variables. Here is an example of what happens when variables are quoted:

```
(defun intro (x y) (list 'x 'this 'is 'y))

(intro 'stanley 'livingstone)  ⇒  (x this is y)
```

The fourth way to misdefine a function is to *not* quote something that should be quoted. In the INTRO function, the symbols X and Y are variables but THIS and IS are not. If we don't quote THIS and IS, an unassigned variable error results.

```
(defun intro (x y) (list x this is y))

(intro 'stanley 'livingstone)
  ⇒    Error! THIS unassigned variable.
```

3.12 MORE ABOUT VARIABLES

In Lisp, a function creates variables automatically when it is is invoked; they (usually) go away when the function returns. Consider the DOUBLE function, which creates a variable named N every time we call it:

```
(defun double (n) (* n 2))
```

Outside of DOUBLE, the symbol N refers to the *global* variable named N. The global variable N has not been assigned any value, so evaluating N results in an error.

n   ⇒   *Error! N unassigned variable.*

Suppose we evaluate (DOUBLE 3). Inside DOUBLE, the symbol N refers to a newly created variable that holds the input to DOUBLE, not the global variable N. The evaltrace diagram below illustrates this.

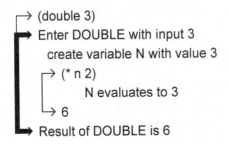

If we call DOUBLE again, for example, (DOUBLE 8), a brand-new variable named N will be created with a value of 8. Outside of DOUBLE the name N still refers to the global variable N, which still has no value.

Now let's try an example with two variables. Here is a definition for QUADRUPLE in terms of DOUBLE:

```
(defun quadruple (n) (double (double n)))
```

Both DOUBLE and QUADRUPLE call their input N. Suppose we evaluate the expression (QUADRUPLE 5) as in the diagram on the next page. When we enter QUADRUPLE, Lisp creates a new variable N with value 5 and evaluates the expression (DOUBLE (DOUBLE N)). What happens when we call DOUBLE with input 5? DOUBLE creates its own variable N, bound to *its own* input, which is 5. The body of DOUBLE evaluates to 10. Now we have evaluated (DOUBLE N), so we can use that result to evaluate (DOUBLE (DOUBLE N)). DOUBLE is called again, this time with input 10, so it creates yet another variable named N, binds it to 10, and evaluates (* N 2). After DOUBLE returns 20, QUADRUPLE returns 20 as its result, and we end up back at top level again, where the name N refers to the global variable N, still with no value assigned.

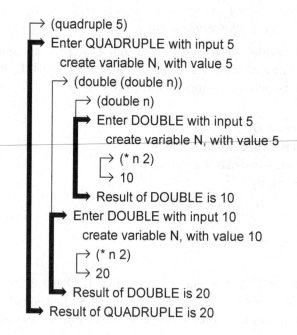

```
→ (quadruple 5)
↪ Enter QUADRUPLE with input 5
    create variable N, with value 5
    → (double (double n))
        → (double n)
        ↪ Enter DOUBLE with input 5
            create variable N, with value 5
            → (* n 2)
            ↳ 10
        ↪ Result of DOUBLE is 10
    ↪ Enter DOUBLE with input 10
        create variable N, with value 10
        → (* n 2)
        ↳ 20
    ↪ Result of DOUBLE is 20
↪ Result of QUADRUPLE is 20
```

EXERCISES

3.15. Consider the following function, paying close attention to the quotes:

```
(defun scrabble (word)
  (list word 'is 'a 'word))
```

The symbol WORD is used two different ways in this function. What are they? What is the result of (SCRABBLE 'AARDVARK)? What is the result of (SCRABBLE 'WORD)?

3.16. Here's a real confuser:

```
(defun stooge (larry moe curly)
  (list larry (list 'moe curly) curly 'larry))
```

What does the following evaluate to? It will help to write down what value each variable is bound to and, of course, mind the quotes!

```
(stooge 'moe 'curly 'larry)
```

3.17. Why can't the special symbols T or NIL be used as variables in a function definition? (Consider the evaluation rule for T and NIL versus the rule for evaluating ordinary symbols.)

SUMMARY

In this chapter we learned EVAL notation, which allows expressions to be represented as lists. Lists are interpreted by the EVAL function according to a built-in set of evaluation rules. The evaluation rules we learned were:

- Numbers are self-evaluating, meaning they evaluate to themselves. So do T and NIL.

- When evaluating a list, the first element specifies a function to call, and the remaining elements specify its arguments. The arguments are evaluated from left to right to derive the inputs that are passed to the function.

- Symbols appearing anywhere other than the first element of a list are interpreted as variable references. A symbol evaluates to the value of the variable it names. Exactly which variable a symbol is referring to depends on the context in which the symbol appears. Variables that haven't been assigned values cause ''unassigned variable'' errors when the symbol is evaluated.

- A quoted list or symbol evaluates to itself, without the quote.

A list of form (DEFUN function-name (argument-list) function-body) defines a function. DEFUN is a special kind of function; its inputs do not have to be quoted. A function's argument list is a list of symbols giving names to the function's inputs. Inside the body of the function, the variables that hold the function's inputs can be referred to by these symbols.

REVIEW EXERCISES

3.18. Name two advantages of EVAL notation over box notation.

3.19. Evaluate each of the following lists. If the list causes an error, tell what the error is. Otherwise, write the result of the evaluation.

```
(cons 'grapes '(of wrath))

(list t 'is 'not nil)

(first '(list moose goose))

(first (list 'moose 'goose))

(cons 'home ('sweet 'home))
```

3.20. Here is a mystery function:

```
(defun mystery (x)
   (list (second x) (first x)))
```

What result or error is produced by evaluating each of the following?

```
(mystery '(dancing bear))
```

```
(mystery 'dancing 'bear)
```

```
(mystery '(zowie))
```

```
(mystery (list 'first 'second))
```

3.21. What is wrong with each of the following function definitions?

```
(defun speak (x y) (list 'all 'x 'is 'y))
```

```
(defun speak (x) (y) (list 'all x 'is y))
```

```
(defun speak ((x) (y)) (list all 'x is 'y))
```

FUNCTIONS COVERED IN THIS CHAPTER

The evaluator: EVAL.

Macro function for defining new functions: DEFUN.

Lisp on the Computer

Congratulations! Having made it successfully through all the pencil-and-paper work, it's time for you to learn how to use Lisp on a real computer. Unfortunately, I can't give you a detailed introduction; there are too many types of computers—and too many implementations of Common Lisp—for that to be practical. You might want to spend a few minutes glancing through the user's manuals for the computer and Lisp implementation you'll be using. A better approach would be to talk to someone who is already familiar with your machine.

3.13 RUNNING LISP

The first thing you need to find out is how to start up Lisp on your computer. If you're lucky you can just type "lisp" and hit the Return key, but you might have to type something more complicated. When Lisp starts up it prints a greeting message. Each implementation has its own style of greeting, but a typical message looks something like this:

```
CMU Common Lisp M2.8 (29-Mar-89)
Hemlock M3.0 (29-Mar-89), Compiler M1.7 (29-Mar-89)
Send bug reports and questions to Gripe.

>
```

The ">" character that appears after the greeting is called a **top-level prompt**. It indicates that Lisp is waiting for you to type something. Some Lisps use a different prompt character; many use "*" (an asterisk).

The next thing you need to find out is which control characters your Lisp uses, specifically:

- How do you delete a character: by pressing Delete, Backspace, or some other key?

- How do you throw away a line of input so you can start over? In some Lisps you can discard a line before hitting Return by typing a Control-U. (While holding down the Control key, press the "U" key.) Other Lisps use a different character.

- What is the "abort" character that gets you back to the top-level prompt? Many Lisps use Control-G or Control-C for this purpose.

While we're on the subject of special characters, remember that computers always provide separate keys for the letter "O" and the digit "0," and for the letter "l" and the digit "1." On conventional typewriters it's fine to type "O" for "0" or "l" for "1," but when you talk to a computer you must be sure to use the correct character for what you mean.

Finally, you need to find out how to get out of Lisp when you're done. Most Lisps require you to type something like (QUIT) or (EXIT) to leave. Sometimes an end-of-file character like Control-D will also work.

3.14 THE READ-EVAL-PRINT LOOP

A computer running Lisp behaves a lot like a pocket calculator. It reads an expression that you type on the keyboard, evaluates it (using EVAL), and prints the result on the screen. Then it prints another prompt and waits for you to type the next expression. This process is called a **read-eval-print loop**.

Here is a sample dialog with a computer in which I define a function and then use that function. In this example, what I type appears after the ">" in lowercase; the computer's response is in uppercase. Not all Lisps follow this convention, but many do.

```
> (defun square (n) (* n n))          First I define SQUARE.
SQUARE                                Computer accepts my definition.

> (square 4)                          Try to square 4.
16                                    Computer prints the answer.

> (square 5)                          Try squaring another number.
25                                    It works just fine.

> (square 123456789)                  Square a big number...
15241578750190521                     and get a really big result.
```

3.15 RECOVERING FROM ERRORS

A very important thing to learn at this point is how to recover from errors. First let's consider typing errors. If after entering a long expression I realize I've made a typing error near the beginning, I may want to throw away the entire expression and start over. In my Lisp, the way to do that is to type Control-G to get back to the top-level prompt. Here's an example:

```
> (defun add87 ((n))                  Too many parens around the N.
    (+ n   ^G                         So I hit Control-G to abort.
Aborted.

> (defun add87 (n)                    This time I typed it correctly.
    (+ n 87))
ADD87
```

A more common problem is an expression that is typed correctly but results in an evaluation error. Trying to add a number and a symbol is an example. When an evaluation error occurs, Lisp prints an error message and puts you in a different kind of input loop. Instead of talking to the top-level read-eval-print loop, you are now talking to the **debugger's** read-eval-print loop. We'll learn how to use the debugger in Chapter 8. For now, all you need to know is how to get out of the debugger and back to top level. In my Lisp, Control-G is the abort character that gets me out of the debugger and back to top level.

```
> (+ 1 'foo)                    This expression causes an error.
Error in function +.            Lisp complains.
Wrong type argument, FOO,
 should have been of type NUMBER.


Debug   (type H for help)       I land in the debugger.
0] ^G                           Type Control-G to get out.
Aborted.


>                               Back at top level again.
```

If you define a function in Lisp and it doesn't work, you can redefine it and try again. You can redefine a function as often as you like; only the last definition is retained. The following example illustrates this and also shows that you can hit Return at any point in an expression with no ill effect. This is because expressions are lists; their spacing and indentation is arbitrary.

```
> (defun intro (x y)            INTRO misdefined! No quotes.
    (list x this is y))
INTRO


> (intro 'stanley 'livingstone) Testing the INTRO function.
Error in function INTRO.
Unassigned variable: THIS.


Debug   (type H for help)       I land in the debugger again.
0] ^G                           Type Control-G to get out.
Aborted.


> (defun intro (x y)            Redefine INTRO correctly.
    (list x 'this 'is y))
INTRO


> (intro 'stanley 'livingstone) Test it again.
(STANLEY THIS IS LIVINGSTONE)   Now it works.
```

Be sure you don't use names like CONS, +, or LIST for your own functions; in Lisp these are the names of built-in functions. Redefining these functions may cause a ''fatal'' error, in which case you will have to leave Lisp and start it up again, and any functions you defined previously will be lost.

Lisp Toolkit: ED

The Lisp Toolkit sections appearing in this and subsequent chapters will introduce you to the important tools of the Lisp programming environment. Some of these tools, such as language-specific editors, program formatters, and source-level debuggers, are available today for other languages, but they first appeared in Lisp. Other tools remain unique to Lisp, and two of them, SDRAW and DTRACE, are unique to this book. The source listings for both appear in an appendix.

The tool we will cover first is the Lisp editor. The Common Lisp standard does not specify what sort of editor should be provided with a Lisp implementation, so I can't tell you exactly how your editor works. But I can tell you something about Lisp editors in general, why they're different from ordinary text editors, and why you ought to take the time to learn to use whatever editor your Lisp provides.

> *The most frequently occurring errors in LISP are parenthetical errors. It is thus almost imperative to employ some sort of counting or pairing device to check parentheses every time that a function is changed.*
> — Elaine Gord, ''Notes on the debugging of LISP programs,'' 1964.

The above quote was written 25 years ago, when Lisp programs were typed on punched cards. Today, of course, we use interactive editors. Lisp editors are not ordinary text editors: They ''understand'' the syntax of Lisp programs. On my machine, whenever I type a right parenthesis, the editor flashes the corresponding left parenthesis for me. This keeps me from making a ''parenthetical error'' when entering Lisp expressions. Another one of my editor's jobs is to automatically indent every line as I type it. If a function definition takes several lines, it will be indented in a neat and orderly format that is easy to read.

Some of the earliest Lisp books were written before anyone thought of systematically indenting programs to make them readable. A program that would have been written this way back then:

```
(defun long-function (some-list) (cons
(third some-list) (list (second some-list)
(fourth some-list) (first some-list))))
```

would today be automatically indented to look like this:

```
(defun long-function (some-list)
  (cons (third some-list)
        (list (second some-list)
              (fourth some-list)
              (first some-list)))))
```

There are two more things a good Lisp editor provides. One is an easy way to evaluate expressions while editing. You can position the cursor (or mouse) on a function definition, hit a few keys, and that function definition will be evaluated without ever leaving the editor. The second thing a good editor provides is rapid access to online documentation. If I want to see the documentation for any Lisp function or variable, I can call it up with just a few keystrokes. The editor also provides online documentation about itself.

The Common Lisp standard specifies the interface between a Lisp implementation and the editor it provides. The interface is a function called ED. Typing (ED) when at the top-level read-eval-print loop causes you to enter the editor, but many Lisps also provide faster ways, such as by typing a character like Control-E.

It is possible to supply arguments to ED to cause it to edit a particular function or file of functions, but we won't go into that here. It's usually easier to just enter the editor first, then use the editor's commands to call up whatever it is on which you wish to work.

Keyboard Exercise

Keyboard exercises are modest programming projects you can solve while sitting at a computer. (However, this first keyboard exercise is just a collection of small unrelated problems, since we haven't covered enough of Lisp yet to do anything more ambitious.) Before attempting a keyboard exercise you should have a firm understanding of the material in that chapter and be able to handle the regular exercises included there.

EXERCISE

3.22. The exercises below may be done in any order. What's most important is that you get comfortable with using the computer. You don't have to solve all of these problems; feel free to experiment and improvise on your own if you like.

a. Find out how to run Lisp on your computer, and start it up.

b. For each following expression, write down what you think it evaluates to or what kind of error it will cause. Then try it on the computer and see.

```
(+ 3 5)

(3 + 5)

(+ 3 (5 6))

(+ 3 (* 5 6))

'(morning noon night)

('morning 'noon 'night)

(list 'morning 'noon 'night)

(car nil)

(+ 3 foo)

(+ 3 'foo)
```

c. Here is an example of the function MYFUN, a strange function of two inputs.

```
(myfun 'alpha 'beta)  ⇒  ((ALPHA) BETA)
```

Write MYFUN. Test your function to make certain it works correctly.

d. Write a predicate FIRSTP that returns T if its first argument (a symbol) is equal to the first element of its second argument (a list). That is, (FIRSTP 'FOO '(FOO BAR BAZ)) should return T. (FIRSTP 'BOING '(FOO BAR BAZ)) should return NIL.

e. Write a function MID-ADD1 that adds 1 to the middle element of a three-element list. For example, (MID-ADD1 '(TAKE 2 COOKIES)) should return the list (TAKE 3 COOKIES). *Note:* You are *not* allowed to make MID-ADD1 a function of three inputs. It has to take a single input that is a list of three elements.

f. Write a function F-TO-C that converts a temperature from Fahrenheit to Celsius. The formula for doing the conversion is: Celsius temperature = [5×(Fahrenheit temperature - 32)]/9. To go in the opposite direction, the formula is: Fahrenheit temperature = (9/5 × Celsius temperature) + 32.

g. What is wrong with this function? What does (FOO 5) do?

```
(defun foo (x) (+ 1 (zerop x)))
```

3 Advanced Topics

3.16 FUNCTIONS OF NO ARGUMENTS

Suppose we wanted to write a function that multiplies 85 by 97. Notice that this function requires no inputs; it does its computation using only prespecified constants. Since the function doesn't take any inputs, when we write its definition, it will have an *empty* argument list. The empty list, of course, is NIL. Let's define this function under the name TEST:

```
(defun test () (* 85 97))
```

After doing this, we see that

```
(test) ⇒ 8245
```

```
(test 1) ⇒ Error! Too many arguments.
```

TEST is a function, so we must put parentheses around it to call it. If we omit them, the symbol TEST is interpreted as a reference to a variable.

`test` ⇒ *Error! TEST unassigned variable.*

3.17 THE QUOTE SPECIAL FUNCTION

QUOTE is a special function: Its input does not get evaluated. The QUOTE special function simply returns its input. For example:

`(quote foo)` ⇒ `foo`

`(quote (hello world))` ⇒ `(hello world)`

Early versions of Lisp used QUOTE instead of an apostrophe to indicate that something shouldn't be evaluated. That is, where we would write

`(cons 'up '(down sideways))`

old-style Lisp programmers would write

`(cons (quote up) (quote (down sideways)))`

Modern Lisp systems use the apostrophe as shorthand for QUOTE. Internally, however, they convert the apostrophe to QUOTE. We can demonstrate that this happens by using multiple quotes. The first quote is stripped away by the evaluation process, but any extra quotes remain.

`'foo` ⇒ `foo`

`''foo` ⇒ `'foo` *also written* `(quote foo)`

`(list 'quote 'foo)` ⇒ `(quote foo)` *also written* `'foo`

`(first ''foo)` ⇒ `quote`

`(rest ''foo)` ⇒ `(foo)`

`(length ''foo)` ⇒ `2`

Depending on the version of Lisp your computer runs, you may occasionally see QUOTE written out instead of in its shorthand form, the apostrophe.

3.18 INTERNAL STRUCTURE OF SYMBOLS

So far in this book we have been drawing symbols by writing their names. But symbols in Common Lisp are actually composite objects, meaning they have several parts to them. Conceptually, a symbol is a block of five pointers, one of which points to the representation of the symbol's name. The others will be defined later. The internal structure of the symbol FRED looks like this:

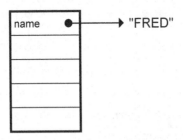

The "FRED" appearing above in quotation marks is called a **string**. Strings are sequences of characters; they will be covered more fully in Chapter 9. For now it suffices to note that strings are used to store the names of symbols; a symbol and its name are actually two different things.

Some symbols, like CONS or +, are used to name built-in Lisp functions. The symbol CONS has a pointer in its **function cell** to a "compiled code object" that represents the machine language instructions for creating new cons cells.

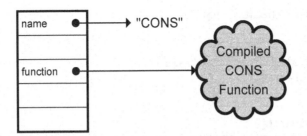

When we draw Lisp expressions such as (EQUAL 3 5) as cons cell chains, we usually write just the name of the symbol instead of showing its internal structure:

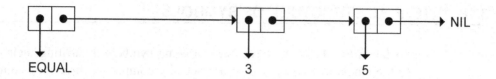

EQUAL   3   5

But if we choose we can show more detail, in which case the expression (EQUAL 3 5) looks like this:

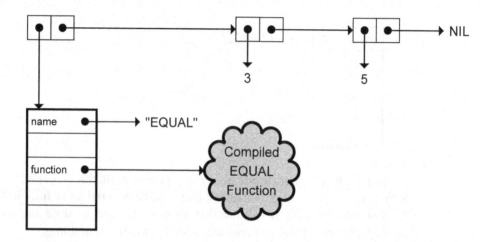

We can extract the various components of a symbol using built-in Common Lisp functions like SYMBOL-NAME and SYMBOL-FUNCTION. The following dialog illustrates this; you'll see something slightly different if you try it on your computer, but the basic idea is the same.

```
> (symbol-name 'equal)
"EQUAL"

> (symbol-function 'equal)
#<Compiled EQUAL function {60463B0}>
```

3.19 LAMBDA NOTATION

Lambda notation was created by Alonzo Church, a mathematician at Princeton University. Church wanted a clear, unambiguous way to describe functions, their inputs, and the computations they perform. In lambda notation, a function that adds 3 to a number would be written as shown below; the λ is the

Greek letter lambda: λx.(3+x).

John McCarthy, the originator of Lisp, was a student of Church. He adopted Church's notation for specifying functions. The Lisp equivalent of the unnamed function λx.(3+x) is the list

```
(lambda (x) (+ 3 x))
```

A function $f(x,y) = 3x+y^2$ would be written $\lambda(x,y).(3x+y^2)$ in lambda notation. In Lisp it is written

```
(lambda (x y) (+ (* 3 x) (* y y)))
```

As you can see, the syntax of lambda expressions in Lisp is similar to that of Church's notation, and even more similar to DEFUN. But unlike DEFUN, LAMBDA is not a function; it is a marker treated specially by EVAL. We'll learn more about lambda expressions in chapter 7.

DEFUN's job is to associate names with functions. When typing in a new function definition, such as for HALF, there are two kinds of naming going on. The string "HALF" names the symbol, and the symbol HALF names the function. In the diagram below, you can see the name cell of HALF pointing to the string "HALF". Its function cell points to a **function object** that is the *real* function. Exactly what this function object looks like depends on which implementation of Common Lisp you're using, but as the diagram indicates, there's probably a lambda expression in there somewhere.

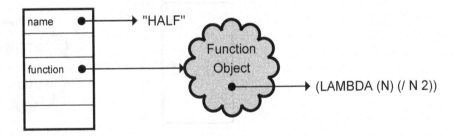

Of course, the lambda expression is just a list constructed out of cons cells. And each of the symbols in the lambda expression, such as N and /, is really a block of five pointers. Since the symbol / names the division function, it contains a pointer to a built-in function object for performing division. So, indirectly, HALF points to the built-in division function. Figure 3-1 shows these details.

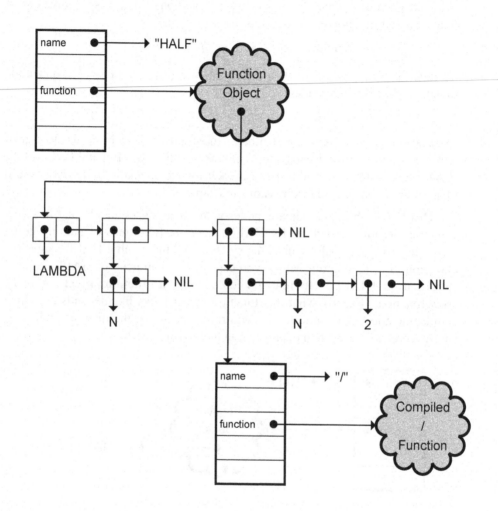

Figure 3-1 The internal representation of HALF.

EXERCISE

3.23. Write each of the following functions in Church's lambda notation: DOUBLE, SQUARE, ONEMOREP.

3.20 SCOPE OF VARIABLES

The **scope** of a variable is the region in which it may be referenced. For example, the variable N that holds the input to HALF has scope limited to the body of HALF. Another way to express this is to say that the variable N is **local** to HALF. Global variables have unbounded scope; they may be referenced anywhere.

In an evaltrace diagram, the scope of a local variable is delimited by the thick arrow containing the creation of that variable. Outside the thick arrow the variable cannot be referenced. The following program illustrates this.

```
(defun parent (n)
  (child (+ n 2)))

(defun child (p)
  (list n p))
```

This program is in error. PARENT calls CHILD after creating a local variable N. Let's see where the problem lies:

Thick arrows in evaltrace diagrams depict scope boundaries. The scope of PARENT's N is limited to the body of PARENT. Inside the body of CHILD there is a reference to N. But there is no N local to CHILD, and since the body of CHILD is surrounded by a thick arrow, we cannot refer to the N in PARENT from there. So the N appearing in the body of CHILD is interpreted as a reference to the global N, which has not been assigned a value. Hence we get an unassigned variable error.

EXERCISE

3.24. Assume we have defined the following functions:

```
(defun alpha (x)
  (bravo (+ x 2) (charlie x 1)))

(defun bravo (y z) (* y z))

(defun charlie (y x) (- y x))
```

Suppose we now evaluate (ALPHA 3). Show the resulting creation and use of variables X, Y, and Z by drawing an evaltrace diagram.

3.21 EVAL AND APPLY

EVAL is a Lisp primitive function. Each use of EVAL gives one level of evaluation.

```
'(+ 2 2) ⇒ (+ 2 2)

(eval '(+ 2 2)) ⇒ 4


'''boing ⇒ ''boing

(eval '''boing) ⇒ 'boing

(eval (eval '''boing) ⇒ boing

(eval (eval (eval '''boing))) ⇒
  Error! BOING unassigned variable.


'(list '* 9 6)) ⇒ (list '* 9 6)

(eval '(list '* 9 6)) ⇒ (* 9 6)

(eval (eval '(list '* 9 6))) ⇒ 54
```

We won't use EVAL explicitly in any of the programs we write, but we make implicit use of it all the time. You can think of the computer as a physical manifestation of EVAL. When it runs Lisp, everything you type is evaluated.

APPLY is also a Lisp primitive function. APPLY takes a function and a list of objects as input. It invokes the specified function with those objects as its inputs. The first argument to APPLY should be quoted with #' rather than an ordinary quote; #' is the proper way to quote functions supplied as inputs to other functions. This will be explained in more detail in Chapter 7.

```
(apply #'+ '(2 3)) ⇒ 5

(apply #'equal '(12 17)) ⇒ nil
```

The objects APPLY passes to the function are *not* evaluated first. In the following example, the objects are a symbol and a list. Evaluating either the symbol AS or the list (YOU LIKE IT) would cause an error.

```
(apply #'cons '(as (you like it)))
   ⇒ (as you like it)
```

EVAL and APPLY are related to each other. A popular exercise in more advanced Lisp texts involves writing each function in terms of the other.

EXERCISE

3.25. What do each of the following expressions evaluate to?

```
(list 'cons t nil)

(eval (list 'cons t nil))

(eval (eval (list 'cons t nil)))

(apply #'cons '(t nil))

(eval nil)

(list 'eval nil)

(eval (list 'eval nil))
```

FUNCTIONS COVERED IN ADVANCED TOPICS

EVAL-related function: APPLY.

EVAL (used explicitly).

Special function: QUOTE.

4

Conditionals

4.1 INTRODUCTION

Decision making is a fundamental part of computing; all nontrivial programs make decisions. In this chapter we will study some special decision-making functions, called **conditionals**, that choose their result from among a set of alternatives based on the value of one or more **predicate expressions**. (A predicate expression is an expression whose value is interpreted as either "true" or "false.")

Conditionals allow functions to vary their behavior for different sorts of inputs. Since we can construct our own predicate expressions to control these conditionals, we can write functions that make arbitrarily complex decisions.

4.2 THE IF SPECIAL FUNCTION

IF is the simplest Lisp conditional. Conditionals are always macros or special functions,[*] so their arguments do not get evaluated automatically. DEFUN and QUOTE are two other function we've studied with this property. Ordinary functions, like + and CONS, always evaluate their arguments.

---

[*]This terminology was suggested by Robert Wilensky. The distinction between "macro" functions and "special" functions is explained in Chapter 14; for now you can think of them as the same.

The IF special function takes three arguments: a **test**, a **true-part**, and a **false-part**. If the test is true, IF returns the value of the true-part. If the test is false, it skips the true-part and instead returns the value of the false-part. Here are some examples.

```
(if (oddp 1) 'odd 'even)  ⇒  odd

(if (oddp 2) 'odd 'even)  ⇒  even

(if t 'test-was-true 'test-was-false)  ⇒
   test-was-true

(if nil 'test-was-true 'test-was-false)  ⇒
   test-was-false

(if (symbolp 'foo) (* 5 5) (+ 5 5))  ⇒  25

(if (symbolp 1) (* 5 5) (+ 5 5))  ⇒  10
```

Let's use IF to construct a function that takes the absolute value of a number. Absolute values are always nonnegative. For negative numbers the absolute value is the negation of the number; for positive numbers and zero the absolute value is the number itself. This leads to a simple definition for MY-ABS, our absolute value function. (We call the function MY-ABS rather than ABS because there is already an ABS function built in to Common Lisp; we don't want to interfere with any of Lisp's built-in functions.)

```
(defun my-abs (x)
  (if (< x 0) (- x) x))
```

The test part of the IF is the expression (< X 0). If the test evaluates to true, the true-part, (- X), will be evaluated and will return the negation of X. If the test evaluates to false, meaning X is zero or positive, the false-part of the IF will be evaluated. The false-part is just X, so the input to MY-ABS will be returned unchanged in this case. Here is how you should be reading the definition of MY-ABS: "DEFUN MY-ABS of X: IF (< X 0) then minus X else X." The words "then" and "else" don't actually appear in the function, but mentally inserting them can help to clarify the function in your mind.

```
> (my-abs -5)      True-part takes the negation.
5

> (my-abs 5)       False-part returns the number unchanged.
5
```

Here's another simple decision-making function. SYMBOL-TEST returns a message telling whether or not its input is a symbol.

```
(defun symbol-test (x)
   (if (symbolp x) (list 'yes x 'is 'a 'symbol)
       (list 'no x 'is 'not 'a 'symbol)))
```

When you read this function definition to yourself, you should read the IF part as "If SYMBOLP of X then...else...."

```
> (symbol-test 'rutabaga)          Evaluate true-part.
(YES RUTABAGA IS A SYMBOL)


> (symbol-test 12345)              Evaluate false-part.
(NO 12345 IS NOT A SYMBOL)
```

IF can be given two inputs instead of three, in which case it behaves as if its third input (the false-part) were the symbol NIL.

```
(if t 'happy)   ⇒   happy

(if nil 'happy)   ⇒   nil
```

EXERCISES

4.1. Write a function MAKE-EVEN that makes an odd number even by adding one to it. If the input to MAKE-EVEN is already even, it should be returned unchanged.

4.2. Write a function FURTHER that makes a positive number larger by adding one to it, and a negative number smaller by subtracting one from it. What does your function do if given the number 0?

4.3. Recall the primitive function NOT: It returns NIL for a true input and T for a false one. Suppose Lisp didn't have a NOT primitive. Show how to write NOT using just IF and constants (no other functions). Call your function MY-NOT.

4.4. Write a function ORDERED that takes two numbers as input and makes a list of them in ascending order. (ORDERED 3 4) should return the list (3 4). (ORDERED 4 3) should also return (3 4), in other words, the first and second inputs should appear in reverse order when the first is greater than the second.

4.3 THE COND MACRO

COND is the classic Lisp conditional. Its input consists of any number of test-and-consequent **clauses**. The general form of a COND expression will be described in Chapter 5, but a slightly simplified form is:

```
(COND (test-1 consequent-1)
      (test-2 consequent-2)
      (test-3 consequent-3)
           ....
      (test-n consequent-n))
```

Here is how COND works: It goes through the clauses sequentially. If the test part of a clause evaluates to *true*, COND evaluates the consequent part and returns its value; it does not consider any more clauses. If the test evaluates to *false*, COND skips the consequent part and examines the next clause. If it goes through all the clauses without finding any whose test is true, CONS returns NIL.

Let's use COND to write a function COMPARE that compares two numbers. If the numbers are equal, COMPARE will say ''numbers are the same''; if the first number is less than the second, it will say ''first is smaller''; if the first number is greater than the second, it will say ''first is bigger.'' Each case is handled by a separate COND clause.

```
(defun compare (x y)
  (cond ((equal x y) 'numbers-are-the-same)
        ((< x y) 'first-is-smaller)
        ((> x y) 'first-is-bigger)))
```

Take a closer look at the COND. It is a four-element list, where the first element is the symbol COND and the remaining three elements are test-and-consequent clauses. The first clause is a two-element list whose first element is the expression (EQUAL X Y). This is the test part of the clause. The second element, the consequent part, is the quoted symbol 'NUMBERS-ARE-THE-SAME.

Here are some examples of the COMPARE function:

```
(compare 3 5)   ⟹   first-is-smaller

(compare 7 2)   ⟹   first-is-bigger

(compare 4 4)   ⟹   numbers-are-the-same
```

EXERCISE

4.5. For each of the following calls to COMPARE, write "1," "2," or "3" to indicate which clause of the COND will have a predicate that evaluates to true.

    _____   (compare 9 1)

    _____   (compare (+ 2 2) 5)

    _____   (compare 6 (* 2 3))

COND and IF are similar functions. COND may appear more versatile since it accepts any number of clauses, but there is a way to do the same thing with nested IFs. This is explained later in the chapter.

4.4 USING T AS A TEST

One of the standard tricks for using COND is to include a clause of form

```
(T consequent)
```

The test T is always true, so if COND ever reaches this clause, it is guaranteed to evaluate the consequent. We put this clause at the very end so that it will be reached only if all the preceding clauses' tests fail. *Example:* The following function returns the country in which a given city is. If the function doesn't know a particular city, it returns the symbol UNKNOWN.

```
(defun where-is (x)
  (cond ((equal x 'paris) 'france)
        ((equal x 'london) 'england)
        ((equal x 'beijing) 'china)
        (t 'unknown)))
```

Note that the last clause of the COND begins with T. If none of the preceding clauses have tests that return true, the last clause will be reached and the function will return UNKNOWN.

(where-is 'london) $\Rightarrow$ england

(where-is 'beijing) $\Rightarrow$ china

(where-is 'hackensack) $\Rightarrow$ unknown

Recall that the general form of an IF expression is

```
(IF test true-part false-part)
```

We can translate any IF expression into a COND expression using two clauses:

```
(COND (test true-part)
      (T false-part))
```

EXERCISE

4.6. Write a version of the absolute value function MY-ABS using COND instead of IF.

4.5 TWO MORE EXAMPLES OF COND

Here is another function, called EMPHASIZE, that changes the first word of a phrase from ''good'' to ''great,'' or from ''bad'' to ''awful,'' and returns the modified phrase:

```
(defun emphasize (x)
  (cond ((equal (first x) 'good) (cons 'great (rest x)))
        ((equal (first x) 'bad) (cons 'awful (rest x)))))
```

Let's take as an example the phrase (GOOD MYSTERY STORY). What happens inside EMPHASIZE? The variable X is assigned the value (GOOD MYSTERY STORY), and COND starts going through the test-and-consequent clauses. The first one is:

```
((equal (first x) 'good) (cons 'great (rest x)))
```

Since (FIRST X) evaluates to GOOD, the test part of this clause is true. The consequent part then constructs a new list from the symbol GREAT and the REST of the input, and that is what the function returns:

```
(emphasize '(good mystery story))
   ⇒  (great mystery story)
```

Now suppose we try to emphasize (MEDIOCRE MYSTERY STORY). The first clause compares MEDIOCRE to GOOD and returns NIL. The next compares MEDIOCRE to BAD and also returns NIL. Now COND has run out of clauses, so it returns NIL. Therefore, NIL is the result of the EMPHASIZE function:

```
(emphasize '(mediocre mystery story))  ⇒  nil
```

What if we want EMPHASIZE to return the original input instead of NIL when it can't figure out how to emphasize it? We simply use the T-as-test trick, demonstrated in the function EMPHASIZE2:

```
(defun emphasize2 (x)
  (cond ((equal (first x) 'good) (cons 'great (rest x)))
        ((equal (first x) 'bad) (cons 'awful (rest x)))
        (t x)))
```

If the COND reaches the last clause, the test T is guaranteed to evaluate to true and the input, X, is returned.

```
(emphasize2 '(good day))  ⇒  (great day)

(emphasize2 '(bad day))  ⇒  (awful day)

(emphasize2 '(long day))  ⇒  (long day)
```

Here is a function COMPUTE that takes three inputs. If the first input is the symbol SUM-OF, the function returns the sum of the second and third inputs. If it is the symbol PRODUCT-OF, the function returns the product of the second and third inputs. Otherwise it returns the list (THAT DOES NOT COMPUTE).

```
(defun compute (op x y)
  (cond ((equal op 'sum-of) (+ x y))
        ((equal op 'product-of) (* x y))
        (t '(that does not compute))))
```

Here are some examples of the COMPUTE function:

```
(compute 'sum-of 3 7)  ⇒  10

(compute 'product-of 2 4)  ⇒  8

(compute 'zorch-of 3 1)
    ⇒  (that does not compute)
```

4.6 COND AND PARENTHESIS ERRORS

Parenthesis errors can play havoc with COND expressions. Most COND clauses begin with exactly two parentheses. The first marks the beginning of the clause, and the second marks the beginning of the clause's test. For example, in the WHERE-IS function, the test part of the first clause is the expression

```
(EQUAL X 'PARIS)
```

so the clause itself looks like

```
((EQUAL X 'PARIS) . . .)
```

If the test part of a clause is just a symbol, not a call to a function, then the clause should begin with a single parenthesis. Notice that in WHERE-IS the clause with T as the test begins with only one parenthesis.

Here are two of the more common parenthesis errors made with COND. First, suppose we leave a parenthesis out of a COND clause. What would happen?

```
(cond (equal x 'paris 'france)
      (. . .)
      (. . .)
      (t 'unknown))
```

The first clause of the COND starts with only one left parenthesis instead of two. As a result, the test part of this clause is just the symbol EQUAL. When the test is evaluated, EQUAL will cause an unassigned variable error.

On the other hand, consider what happens when too many parentheses are used:

```
(cond ((. . .) 'france)
      ((. . .) 'england)
      ((. . .) 'china)
      ((t 'unknown)))
```

If X has the value HACKENSACK, we will reach the fourth COND clause. Due to the presence of an extra pair of parentheses in this clause, the test is (T 'UNKNOWN) instead of simply T. T is not a function, so this test will generate an undefined function error.

EXERCISES

4.7. For each of the following COND expressions, tell whether the parenthesization is correct or incorrect. If incorrect, explain where the error lies.

```
(cond (symbolp x) 'symbol
      (t 'not-a-symbol))
```

```
(cond ((symbolp x) 'symbol)
      (t 'not-a-symbol))
```

```
(cond ((symbolp x) ('symbol))
      (t 'not-a-symbol))

(cond ((symbolp x) 'symbol)
      ((t 'not-a-symbol)))
```

4.8. Write EMPHASIZE3, which is like EMPHASIZE2 but adds the symbol VERY onto the list if it doesn't know how to emphasize it. For example, EMPHASIZE3 of (LONG DAY) should produce (VERY LONG DAY). What does EMPHASIZE3 of (VERY LONG DAY) produce?

4.9. Type in the following suspicious function definition:

```
(defun make-odd (x)
  (cond (t x)
        ((not (oddp x)) (+ x 1))))
```

What is wrong with this function? Try out the function on the numbers 3, 4, and -2. Rewrite it so it works correctly.

4.10. Write a function CONSTRAIN that takes three inputs called X, MAX, and MIN. If X is less than MIN, it should return MIN; if X is greater than MAX, it should return MAX. Otherwise, since X is between MIN and MAX, it should return X. (CONSTRAIN 3 -50 50) should return 3. (CONSTRAIN 92 -50 50) should return 50. Write one version using COND and another using nested IFs.

4.11. Write a function FIRSTZERO that takes a list of three numbers as input and returns a word (one of "first," "second," "third," or "none") indicating where the first zero appears in the list. Example: (FIRSTZERO '(3 0 4)) should return SECOND. What happens if you try to call FIRSTZERO with three separate numbers instead of a list of three numbers, as in (FIRSTZERO 3 0 4)?

4.12. Write a function CYCLE that cyclically counts from 1 to 99. CYCLE called with an input of 1 should return 2, with an input of 2 should return 3, with an input of 3 should return 4, and so on. With an input of 99, CYCLE should return 1. That's the cyclical part. Do not try to solve this with 99 COND clauses!

4.13. Write a function HOWCOMPUTE that is the inverse of the COMPUTE function described previously. HOWCOMPUTE takes three numbers as input and figures out what operation would produce the third from the first two. (HOWCOMPUTE 3 4 7) should return SUM-OF.

(HOWCOMPUTE 3 4 12) should return PRODUCT-OF. HOWCOMPUTE should return the list (BEATS ME) if it can't find a relationship between the first two inputs and the third. Suggest some ways to extend HOWCOMPUTE.

4.7 THE AND AND OR MACROS

We will often need to construct complex predicates from simple ones. The AND and OR macros make this possible. Before giving the precise rules for evaluating AND and OR, let's just look at an example. Suppose we want a predicate for small (no more than two digit) positive odd numbers. We can use AND to express this conjunction of simple conditions:

```
(defun small-positive-oddp (x)
  (and (< x 100)
       (> x 0)
       (oddp x)))
```

Or suppose we want a function GTEST that takes two numbers as input and returns T if either the first is greater than the second or one of them is zero. These conditions form a disjunctive set; only one need be true for GTEST to return T. OR is used for disjunctions.

```
(defun gtest (x y)
  (or (> x y)
      (zerop x)
      (zerop y)))
```

Like COND, AND and OR are macros: they can accept any number of clauses, and they do not evaluate their arguments first. For AND and OR, however, the clauses are simply tests, not test-and-consequent pairs.

4.8 EVALUATING AND AND OR

AND and OR have slightly different meanings in Lisp than they do in logic or in English. The precise rule for evaluating AND is: Evaluate the clauses one at a time. If a clause returns NIL, stop and return NIL; otherwise go on to the next one. If all the clauses yield non-NIL results, return the value of the last clause. Examples:

```
(and nil t t)   ⇒   nil
```

```
(and 'george nil 'harry)  ⇒  nil
```

```
(and 'george 'fred 'harry)  ⇒  harry
```

```
(and 1 2 3 4 5)  ⇒  5
```

The rule for evaluating OR is: Evaluate the clauses one at a time. If a clause returns something other than NIL, stop and return that value; otherwise go on to the next clause, or return NIL if none are left.

```
(or nil t t)  ⇒  t
```

```
(or 'george nil 'harry)  ⇒  george
```

```
(or 'george 'fred 'harry)  ⇒  george
```

```
(or nil 'fred 'harry)  ⇒  fred
```

EXERCISE

4.14. What results do the following expressions produce? Read the evaluation rules for AND and OR carefully before answering.

```
(and 'fee 'fie 'foe)
```

```
(or 'fee 'fie 'foe)
```

```
(or nil 'foe nil)
```

```
(and 'fee 'fie nil)
```

```
(and (equal 'abc 'abc) 'yes)
```

```
(or (equal 'abc 'abc) 'yes)
```

4.9 BUILDING COMPLEX PREDICATES

The HOW-ALIKE function compares two numbers several different ways to see in what way they are similar. It uses AND to construct complex predicates as part of a COND clause:

```
(defun how-alike (a b)
  (cond ((equal a b) 'the-same)
        ((and (oddp a) (oddp b)) 'both-odd)
        ((and (not (oddp a)) (not (oddp b)))
         'both-even)
        ((and (< a 0) (< b 0)) 'both-negative)
        (t 'not-alike)))
```

(how-alike 7 7)   ⇒   the-same

(how-alike 3 5)   ⇒   both-odd

(how-alike -2 -3)   ⇒   both-negative

(how-alike 5 8)   ⇒   not-alike

The SAME-SIGN predicate uses a combination of AND and OR to test if its two inputs have the same sign:

```
(defun same-sign (x y)
  (or (and (zerop x) (zerop y))
      (and (< x 0) (< y 0))
      (and (> x 0) (> y 0))))
```

SAME-SIGN returns T if any of the inputs to OR returns T. Each of these inputs is an AND expression. The first one tests whether X is zero and Y is zero, the second tests whether X is negative and Y is negative, and the third tests whether X is positive and Y is positive. Examples:

(same-sign 0 0)   ⇒   t

(same-sign -3 -4)   ⇒   t

(same-sign 3 4)   ⇒   t

(same-sign -3 4)   ⇒   nil

EXERCISES

4.15. Write a predicate called GEQ that returns T if its first input is greater than or equal to its second input.

4.16. Write a function that squares a number if it is odd and positive, doubles it if it is odd and negative, and otherwise divides the number by 2.

4.17. Write a predicate that returns T if the first input is either BOY or GIRL and the second input is CHILD, or the first input is either MAN or WOMAN and the second input is ADULT.

4.18. Write a function to act as referee in the Rock-Scissors-Paper game. In this game, each player picks one of Rock, Scissors, or Paper, and then both players tell what they picked. Rock "breaks" Scissors, so if the first player picks Rock and the second picks Scissors, the first player wins. Scissors "cuts" Paper, and Paper "covers" Rock. If both players pick the same thing, it's a tie. The function PLAY should take two inputs, each of which is either ROCK, SCISSORS, or PAPER, and return one of the symbols FIRST-WINS, SECOND-WINS, or TIE. Examples: (PLAY 'ROCK 'SCISSORS) should return FIRST-WINS. (PLAY 'PAPER 'SCISSORS) should return SECOND-WINS.

4.10 WHY AND AND OR ARE CONDITIONALS

Why are AND and OR classed as conditionals instead of regular functions? The reason is that they are not required to evaluate every clause. If any clause of an AND returns NIL, or any clause of an OR returns non-NIL, none of the succeeding clauses get evaluated. This property can be valuable, because we may need to halt evaluation to avoid errors that would otherwise occur. For example, consider the POSNUMP predicate:

```
(defun posnump (x)
  (and (numberp x) (plusp x)))
```

POSNUMP returns T if its input is a number and is positive. The built-in PLUSP predicate can be used to tell if a number is positive, but if PLUSP is used on something other than a number, it signals a "wrong type input" error, so it is important to make sure that the input to POSNUMP is a number *before* invoking PLUSP. If the input isn't a number, we must not call PLUSP.

Here is an incorrect version of POSNUMP:

```
(defun faulty-posnump (x)
  (and (plusp x) (numberp x)))
```

If FAULTY-POSNUMP is called on the symbol FRED instead of a number, the first thing it does is check if FRED is greater than 0, which causes a wrong type input error. However, if the regular POSNUMP function is called with input FRED, the NUMBERP predicate returns NIL, so AND returns NIL *without ever calling* PLUSP.

4.11 CONDITIONALS ARE INTERCHANGEABLE

Functions that use AND and OR can also be implemented using COND or IF, and vice versa. Recall the definition of POSNUMP:

```
(defun posnump (x)
  (and (numberp x) (> x 0)))
```

Here is a version of POSNUMP written with IF instead of AND:

```
(defun posnump-2 (x)
  (if (numberp x) (> x 0) nil))
```

This version of POSNUMP tests for a number first, and if the condition succeeds, the true-part of the IF evaluates (> X 0). If the number test fails, the false-part of the IF is NIL. Trace the evaluation of the function on paper with inputs like FRED, 7, and -2 to better understand how it works. Here is another version of POSNUMP, this time using COND:

```
(defun posnump-3 (x)
  (cond ((numberp x) (> x 0))
        (t nil)))
```

Let's look at another use of conditionals. This is the original version of WHERE-IS, using COND:

```
(defun where-is (x)
  (cond ((equal x 'paris) 'france)
        ((equal x 'london) 'england)
        ((equal x 'beijing) 'china)
        (t 'unknown)))
```

This COND has four clauses. We can write WHERE-IS using IF instead of COND by putting three IFs together. Such a construct is called a **nested if**.

```
(defun where-is-2 (x)
  (if (equal x 'paris) 'france
      (if (equal x 'london) 'england
          (if (equal x 'beijing) 'china
              'unknown))))
```

Suppose we call WHERE-IS-2 with the input BEIJING. As the evaltrace shows, the local variable X is assigned the value BEIJING, and the body is evaluated. The body of WHERE-IS-2 is a single IF whose test checks if X is equal to PARIS. It is not, so the IF evaluates its false-part. The false-part is also an IF, and this IF's test checks whether X is equal to LONDON. It is not, so the IF evaluates its own false-part—yet another IF. This third IF tests if X is equal to BEIJING, which it is, so its true part evaluates to CHINA. The third IF returns CHINA, which is now the value of the false-part of the second IF so it returns CHINA, which is now the value of the false-part of the first IF so it returns CHINA as well. The result of (WHERE-IS-2 'BEIJING) is CHINA.

```
┌→ (where-is-2 'beijing)
│ ┌→ Enter WHERE-IS-2 with input BEIJING
│ │   create variable X, with value BEIJING
│ │   ┌→ (if (equal x 'paris) ...)
│ │   │   ┌→ (equal x 'paris)
│ │   │   └→ NIL
│ │   │   ┌→ (if (equal x 'london) ...)
│ │   │   │   ┌→ (equal x 'london)
│ │   │   │   └→ NIL
│ │   │   │   ┌→ (if (equal x 'beijing) 'china 'unknown)
│ │   │   │   │   ┌→ (equal x 'beijing)
│ │   │   │   │   └→ T
│ │   │   │   │   ┌→ 'china
│ │   │   │   │   └→ CHINA
│ │   │   │   └→ CHINA
│ │   │   └→ CHINA
│ │   └→ CHINA
│ └→ Result of WHERE-IS-2 is CHINA
```

We can write another version of WHERE-IS using AND and OR. This version employs a simple two-level scheme rather than the more complex nesting required for IF.

```
(defun where-is-3 (x)
  (or (and (equal x 'paris) 'france)
      (and (equal x 'london) 'england)
      (and (equal x 'beijing) 'china)
      'unknown)))
```

Let's evaluate (WHERE-IS-3 'LONDON). X is bound to LONDON, and OR starts going through its clauses looking for one that isn't NIL. The first clause is an AND expression; AND evaluates (EQUAL X 'PARIS) and gets a NIL result, so AND gives up and returns NIL. OR moves on to its second clause. This is also an AND expression; (EQUAL X 'LONDON) returns T, so

AND moves on to its next clause. 'ENGLAND evaluates to ENGLAND; AND has run out of clauses, so it returns the value of the last one. Since OR has found a non-NIL clause, OR now returns ENGLAND.

Since IF, COND, and AND/OR are interchangeable conditionals, you may wonder why Lisp has more than one. It's a matter of convenience. IF is the easiest to use for simple functions like absolute value. AND and OR are good for writing complex predicates like SMALL-POSITIVE-ODDP. COND is easiest to use when there are many tests, as in WHERE-IS and HOW-ALIKE. Choosing the right conditional for the job is part of the art of programming.

EXERCISES

4.19. Show how to write the expression (AND X Y Z W) using COND instead of AND. Then show how to write it using nested IFs instead of AND.

4.20. Write a version of the COMPARE function using IF instead of COND. Also write a version using AND and OR.

4.21. Write versions of the GTEST function using IF and COND.

4.22. Use COND to write a predicate BOILINGP that takes two inputs, TEMP and SCALE, and returns T if the temperature is above the boiling point of water on the specified scale. If the scale is FAHRENHEIT, the boiling point is 212 degrees; if CELSIUS, the boiling point is 100 degrees. Also write versions using IF and AND/OR instead of COND.

4.23. The WHERE-IS function has four COND clauses, so WHERE-IS-2 needs three nested IFs. Suppose WHERE-IS had eight COND clauses. How many IFs would WHERE-IS-2 need? How many ORs would WHERE-IS-3 need? How many ANDs would it need?

SUMMARY

Conditionals allow the computer to make decisions that control its behavior. IF is a simple conditional; its syntax is (IF condition true-part false-part). COND, the most general conditional, takes a set of test-and-consequent clauses as input and evaluates the tests one at a time until it finds a true one. It then returns the value of the consequent of that clause. If none of the tests are true, COND returns NIL.

AND and OR are also conditionals. AND evaluates clauses one at a time until one of them returns NIL, which AND then returns. If all the clauses evaluate to true, AND returns the value of the last one. OR evaluates clauses until a non-NIL value is found, and returns that value. If all the clauses evaluate to NIL, OR returns NIL. AND and OR aren't considered predicates because they're not ordinary functions.

A useful programming trick when writing COND expressions is to place a list of form (T consequent) as the final clause of the COND. Since the test T is always true, the clause serves as a kind of catchall case that will be evaluated when the tests of all the preceding clauses are false.

An important feature of conditionals is their ability to not evaluate all of their inputs. This lets us prevent errors by protecting a sensitive expression with predicate expressions that can cause evaluation to stop. Conditionals can do this because they are either macros or special functions, not ordinary functions.

REVIEW EXERCISES

4.24. Why are conditionals important?

4.25. What does IF do if given two inputs instead of three?

4.26. COND can accept any number of clauses, but IF takes at most three inputs. How is it then that any function involving COND can be rewritten to use IF instead?

4.27. What does COND return if given *no* clauses, in other words, what does (COND) evaluate to?

4.28. We can usually rewrite an IF as a combination of AND plus OR by following this simple scheme: Replace (IF *test true-part false-part*) with the equivalent expression (OR (AND *test true-part*) *false-part*). But this scheme fails for the expression (IF (ODDP 5) (EVENP 7) 'FOO). Why does it fail? Suggest a more sophisticated way to rewrite IF as a combination of ANDs and ORs that does not fail.

FUNCTIONS COVERED IN THIS CHAPTER

Conditionals: IF, COND, AND, OR.

Predicate: PLUSP.

Lisp Toolkit: STEP

STEP is a tool that lets you interactively step through the evaluation of a Lisp expression so you can see everything that takes place. It is mostly used for **debugging** (finding and eliminating errors in programs),** but it can also be useful for learning about new special functions like conditionals.

Each implementation of Common Lisp provides its own version of this tool; only the name has been standardized. Most steppers accept one-letter commands telling them what to do at each iteration, such as continue stepping, proceed with the evaluation without stepping, enter the debugger, and so forth. Steppers are supposed to respond to a "?" by printing a list of commands they understand. In this book we will use just one command, "n," to go to the next step of the evaluation.

Because STEP is a macro, its input should not be quoted. Here is an example of the use of STEP.

```
> (step (if (oddp 5) 'yes 'no))
  (IF (ODDP 5) 'YES 'NO) : n      Stepping through the IF...
    (ODDP 5) : n                       The test is (ODDP 5).
      5 = 5                              5 evaluates to itself.
    T                                  ODDP returns T.
    'YES = YES                         The true-part is 'YES.
  YES                                  The IF returns YES.
YES
```

Here is a more detailed example using MY-ABS, our own version of the absolute value function. The BLOCK special function that shows up in the step output can be ignored. Some Lisp implementations put a BLOCK form around the body of every function definition; in other implementations this form is implicit and does not show up in the stepper.

**The term "debugging" arose from an incident in the early days of computing, when computers were built from electromechanical switches called relays. Erroneous behavior in one machine was found to be due to a moth having gotten stuck in one of the relays, preventing it from making a good electrical connection. Removal of the "bug" fixed the problem.

```
> (defun my-abs (x)
    (if (< x 0) (- x) x))
MY-ABS

> (step (my-abs -5))
  (MY-ABS -5) : n                        Call MY-ABS with input -5.
   -5 = -5
   (BLOCK MY-ABS (IF (< X 0) (- X) X)) : n
    (IF (< X 0) (- X) X) : n             Stepping through the IF.
     (< X 0) : n                         The test is (< X 0).
      X = -5                             X is -5.
      0 = 0                              0 evaluates to itself.
      T                                  The < pred. returns T.
     (- X) : n                           The true-part is (- X).
      X = -5                             X is -5.
      5                                  The - function returns 5.
     5                                   The IF returns 5.
    5                                    The BLOCK returns 5.
   5                                     MY-ABS returns 5.
  5
```

The output of STEP is similar to an evaltrace diagram, without the arrows. Here is an evaltrace diagram of (MY-ABS -5) for comparison.

4 Advanced Topics

4.12 BOOLEAN FUNCTIONS

Boolean functions are functions whose inputs and outputs are truth values, meaning T or NIL. We have already encountered boolean functions under the name **truth functions** in previous chapters. The term "boolean" comes from George Boole, a nineteenth century English mathematician. Boolean logic is used today to describe the behavior of most computer circuits.

Yet another name for boolean functions is **logical functions**, since they use the logical values *true* and *false*. Let's define a two-input LOGICAL-AND function:

```
(defun logical-and (x y) (and x y t))
```

This ordinary function differs from the AND macro in several respects. First, as already noted, it must be given exactly two inputs. This is a minor point because we can always nest or cascade several of them to handle more inputs. Second, LOGICAL-AND returns only the logical values T or NIL, nothing else.

```
(logical-and 'tweet 'woof)  ⇒  t
```

```
(and 'tweet 'woof)  ⇒  woof
```

Most important of all is the fact that LOGICAL-AND is not a macro: It cannot control whether or not its arguments get evaluated. In the following example, the expression (ODDP 'FRED) causes an error for LOGICAL-AND but not for AND, because AND never evaluates the second clause.

```
(and (numberp 'fred) (oddp 'fred))  ⇒ nil
```

```
(logical-and (numberp 'fred) (oddp 'fred))
    ⇒   Error! FRED wrong type input to ODDP.
```

Boolean functions are simpler than conditionals. Boolean functions in Lisp correspond to boolean circuits in electronics: They are the primitive logical operations from which computer circuitry is built.

EXERCISES

4.29. Write versions of LOGICAL-AND using IF and COND instead of AND.

4.30. Write LOGICAL-OR. Make sure it returns only T or NIL for its result.

4.31. Is NOT a conditional? Is it a boolean function? Do you need to write a LOGICAL-NOT function?

4.13 TRUTH TABLES

Truth tables are a convenient way of describing boolean functions. To describe a function with a truth table, we simply consider in turn every possible combination of T and NIL as inputs, and write down the result the function should produce. Here is the truth table for NOT:

x	(NOT x)
T	NIL
NIL	T

Here is the truth table for LOGICAL-AND. Since this function takes two inputs, each of which has two possible values, the table has $2^2 = 4$ lines.

x	y	(LOGICAL-AND x y)
T	T	T
T	NIL	NIL
NIL	T	NIL
NIL	NIL	NIL

EXERCISES

4.32. Construct a truth table for LOGICAL-OR.

4.33. Imagine a LOGICAL-IF function that works like IF does, except it always takes exactly three inputs, and its outputs are limited to T or NIL. How many lines are in its truth table?

4.34. Write down the truth table for LOGICAL-IF.

4.14 DEMORGAN'S THEOREM

DeMorgan's Theorem concerns the interchangeability of AND and OR. If you have one of these functions plus NOT you can always build the other. Here is DeMorgan's Theorem stated two different ways:

```
(and x y)  =  (not (or (not x) (not y)))
```

```
(or x y)  =  (not (and (not x) (not y)))
```

These equations look pretty tricky, so let me also state them in English. The first equation says that if X and Y are true, then neither is X false nor is Y false. The second equation says that if either X or Y is true, then X and Y can't both be false. The English versions sound obvious, but do you believe the equations? Let's test them out.

```
(defun demorgan-and (x y)
  (not (or (not x) (not y))))
```

```
(defun demorgan-or (x y)
  (not (and (not x) (not y))))
```

```
(logical-and t t)  ⇒  t
```

```
(demorgan-and t t)  ⇒  t
```

```
(logical-and t nil)  ⇒  nil
```

```
(demorgan-and t nil)  ⇒  nil
```

```
(logical-or t nil)  ⇒  t
```

```
(demorgan-or t nil)  ⇒  t
```

```
(logical-or nil nil)  ⇒  nil
```

```
(demorgan-or nil nil)  ⇒  nil
```

That was not a complete test of the equations; you are welcome to test out the remaining cases yourself.

DeMorgan's Theorem proved the interchangeability of the *logical* AND and OR functions. Does it hold for Lisp's *conditional* AND and OR functions as well? Not exactly. The use of double NOTs means that arbitrary true inputs like FOO will be changed to the canonical true value T on output, so in this sense DeMorgan's Theorem doesn't hold.

```
(and 'foo 'bar)  ⇒  bar

(not (or (not 'foo) (not 'bar)))  ⇒  t
```

However, DeMorgan's Theorem does preserve the conditional property of AND and OR. That is, clauses that (AND X Y) would evaluate would also be evaluated by (NOT (OR (NOT X) (NOT Y))), and clauses that AND would not evaluate would not be evaluated by the other expression. Example:

```
(and (numberp 'fred) (plusp 'fred))  ⇒  nil

(not (or (not (numberp 'fred))
         (not (plusp 'fred))))       ⇒  nil
```

DeMorgan's Theorem is especial ly useful for simplifying expressions involving complex combinations of predicates. Consider this function:

```
(defun complicated-predicate (x y)
  (not (and (evenp x) (evenp y))))
```

The body can be converted to an OR by writing:

```
(or (not (evenp x)) (not (evenp y)))
```

Since EVENP is the opposite of ODDP, we derive:

```
(defun simplified-predicate (x y)
  (or (oddp x) (oddp y)))
```

EXERCISES

4.35. Write down the DeMorgan equations for the three-input versions of AND and OR.

4.36. The NAND function (NAND is short for Not AND) is very commonly found in computer circuitry. Here is a definition of NAND. Write down its truth table.

```
(defun nand (x y) (not (and x y)))
```

4.37. NAND is called a **logically complete** function because we can construct all other boolean functions from various combinations of NAND. For example, here is a version of NOT called NOT2 contructed from NAND:

```
(defun not2 (x) (nand x x))
```

Construct versions of LOGICAL-AND and LOGICAL-OR by putting together NANDs. You will have to use more than one NAND in each case.

4.38. Consider the NOR function (short for Not OR). Can you write versions of NOT, LOGICAL-AND, NAND, and LOGICAL-OR by putting NORs together?

4.39. Is LOGICAL-AND logically complete the way NAND and NOR are?

5

Variables and Side Effects

5.1 INTRODUCTION

This chapter will give you a better understanding of the different kinds of variables that may appear in Lisp programs, how variables are created, and how their values may change over time. Common Lisp is more sophisticated in this regard than earlier Lisp dialects. We will also talk about **side effects**, which are actions a function takes other than returning a value. Changing the value of a variable is one kind of side effect.

5.2 LOCAL AND GLOBAL VARIABLES

Every variable has a **scope**, which is the region in which it can be referenced. So far the only variables we've seen are the ones that appear in a function's argument list. Since their scope is restricted to the body of the function, they are called **local variables**. Consider this example:

```
(defun double (n) (* n 2))
```

Every time we call the DOUBLE function, a new local variable named N is created. Inside the body of DOUBLE, the name N refers to that variable. Outside of DOUBLE, we cannot refer to the variable at all because we are outside its scope. In other words, the name N has a different meaning outside of DOUBLE than inside.

```
(defun double (n) (* n 2))

(double 5)  ⇒  10
```

n   ⇒   *Error! N unassigned variable.*

The unassigned variable N referred to in the error message above is not the local N created by DOUBLE. It is another variable, one that is not local to any specific function. For this reason it is known as a **global variable**. Because the global variable N initially has no value (is "unbound," in older terminology), we get an unassigned variable error when we type N at the top-level read-eval-print loop. If we look at the evaltrace of (DOUBLE 5), the distinction between the two meanings of N becomes apparent:

the global variable N has no value

```
 ┌─► (double 5)
 │   Enter DOUBLE with input 5
 │     create (local) variable N, with value 5
 │      ┌─► (* n 2)
 │      │      N evaluates to 5
 │      └─► 10
 └─► Result of DOUBLE is 10
```

the global variable N still has no value

There can be only one global variable named N, but there can be many local variables with this name because each resides in a different lexical context.

5.3 SETF ASSIGNS A VALUE TO A VARIABLE

The SETF macro function assigns a value to a variable. If the variable already has a value, the new value replaces the old one. Here is an example of SETF assigning a value to a global variable, and later changing its value.

```
> vowels                        VOWELS initially has no value.
Error: VOWELS unassigned variable.

> (setf vowels '(a e i o u))    SETF gives VOWELS a
(A E I O U)                     value.
```

```
> (length vowels)
5
```
*Now we can use VOWELS
in Lisp expressions.*

```
> (rest vowels)
(E I O U)
```

```
> vowels
(A E I O U)
```
Its value is unchanged.

```
> (setf vowels
    '(a e i o u and sometimes y))
(A E I O U AND SOMETIMES Y)
```
*Give VOWELS a new
value.*

```
> (rest (rest vowels))
(I O U AND SOMETIMES Y)
```
Use the new value.

The first argument to SETF is the name of a variable; SETF does not evaluate this argument. (It can do this because it is a macro function.) The second argument is the value to which the variable is set; this argument is evaluated. The value returned by SETF is the value to which it set the variable. Note: some Common Lisp implementations complain if you define a global variable using SETF without first "declaring" it. If your implementation gives a warning, you can solve the problem using DEFVAR to properly declare the variable first, like this:

```
> (defvar trowels)
TROWELS
```

```
> (setf trowels '(gardening pointing finishing))
(GARDENING POINTING FINISHING)
```

Global variables are useful for holding on to values so we don't have to continually retype them. Example:

```
> (setf long-list '(a b c d e f g h i))
(A B C D E F G H I)
```

```
> (setf head (first long-list))
A
```

```
> (setf tail (rest long-list))
(B C D E F G H I)
```

```
> (cons head tail)
(A B C D E F G H I)
```

```
> (equal long-list (cons head tail))
T

> (list head tail)
(A (B C D E F G H I))
```

HEAD, TAIL, and LONG-LIST are all global variables.

5.4 SIDE EFFECTS

Ordinary functions like CAR and + are useful only because of the values they
return.   Other functions are useful primarily because of their side effects.
SETF's side effect is that it changes the value of a variable. This side effect is
much more important than the value SETF returns. DEFUN is also called
purely for its side effect: It defines a new function. The value returned by
DEFUN is the name of the function it defined.

Another function with a side effect is RANDOM, Common Lisp's random
number generator. (RANDOM *n*) returns a number chosen at random, from
zero up to (but not including) *n*. If *n* is an integer, RANDOM returns an
integer; if it is a floating point number, RANDOM returns a floating point
number.

```
> (list (random 5)(random 5))
(3 1)

> (list (random 5.0)(random 5.0))
(2.32459 4.94179)
```

RANDOM's side effect is hidden from the user. It changes the values of
some variables inside the random number generator, allowing it to produce a
different random number each time it is called.

The SETF function can change the value of any variable, local or global.
In this book we will use SETF only on global variables, because it is good
programming style to avoid changing the values of local variables. But just to
show that it can be done, here is an example where a function changes the
value of a local variable, P. Notice that this function has two **forms**
(expressions) in its body. When a function body contains more than one form,
it evaluates all of them and returns the value of the last one.

```
(defun poor-style (p)
  (setf p (+ p 5))
  (list 'result 'is p))
```

```
> (poor-style 8)
(RESULT IS 13)

> p
Error! P unassigned variable
```

Inside POOR-STYLE the symbol P refers to a local variable, so SETF changes the value of this local variable. The global variable P is unaffected by the SETF. In evaltrace notation, the assignment is shown as a side effect of the SETF form nested within the body of POOR-STYLE. You can also see that the result of this form is not returned by POOR-STYLE, because it is not the last form in the body.

```
┌─→ (poor-style 8)
├─→ Enter POOR-STYLE with input 8
│      create variable P, with value 8
│      ┌─→ (setf p (+ p 5))
│      │      ┌─→ (+ p 5)
│      │      │       P evaluates to 8
│      │      └─→ 13
│      │         set P to 13
│      └─→ 13
│      ┌─→ (list 'result 'is p)
│      │       P evaluates to 13
│      └─→ (RESULT IS 13)
└─→ Result of POOR-STYLE is (RESULT IS 13)
```

5.5 THE LET SPECIAL FUNCTION

So far, the only local variables we've seen have been those created by calling user-defined functions, such as DOUBLE or AVERAGE. Another way to create a local variable is with the LET special function. For example, since the average of two numbers is half their sum, we might want to use a local variable called SUM inside our AVERAGE function. We can use LET to create this local variable and give it the desired initial value. Then, in the body of the LET form, we can compute the average.

```
(defun average (x y)
  (let ((sum (+ x y)))
    (list x y 'average 'is (/ sum 2.0))))
```

```
> (average 3 7)
(3 7 AVERAGE IS 5.0)
```

The right way to read a LET form such as

```
(let ((x 2)
      (y 'aardvark))
  (list x y))
```

is to say "Let X be 2, and Y be AARDVARK; return (LIST X Y)." The general syntax of LET is:

```
(LET ((var-1 value-1)
      (var-2 value-2)
      ...
      (var-n value-n))
  body)
```

The first argument to LET is a list of variable-value pairs. The n value forms are evaluated, then n local variables are created to hold the results, finally the forms in the body of the LET are evaluated. Here is an evaltrace of the call to AVERAGE.

```
┌→ (average 3 7)
├→ Enter AVERAGE with inputs 3 and 7
│     create variable X, with value 3
│     create variable Y, with value 7
│  ┌→ (let ...)
│  │  ┌→ (+ x y)
│  │  └→ 10
│  ┌→ Enter LET body
│  │     create variable SUM, with value 10
│  │  ┌→ (list x y 'average 'is (/ sum 2))
│  │  │     X evaluates to 3
│  │  │     Y evaluates to 7
│  │  │  ┌→ (/ sum 2.0)
│  │  │  │     SUM evaluates to 10
│  │  │  └→ 5.0
│  │  └→ (3 7 AVERAGE IS 5.0)
│  └→ Result of LET is (3 7 AVERAGE IS 5.0)
└→ Result of AVERAGE is (3 7 AVERAGE IS 5.0)
```

Let's focus on what goes on inside the body of the LET. The inner thick arrow with the hollow shaft, which marks the LET body in the evaltrace diagram, indicates that the LET creates its own lexical context *within* the lexical context of AVERAGE. When evaluating the LET body, EVAL can see through the hollow shaft to the local variables X and Y that AVERAGE created. If the LET's arrow had been solid like AVERAGE's instead of hollow it would be a scoping boundary: EVAL would not be able to see through it when searching for variables. In that case, when evaluating the expression (LIST X Y *etc.*) in the LET body, EVAL would hit the boundary and immediately jump to the global lexical context to look for a *global* variable named X or Y. This would obviously not produce the intended result; it would probably cause an unassigned variable error.

Here is an example of using LET to create two local variables at once.

```
(defun switch-billing (x)
  (let ((star (first x))
        (co-star (third x)))
    (list co-star 'accompanied 'by star)))
```

```
> (switch-billing '(fred and ginger))
(GINGER ACCOMPANIED BY FRED)
```

Here is an evaltrace showing exactly how LET creates the local variables STAR and CO-STAR. Note that the two value forms, (FIRST X) and (THIRD X), are both evaluated before any local variables are created.

```
→ (switch-billing '(fred and ginger))
→ Enter SWITCH-BILLING with input (FRED AND GINGER)
    create variable X, with value (FRED AND GINGER)
    → (let ...)
        → (first x)
        → FRED
        → (third x)
        → GINGER
    → Enter LET body
        create variable STAR, with value FRED
        create variable CO-STAR, with value GINGER
        → (list co-star 'accompanied 'by star)
        → (GINGER ACCOMPANIED BY FRED)
    → Result of LET is (GINGER ACCOMPANIED BY FRED)
→ Result of SWITCH-BILLING is (GINGER ACCOMPANIED BY FRED)
```

EXERCISE

5.1. Rewrite function POOR-STYLE to create a new local variable Q using LET, instead of using SETF to change P. Call your new function GOOD-STYLE.

5.6 THE LET* SPECIAL FUNCTION

The LET* special function is similar to LET, except it creates the local variables one at a time instead of all at once. Therefore, the first local variable forms part of the lexical context in which the value of the second variable is computed, and so on. This way of creating local variables is useful when one wants to assign names to several intermediate steps in a long computation.

For example, suppose we want a function that computes the percent change in the price of widgets given the old and new prices as input. Our function must compute the difference between the two prices, then divide this difference by the old price to get the proportional change in price, and then multiply that by 100 to get the percent change. We can use local variables named DIFF, PROPORTION, and PERCENTAGE to hold these values. We use LET* instead of LET because these variables must be created one at a time, since each depends on its predecessor.

```
(defun price-change (old new)
  (let* ((diff (- new old))
         (proportion (/ diff old))
         (percentage (* proportion 100.0)))
    (list 'widgets 'changed 'by percentage
          'percent)))

> (price-change 1.25 1.35)
(WIDGETS CHANGED BY 8.0 PERCENT)
```

An evaltrace of PRICE-CHANGE shows how LET* creates its local variables. Notice that the expression (- NEW OLD) occurs in the lexical context containing just the local variables NEW and OLD. The expression (/ DIFF OLD) occurs in a nested lexical context in which the local variable DIFF is also defined. And the expression (* PROPORTION 100.0) occurs in a more deeply nested context, containing OLD, NEW, DIFF, and PROPORTION. The body of the LET* form is evaluated in a context containing all these variables plus PERCENTAGE.

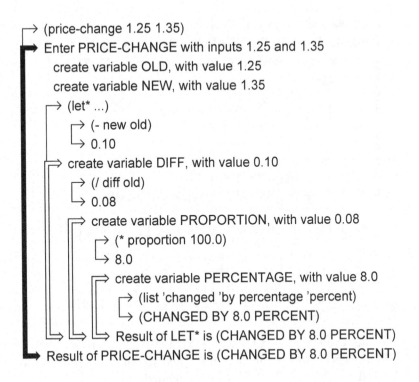

```
→ (price-change 1.25 1.35)
→ Enter PRICE-CHANGE with inputs 1.25 and 1.35
     create variable OLD, with value 1.25
     create variable NEW, with value 1.35
   → (let* ...)
      → (- new old)
      → 0.10
   → create variable DIFF, with value 0.10
      → (/ diff old)
      → 0.08
      → create variable PROPORTION, with value 0.08
         → (* proportion 100.0)
         → 8.0
      → create variable PERCENTAGE, with value 8.0
         → (list 'changed 'by percentage 'percent)
         → (CHANGED BY 8.0 PERCENT)
      → Result of LET* is (CHANGED BY 8.0 PERCENT)
→ Result of PRICE-CHANGE is (CHANGED BY 8.0 PERCENT)
```

A common programming error is to use LET when LET* is required. Consider the following FAULTY-SIZE-RANGE function. It uses MAX and MIN to find the largest and smallest of a group of numbers. MAX and MIN are built in to Common Lisp; they both accept one or more inputs. The extra 1.0 argument to / is used to force the result to be a floating point number rather than a ratio.

```
(defun faulty-size-range (x y z)
  (let ((biggest (max x y z))
        (smallest (min x y z))
        (r (/ biggest smallest 1.0)))
    (list 'factor 'of r)))

> (faulty-size-range 35 87 4)
Error in function SIZE-RANGE:
  BIGGEST unassigned variable.
```

The problem is that the expression (/ BIGGEST SMALLEST 1.0) is being evaluated in a lexical context that does not include these variables. Therefore the symbol BIGGEST is interpreted as a reference to a global variable by that name. This is readily apparent in an evaltrace.

```
→ (faulty-size-range 35 87 4)
→ Enter FAULTY-SIZE-RANGE with inputs 35, 87, and 4
    create variables X, Y, and Z, with values 35, 87, and 4
    → (let ...)
        → (max x y z)
        → 87
        → (min x y z)
        → 4
        → (/ biggest smallest 1.0)
        → Error! BIGGEST unassigned variable.
```

The problem is solved by replacing the LET with a LET*:

```
(defun correct-size-range (x y z)
  (let* ((biggest (max x y z))
         (smallest (min x y z))
         (r (/ biggest smallest 1.0)))
    (list 'factor 'of r)))
```

The evaltrace of CORRECT-SIZE-RANGE shows that (/ BIGGEST SMALLEST 1.0) is evaluated in the lexical context containing local variables BIGGEST and SMALLEST, as we intended.

```
→ (correct-size-range 35 87 4)
→ Enter CORRECT-SIZE-RANGE with inputs 35, 87, and 4
    create variables X, Y, and Z, with values 35, 87, and 4
    → (let* ...)
        → (max x y z)
        → 87
    → create variable BIGGEST, with value 87
        → (min x y z)
        → 4
        → create variable SMALLEST, with value 4
            → (/ biggest smallest 1.0)
            → 21.75
            → create variable R, with value 21.75
                → (list 'factor 'of r)
                → (FACTOR OF 21.75)
            → Result of LET* is (FACTOR OF 21.75)
→ Result of CORRECT-SIZE-RANGE is (FACTOR OF 21.75)
```

Don't be misled by this example into thinking that LET* should always be used in place of LET. There are some situations where LET is the only correct choice, but we won't go into the details here. Stylistically, it is better to use LET than LET* where possible, because this indicates to anyone reading the program that there are no dependencies among the local variables that are being created. Programs with few dependencies are easier to understand.

5.7 SIDE EFFECTS CAN CAUSE BUGS

It is best to avoid side effects in your programs wherever possible. Here is an example where the side effects of RANDOM cause a bug. Suppose we want a function that simulates a coin toss. Most of the time it should return HEADS or TAILS, but once in a great while it should return EDGE, indicating that the coin landed on its edge instead of one of its two faces. Here's how we'll do it: Pick a random number from 0 up to (but not including) 101. If the number is in the range 0 to 49, we'll return HEADS. If it's in the range 51 to 100, we'll return tails. If it is exactly equal to 50, we'll return EDGE.

```
(defun coin-with-bug ()
  (cond ((< (random 101) 50) 'heads)
        ((> (random 101) 50) 'tails)
        ((equal (random 101) 50) 'edge)))

> (coin-with-bug)
HEADS

> (coin-with-bug)
TAILS

> (coin-with-bug)
TAILS

> (coin-with-bug)
NIL
```

Why did the function return NIL? The bug is that we're evaluating the expression (RANDOM 101) as many as three times per function call. Suppose in the first COND clause (RANDOM 101) returns 65; this makes the first test false. In the second COND clause we again evaluate (RANDOM 101); suppose this time it returns 35, which makes the second test false. In the third clause, suppose (RANDOM 101) returns anything other than 50; this makes the third test false. COND has run out of clauses, so it returns NIL.

The fix for this bug is simple: Use LET to hold the value of (RANDOM 101) in a local variable, so we only have to evaluate the expression once. Also, we can omit the EQUAL test, since if the first two tests fail we know that the result must have been exactly equal to 50.[*]

```
(defun fair-coin ()
  (let ((toss (random 101)))
    (cond ((< toss 50) 'heads)
          ((> toss 50) 'tails)
          (t 'edge))))
```

SUMMARY

A variable is global to a function if it was not created by that function. Local variables have scope limited to the form that created them, for example, the variables in a function's argument list are local to that function, and the variables created by LET or LET* are local to their bodies. Global variables are so named because they have global scope; they are not local to any one function.

SETF is a macro function that assigns a value to a variable, or changes the value if it already has one. This side effect, called "assignment," is what makes SETF useful.

When multiple expressions appear in a function body or LET or LET* body, the value of the last expression is returned. The other expressions are only useful for their side effects.

REVIEW EXERCISES

5.2. What is a side effect?

5.3. What is the difference between a local and global variable?

5.4. Why must SETF be a macro function instead of a regular function?

5.5. Are LET and LET* equivalent when you are only creating one local variable?

---

[*]Eliminating the EQUAL test from COIN-WITH-BUG would not have fixed the bug, but it would have made the symptoms more subtle: The value EDGE would be returned roughly 25% of the time instead of only 1%.

FUNCTIONS COVERED IN THIS CHAPTER

Macro function for assignment: SETF.

Special functions for creating local variables: LET, LET*.

Lisp Toolkit: DOCUMENTATION and APROPOS

Most Common Lisp implementations include online documentation for every built-in function and variable. One way to access this documentation is with the DOCUMENTATION function, which returns a **documentation string**.

```
> (documentation 'cons 'function)
"(CONS x y) returns a list with x as the car
and y as the cdr."

> (documentation '*print-length* 'variable)
"*PRINT-LENGTH* determines how many elements to
print on each level of a list.  Unlimited if NIL."
```

Programmers don't use the DOCUMENTATION function very often, though, because there are faster ways to access online documentation via the editor your Lisp provides. On my machine, for example, when I point the mouse at a symbol and press Control-Meta-Shift-S, the documentation for that function or variable is displayed in a pop-up window.

You can include documentation strings in the functions you write, too. They should be placed immediately after the argument list when calling DEFUN.

```
(defun average (x y)
  "Returns the mean (average value) of its two
   inputs."
  (/ (+ x y) 2.0))

> (documentation 'average 'function)
"Returns the mean (average value) of its two
inputs."
```

Providing documentation strings for functions you write is good programming practice. It also helps other people to use your programs, since online documentation is always available whenever they need assistance.

Another way to document a program is by including **comments** in the file. Comments in Lisp programs must be prefaced with a semicolon. Whenever Lisp encounters a semicolon while loading a program, it discards the semicolon and everything to the right of it until the next carriage return. Comments benefit only those humans who take the trouble to examine the program; they are ignored by Lisp and do not form part of the online documentation. But they are useful because they may provide more lengthy information than a documentation string. They may also be more specific, for example, by explaining one or two of the more subtle lines in a function.

By convention, Lisp comments appear in one of three places. Comments appearing to the right of a line begin with one semicolon. Comments inside a function that occupy a line by themselves are preceded by two semicolons. Comments that begin at the left margin, appearing outside a function definition, are preceded by three semicolons. Some Lisp editors indent comments automatically based on the number of semicolons they contain. An example of all three comment styles follows.

```
;;; Function to compute Einstein's E = mc²

(defun einstein (m)
  (let ((c 300000.0)) ; speed of light in km/sec.
    ;; E is energy
    ;; m is mass
    (* m c c)))
```

Another useful source of documentation is APROPOS. It tells you the names of all symbols containing a specified string. For example, suppose you want to find all the built-in functions and variables containing "TOTAL" in their name. You can do this with APROPOS:

```
> (apropos "TOTAL" "USER")
ARRAY-TOTAL-SIZE (function)
ARRAY-TOTAL-SIZE-LIMIT, constant, value: 134217727
```

We see that there is a built-in Common Lisp function called ARRAY-TOTAL-SIZE, and a built-in constant called ARRAY-TOTAL-SIZE-LIMIT. (A constant is a variable whose value you are not allowed to change. PI is also a constant.)

The second argument to APROPOS is called a **package name**. You should always use the string "USER" (all uppercase) for the second argument; otherwise APROPOS may show you lots of implementation-specific Lisp functions in other packages that you don't care to know anything about. Packages are one of the more obscure features of Common Lisp and will not be covered in this book.

Keyboard Exercise

EXERCISE

5.6. This keyboard exercise is about dice. We will start with a function to throw one die and end up with a program to play craps. Be sure to include a documentation string for each function you write.

a. Write a function THROW-DIE that returns a random number from 1 to 6, inclusive. Remember that (RANDOM 6) will pick numbers from 0 to 5. THROW-DIE doesn't need any inputs, so its argument list should be NIL.

b. Write a function THROW-DICE that throws two dice and returns a list of two numbers: the value of the first die and the value of the second. We'll call this list a ''throw.'' For example, (THROW-DICE) might return the throw (3 5), indicating that the first die was a 3 and the second a 5.

c. Throwing two ones is called ''snake eyes''; two sixes is called ''boxcars.'' Write predicates SNAKE-EYES-P and BOXCARS-P that take a throw as input and return T if the throw is equal to (1 1) or (6 6), respectively.

d. In playing craps, the first throw of the dice is crucial. A throw of 7 or 11 is an instant win. A throw of 2, 3, or 12 is an instant loss (American casino rules). Write predicates INSTANT-WIN-P and INSTANT-LOSS-P to detect these conditions. Each should take a throw as input.

e. Write a function SAY-THROW that takes a throw as input and returns either the sum of the two dice or the symbol SNAKE-EYES or BOXCARS if the sum is 2 or 12. (SAY-THROW '(3 4)) should return 7. (SAY-THROW '(6 6)) should return BOXCARS.

f. If you don't win or lose on the first throw of the dice, the value you threw becomes your "point," which will be explained shortly. Write a function (CRAPS) that produces the following sort of behavior. Your solution should make use of the functions you wrote in previous steps.

```
> (craps)
(THROW 1 AND 1 -- SNAKEYES -- YOU LOSE)

> (craps)
(THROW 3 AND 4 -- 7 -- YOU WIN)

> (craps)
(THROW 2 AND 4 -- YOUR POINT IS 6)
```

g. Once a point has been established, you continue throwing the dice until you either win by making the point again or lose by throwing a 7. Write the function TRY-FOR-POINT that simulates this part of the game, as follows:

```
> (try-for-point 6)
(THROW 3 AND 5 -- 8 -- THROW AGAIN)

> (try-for-point 6)
(THROW 5 AND 1 -- 6 -- YOU WIN)

> (craps)
(THROW 3 AND 6 -- YOUR POINT IS 9)

> (try-for-point 9)
(THROW 6 AND 1 -- 7 -- YOU LOSE)
```

5 Advanced Topics

5.8 SYMBOLS AND VALUE CELLS

Recall that internally a symbol is composed of five components. The two we've seen so far are the symbol's name and function cell. A third component of every symbol is the **value cell**. It points to the value of the global variable named by that symbol. For example, if the global variable TOTAL has the value 12, then the internal structure of the symbol TOTAL would look like this:

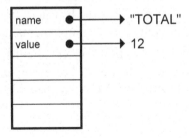

Similarly, if the global variable FISH has the value TROUT, the structure would look like this:

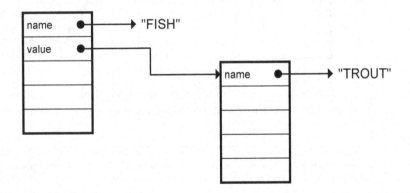

The symbols T and NIL evaluate to themselves because their value cells point to themselves. In other words, T is the name of a global variable whose value happens to be the symbol T; NIL's value is the symbol NIL. The internal structure of these symbols involves a circularity, as shown:

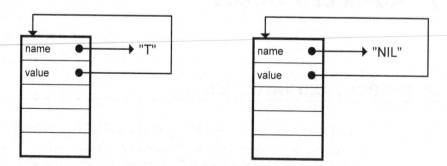

A symbol can be used to name many variables, but only one of these can be global. In other words, only one can exist in the global lexical context. The value cell is reserved for that variable. All the other variables must exist in local contexts, and their values reside someplace other than the symbol's value cell. Common Lisp doesn't specify exactly where the values of local variables are stored; the details are left up to the implementation.

Because symbols have separate function and value cells, we can have a variable and a function with the same name.[**] For example, if we gave the global variable CAR the value ROLLS-ROYCE, the symbol CAR would look like this:

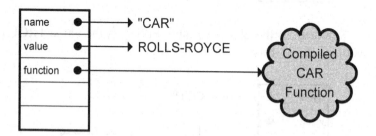

Common Lisp determines whether a symbol refers to a function or a variable based on the context in which it appears. If a symbol appears as the

---

[**]This is not possible in the Scheme dialect of Lisp, which stores functions and variable values in the same cell.

first element of a list that is to be evaluated, it is treated as a function name. In other contexts it is treated as a variable name. So (CAR '(A B C)) calls the CAR function, which returns A. But (LIST 'A 'NEW CAR) references the global variable CAR and produces the result (A NEW ROLLS-ROYCE).

5.9 DISTINGUISHING LOCAL FROM GLOBAL VARIABLES

By now it should be clear that symbols are not variables; they serve as names for variables (and for functions too.) Exactly which variable a symbol refers to depends on the context in which it appears. In the example below, there are two variables named X. The global variable X has the value 57. The variable X that is local to NEWVAR is bound to whatever is the input to NEWVAR.

```
(setf x 57)

(defun newvar (x)
   (list 'value 'of 'x 'is x))

> x
57

> (newvar 'whoopie)
(VALUE OF X IS WHOOPIE)

> x
57
```

Inside NEWVAR the name X refers to the local variable X, which the function created and assigned the value WHOOPIE. Outside the function, X refers to the global variable, whose value is 57. The value cell of the symbol X points to 57 the whole time; NEWVAR's local variable X is stored someplace else. An evaltrace diagram illustrates the relationship between the two Xs:

the global variable X has the value 57

→ (newvar 'whoopie)
→ Enter NEWVAR with input WHOOPIE
    create (local) variable X, with value WHOOPIE
    → (list 'value 'of 'x 'is x)
        X evaluates to WHOOPIE
    → (VALUE OF X IS WHOOPIE)
→ Result of NEWVAR is (VALUE OF X IS WHOOPIE)

the global variable X still has the value 57

The rule for evaluating the symbol X in the body of NEWVAR is to start in the current lexical context and move outward, looking for a variable with the given name. Since there is a variable named X in the current context, its value, WHOOPIE, is used. EVAL never looks at the global variable X.

The rule is actually a little more complex than this. EVAL moves outward from the current lexical context only until it finds a variable with that name or hits a thick line, indicating the end of the lexical environment. In the latter case, it cannot move outward any further; it can only check the global variable with that name. This explains the following example:

```
(setf a 100)

(defun f (a)
  (list a (g (+ a 1))))

(defun g (b)
  (list a b))

> (f 3)
(3 (100 4))
```

In this example, we create a global variable named A with value 100. When we call F, it creates a local variable named A, with value 3, and then calls the function G. G's lexical context is independent of F's. (Every function defined with DEFUN has its own independent lexical context.) In the evaltrace, the thick line denoting the context of G is a barrier: No variables that F creates are visible within G. So in the body of G, since there is no local variable named A, EVAL hits the barrier. The occurrence of A in the body is therefore treated as a reference to the global variable A.

the global variable A has the value 100

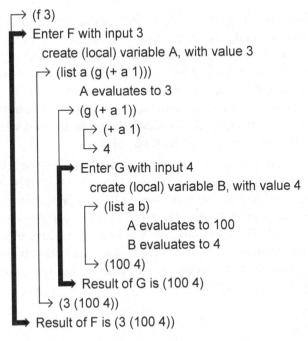

the global variable A still has the value 100

5.10 BINDING, SCOPING, AND ASSIGNMENT

Because Common Lisp evolved from older, less sophisticated Lisp dialects, it has inherited terminology that in a few cases doesn't quite fit. This book strives to use only correct and unambiguous terminology, but for compatibility with other books and the Lisp community at large, I will digress for a section and explain the various uses and misuses of the term ''binding.''

For historical reasons, variables that have values are said to be ''bound,'' and variables with no value are said to be ''unbound.'' While this book talks about ''unassigned variable'' errors, the error message most Lisp implementations produce is ''unbound variable.''

The process of creating a new variable and giving it a value is called ''binding.'' If the variable appears in a function's argument list, it is said to be created by ''lambda binding.'' If it appears in the variable list of a LET or LET* form, it is said to be created by ''LET-binding.'' These uses of ''binding'' are not incorrect today. A Lisp expert might well say that we cured the bug in COIN-WITH-BUG "by LET-binding a variable to the value of (RANDOM 101)."

But old-time Lispers get themselves into terminological trouble when they try to talk about the binding of variables in ways that aren't true for lexically scoped Lisps. While variables are lexically scoped by default, Common Lisp also provides another scoping discipline, called dynamic scoping, which we won't get into until Chapter 14. Dynamic scoping was the default in most earlier Lisp dialects, except for Scheme and T. "Bound" doesn't necesssarily mean "has a value" for dynamically scoped variables, because it is possible for such a variable to be bound but have no value.

Referring to the functions F and G in the preceding section, old-time Lispers would say "the symbol A is bound to 3 by F." This is not proper language if you are speaking about Common Lisp. Symbols are never bound; only variables can be bound. And there is no unique variable named A; there are two. Even while F's local variable A is in existence, the global A can be referenced by functions such as G whose lexical context is outside the body of F. To express the offending phrase in correct Common Lisp, one should say "F binds a local variable A to 3."

6

List Data Structures

6.1 INTRODUCTION

This chapter presents more list-manipulation functions, and shows how lists are used to implement such other data structures as sets, tables, and trees. Common Lisp offers many built-in functions that support these data structures. This is one of the strengths of Lisp compared to other languages. A Lisp programmer can immediately concentrate on the problem he or she wants to solve. A Pascal or C programmer faced with the same problem must first go off and implement parts of a Lisp-like system, such as linked list primitives, symbolic data structures, a storage allocator, and so on, before getting to work on the real problem.

The approach we take to lists in this chapter is somewhat more sophisticated than in Chapter 2. We will discuss not only what various Lisp primitive functions do, but also how they work inside. In preparation for this, you may want to review the discussion of dotted-pair notation in section 2.17. If you haven't been reading the Advanced Topics sections, that's okay; just go back and read section 2.17 now.

6.2 PARENTHESIS NOTATION VS. CONS CELL NOTATION

Writing lists in parenthesis notation is convenient, but it can be misleading. Lists in parenthesis notation appear symmetric: They begin with a left parenthesis and they end with a right one. One might therefore expect the CONS function to treat its arguments symmetrically. If CONS can add a symbol to the front of a list like so:

```
(cons 'w '(x y z))   ⇒   (w x y z)
```

why can't it add a symbol to the end of a list? Beginners who try this are surprised by the result:

```
(cons '(a b c) 'd)   ⇒   ((a b c) . d)
```

There is no reason to view the left end of a list as fundamentally different from the right end if we stick to parenthesis notation. But switching to cons cell notation reveals the crucial difference: Lists are one-way chains of pointers. It is easy to add an element to the *front* of a list because what we're really doing is creating a new cons cell whose cdr points to the existing list. If the inputs to CONS are W and (X Y Z), the result will be a new cell whose car points to W and whose cdr points to the old chain (X Y Z), as shown below. Although we usually display the result as (W X Y Z), we can also write it in dot notation as (W . (X Y Z)).

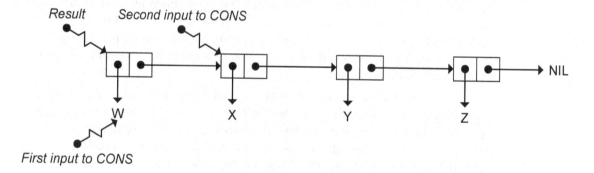

When we cons (A B C) onto D, it's the car of the new cell that points to the old list (A B C); the cdr points to the symbol D. The result is normally written ((A B C) . D), which looks decidedly odd in parenthesis notation. The dot is necessary because the cons cell chain ends in an atom other than NIL. In cons cell notation the structure looks like this:

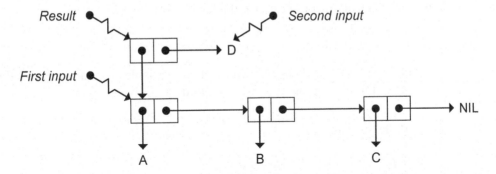

There is no direct way to add an element to the end of a list simply by creating a new cons cell, because the end of the original list already points to NIL.   More sophisticated techniques must be used.   One of these is demonstrated in the next section.

6.3 THE APPEND FUNCTION

APPEND takes two lists as input; it returns a list containing all the elements of the first list followed by all the elements of the second.[*]

```
> (append '(friends romans) '(and countrymen))
(FRIENDS ROMANS AND COUNTRYMEN)

> (append '(l m n o) '(p q r))
(L M N O P Q R)
```

If one of the inputs to APPEND is the empty list, the result will be equal to the other input. Appending NIL to a list is like adding zero to a number.

```
> (append '(april showers) nil)
(APRIL SHOWERS)

> (append nil '(bring may flowers))
(BRING MAY FLOWERS)

> (append nil nil)
NIL
```

---

*Note to instructors: To simplify the upcoming discussion of how APPEND works, we consider only the two-input case. In Common Lisp, APPEND can accept any number of inputs.

APPEND works on nested lists too. It only looks at the top level of each cons cell chain, so it doesn't notice if a list is nested or not.

```
> (append '((a 1) (b 2)) '((c 3) (d 4)))
((A 1) (B 2) (C 3) (D 4))
```

APPEND does not change the value of any variable or modify any existing cons cells. For this reason, it is called a **nondestructive** function.

```
> (setf who '(only the good))
(ONLY THE GOOD)

> (append who '(die young))
(ONLY THE GOOD DIE YOUNG)

> who
(ONLY THE GOOD)                    The value of WHO is unchanged.
```

APPEND may appear to treat its two inputs symmetrically, but this is just an illusion caused by the use of parenthesis notation. APPEND treats its two inputs quite differently. When we append the list (A B C) to the list (D E), APPEND *copies* the first input but not the second. It makes the cdr of the last cell of the copy point to the second input, and returns a pointer to the copy, as shown in Figure 6-1.

This description of how APPEND really works also explains why it is an error for the first input to APPEND to be a non-list, but it's okay if the second input is a non-list.

```
(append 'a '(b c d))    ⇒    Error! A is not a list.

(append '(w x y) 'z)    ⇒    (W X Y . Z)
```

APPEND wants to copy the cons cells that make up its first input. It can't when the first input is A because that isn't a list, so it signals an error. But when we append (W X Y) to Z, APPEND *can* copy its first input and make the cdr of the last cell point to the second input, so it doesn't have to signal an error. In this case the second input is Z rather than a list, so the result looks odd because the cons cell chain doesn't end in NIL.

Let us now return to the problem of adding an element to the end of a list. If we first make a list of the element, we can solve this problem by using APPEND.

```
(append '(a b c) '(d))    ⇒    (A B C D)
```

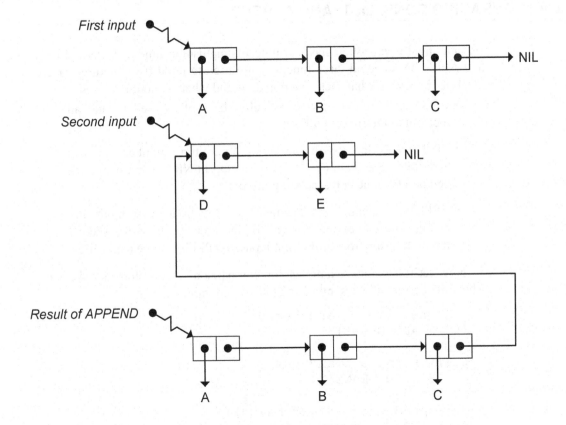

Figure 6-1 Result of appending (A B C) to (D E).

```
(defun add-to-end (x e)
  "Adds element E to the end of list X."
  (append x (list e)))

(add-to-end '(a b c) 'd)  ⇒  (A B C D)
```

6.4 COMPARING CONS, LIST, AND APPEND

Beginninng Lispers often have trouble distinguishing among CONS, LIST, and APPEND, since all three functions are used to build list structures. Here is a brief review of what each function does and when it should be used:

- CONS creates one new cons cell. It is often used to add an element to the front of a list.

- LIST makes new lists by accepting an arbitrary number of inputs and building a chain of cons cells ending in NIL. The car of each cell points to the corresponding input.

- APPEND appends lists together by copying its first input and making the cdr of the last cell of the copy point to the second input. It is an error for the first input to APPEND to be a non-list.

Now let's try some examples for comparison. First, consider the case where the first input is a symbol and the second input a list:

```
> (cons 'rice '(and beans))
(RICE AND BEANS)

> (list 'rice '(and beans))
(RICE (AND BEANS))

> (append 'rice '(and beans))
Error: RICE is not a list.
```

Next, let's see what happens when both inputs are lists:

```
> (cons '(here today) '(gone tomorrow))
((HERE TODAY) GONE TOMORROW)

> (list '(here today) '(gone tomorrow))
((HERE TODAY) (GONE TOMORROW))

> (append '(here today) '(gone tomorrow))
(HERE TODAY GONE TOMORROW)
```

Finally, let's try making the first input a list and the second input a symbol. This is the trickiest case to understand; you must think in terms of cons cells rather than parentheses and dots.

```
> (cons '(eat at) 'joes)
((EAT AT) . JOES)

> (list '(eat at) 'joes)
((EAT AT) JOES)

> (append '(eat at) 'joes)
(EAT AT . JOES)
```

To further develop your intuitions about CONS, LIST, and APPEND, try the above examples using the SDRAW tool described in the Lisp Toolkit section of this chapter. SDRAW draws cons cell diagrams.

6.5 MORE FUNCTIONS ON LISTS

Lisp provides many simple functions for operating on lists. We've already discussed CONS, LIST, APPEND, and LENGTH. Now we will cover REVERSE, NTH, NTHCDR, LAST, and REMOVE. Some of these functions must copy their first input, while others don't have to. See if you can figure out the reason for this.

6.5.1 REVERSE

REVERSE returns the reversal of a list.

```
> (reverse '(one two three four five))
(FIVE FOUR THREE TWO ONE)

> (reverse '(l i v e))
(E V I L)

> (reverse 'live)
Error: Wrong type input.

> (reverse '((my oversight)
             (your blunder)
             (his negligence)))
((HIS NEGLIGENCE) (YOUR BLUNDER) (MY OVERSIGHT))
```

Notice that REVERSE reverses only the *top level* of a list. It does not reverse the individual elements of a list of lists. Another point about REVERSE is that it doesn't work on symbols. REVERSE of the list (L I V E) gives the list (E V I L), but REVERSE of the symbol LIVE gives a wrong-type input error.

Like APPEND, REVERSE is nondestructive. It copies its input rather than modifying it.

```
> (setf vow '(to have and to hold))
(TO HAVE AND TO HOLD)

> (reverse vow)
(HOLD TO AND HAVE TO)

> vow
(TO HAVE AND TO HOLD)
```

We can use REVERSE to add an element to the end of a list, as follows. Suppose we want to add D to the end of the list (A B C). The reverse of (A B C) is (C B A). If we cons D onto that we get (D C B A). Then, reversing the result of CONS gives (A B C D).

```
(defun add-to-end (x y)
    (reverse (cons y (reverse x))))

(add-to-end '(a b c) 'd)  ⇒  (a b c d)
```

Now you know two ways to add an element to the end of a list. The APPEND solution is considered better style than the double REVERSE solution because the latter makes two copies of the list. APPEND is more efficient. Efficiency issues are further discussed in an Advanced Topics section at the end of this chapter.

6.5.2 NTH and NTHCDR

The NTHCDR function returns the *n*th successive cdr of a list. Of course, if we take zero cdrs we are left with the list itself. If we take one too many cdrs, we end up with the atom that terminates the cons cell chain, which usually is NIL.

```
(nthcdr 0 '(a b c))  ⇒  (a b c)

(nthcdr 1 '(a b c))  ⇒  (b c)
```

```
(nthcdr 2 '(a b c))   ⇒   (c)

(nthcdr 3 '(a b c))   ⇒   nil
```

Using inputs greater than 3 does not cause an error; we simply get the same result as for 3. This is one of the consequences of making the cdr of NIL be NIL.

```
(nthcdr 4 '(a b c))   ⇒   nil

(nthcdr 5 '(a b c))   ⇒   nil
```

However, if the list ends in an atom other than NIL, going too far with NTHCDR will cause an error.

```
(nthcdr 2 '(a b c . d))   ⇒   (c . d)

(nthcdr 3 '(a b c . d))   ⇒   d

(nthcdr 4 '(a b c . d))   ⇒   Error! D is not a list.
```

The NTH function takes the CAR of the NTHCDR of a list.

```
(defun nth (n x)
  "Returns the Nth element of the list X,
   counting from 0."
  (car (nthcdr n x)))
```

Since (NTHCDR 0 x) is the list x, (NTH 0 x) is the first element. Therefore, (NTH 1 x) is the second, and so on.

```
(nth 0 '(a b c))   ⇒   a

(nth 1 '(a b c))   ⇒   b

(nth 2 '(a b c))   ⇒   c

(nth 3 '(a b c))   ⇒   nil
```

The convention of numbering things from zero rather than from one is used throughout Common Lisp. You will encounter it again when we discuss arrays in Chapter 13.

EXERCISES

6.1. Why is (NTH 4 '(A B C)) equal to NIL?

6.2. What is the value of (NTH 3 '(A B C . D)), and why?

6.5.3 LAST

LAST returns the last cons cell of a list, in other words, the cell whose car is the list's last element. By definition, the cdr of this cell is an atom; otherwise it wouldn't be the last cell of the list. If the list is empty, LAST just returns NIL.

```
(last '(all is forgiven))  ⇒  (forgiven)

(last nil)  ⇒  nil

(last '(a b c . d))  ⇒  (c . d)

(last 'nevermore)  ⇒  Error! NEVERMORE is not a list.
```

EXERCISES

6.3. What is the value of (LAST '(ROSEBUD)) ?

6.4. What is the value of (LAST '((A B C))), and why?

6.5.4 REMOVE

REMOVE removes an item from a list. Normally it removes all occurrences of the item, although there are ways to tell it to remove only some (see the Advanced Topics section). The result returned by REMOVE is a new list, without the deleted items.

```
(remove 'a '(b a n a n a))  ⇒  (b n n)

(remove 1 '(3 1 4 1 5 9))  ⇒  (3 4 5 9)
```

REMOVE is a nondestructive function. It does not change any variables or cons cells when removing elements from a list. REMOVE builds its result out of fresh cons cells by copying (parts of) the list.

```
> (setf spell '(a b r a c a d a b r a))
(A B R A C A D A B R A)

> (remove 'a spell)
(B R C D B R)

> spell
(A B R A C A D A B R A)
```

The following table should help you remember which functions copy their input and which do not. APPEND, REVERSE, and REMOVE return a new cons cell chain that is not contained in their input, so they must copy their input to produce the new chain. Functions such as NTHCDR, NTH, and LAST return a pointer to some component of their input. They do not need to copy anything because, by definition, the exact object they want to return already exists.

Function	Copies its input?
APPEND	yes *(execept for the last input)*
REVERSE	yes
NTHCDR	no
NTH	no
LAST	no
REMOVE	yes *(only the second input)*

EXERCISES

6.5. Write an expression to set the global variable LINE to the list (ROSES ARE RED). Then write down what each of the following expressions evaluates to:

```
(reverse line)

(first (last line))

(nth 1 line)

(reverse (reverse line))

(append line (list (first line)))

(append (last line) line)

(list (first line) (last line))

(cons (last line) line)

(remove 'are line)

(append line '(violets are blue))
```

6.6. Use the LAST function to write a function called LAST-ELEMENT that returns the last element of a list instead of the last cons cell. Write another version of LAST-ELEMENT using REVERSE instead of LAST. Write another version using NTH and LENGTH.

6.7. Use REVERSE to write a NEXT-TO-LAST function that returns the next-to-last element of a list. Write another version using NTH.

6.8. Write a function MY-BUTLAST that returns a list with the last element removed. (MY-BUTLAST '(ROSES ARE RED)) should return the list (ROSES ARE). (MY-BUTLAST '(G A G A)) should return (G A G).

6.9. What primitive function does the following reduce to?

```
(defun mystery (x) (first (last (reverse x))))
```

6.10. A palindrome is a sequence that reads the same forwards and backwards. The list (A B C D C B A) is a palindrome; (A B C A B C) is not. Write a function PALINDROMEP that returns T if its input is a palindrome.

6.11. Write a function MAKE-PALINDROME that makes a palindrome out of a list, for example, given (YOU AND ME) as input it should return (YOU AND ME ME AND YOU).

6.6 LISTS AS SETS

A set is an unordered collection of items. Each item appears only once in the set. Some typical sets are the set of days of the week, the set of integers (an infinite set), and the set of people in Hackensack, New Jersey, who had spaghetti for dinner last night.

Sets are undoubtedly one of the more useful data structures one can build from lists. The basic set operations are testing if an item is a **member** of a set; taking the **union**, **intersection**, and **set difference** (also called set subtraction) of two sets; and testing if one set is a **subset** of another. The Lisp functions for all these operations are described in the following subsections.

6.6.1 MEMBER

The MEMBER predicate checks whether an item is a member of a list. If the item is found in the list, the sublist beginning with that item is returned. Otherwise NIL is returned. MEMBER never returns T, but by tradition it is counted as a predicate because the value it returns is non-NIL (hence true) if and only if the item is in the list.

> (setf ducks '(huey dewey louie))   *Create a set of ducks.*
(HUEY DEWEY LOUIE)

> (member 'huey ducks)   *Is Huey a duck?*
(HUEY DEWEY LOUIE)   *Non-NIL result: yes.*

> (member 'dewey ducks)   *Is Dewey a duck?*
(DEWEY LOUIE)   *Non-NIL result: yes.*

> (member 'louie ducks)   *Is Louie a duck?*
(LOUIE)   *Non-NIL result: yes.*

> (member 'mickey ducks)   *Is Mickey a duck?*
NIL   *NIL: no.*

In the very first dialect of Lisp, MEMBER returned just T or NIL. But
people decided that having MEMBER return the sublist beginning with the
item sought made it a much more useful function. This extension is consistent
with MEMBER's being a predicate, because the sublist with *zero* elements is
also the only way to say "false."

Here's an example of why it is useful for MEMBER to return a sublist.
The BEFOREP predicate returns a true value if *x* appears earlier than *y* in the
list *l*.

```
(defun beforep (x y l)
  "Returns true if X appears before Y in L"
  (member y (member x l)))

> (beforep 'not 'whom
           '(ask not for whom the bell tolls))
(WHOM THE BELL TOLLS)

> (beforep 'thee 'tolls '(it tolls for thee))
NIL
```

EXERCISE

6.12. Does MEMBER have to copy its input to produce its result? Explain
your reasoning.

6.6.2 INTERSECTION

The INTERSECTION function takes the intersection of two sets and returns a list of items appearing in *both* sets. The exact order in which elements appear in the result is undefined; it may differ from one Lisp implementation to another. Order isn't important for sets anyway; only the elements themselves matter.

```
> (intersection '(fred john mary)
                '(sue mary fred))
(FRED MARY)

> (intersection '(a s d f g)
                '(v w s r a))
(A S)

> (intersection '(foo bar baz)
                '(xam gorp bletch))
NIL
```

If a list contains multiple occurrence of an item, it is not a true set. Common Lisp set functions such as INTERSECTION and UNION can handle lists that are not sets, but whether the result contains duplicates or not is undefined, and may vary across implementations.

EXERCISES

6.13. What is the result of intersecting a set with NIL?

6.14. What is the result of intersecting a set with itself?

6.15. We can use MEMBER to write a predicate that returns a true value if a sentence contains the word "the."

```
(defun contains-the-p (sent)
  (member 'the sent))
```

Suppose we instead want a predicate CONTAINS-ARTICLE-P that returns a true value if a sentence contains any article, such as "the," "a," or "an." Write a version of this predicate using INTERSECTION. Write another version using MEMBER and OR. Could you solve this problem with AND instead of OR?

6.6.3 UNION

The UNION function returns the union of two sets, in other words, a list of items that appear in *either* set. If an item appears in both sets, it will still appear only once in the result. The exact order in of items in the result is undefined (and unimportant) for sets.

```
> (union '(finger hand arm)
         '(toe finger foot leg))
(FINGER HAND ARM TOE FOOT LEG)

> (union '(fred john mary)
         '(sue mary fred))
(FRED JOHN MARY SUE)

> (union '(a s d f g)
         '(v w s r a))
(A S D F G V W R)
```

EXERCISES

6.16. What is the union of a set with NIL?

6.17. What is the union of a set with itself?

6.18. Write a function ADD-VOWELS that takes a set of letters as input and adds the vowels (A E I O U) to the set. For example, calling ADD-VOWELS on the set (X A E Z) should produce the set (X A E Z I O U), except that the exact order of the elements in the result is unimportant.

6.6.4 SET-DIFFERENCE

The SET-DIFFERENCE function performs set subtraction. It returns what is left of the first set when the elements in the second set have been removed. Again, the order of elements in the result is undefined.

```
> (set-difference '(alpha bravo charlie delta)
                  '(bravo charlie))
(ALPHA DELTA)

> (set-difference '(alpha bravo charlie delta)
                  '(echo alpha foxtrot))
(BRAVO CHARLIE DELTA)
```

```
> (set-difference '(alpha bravo) '(bravo alpha))
NIL
```

Unlike UNION and INTERSECTION, SET-DIFFERENCE is not a symmetric function. Switching its first and second inputs usually results in a different set being produced as output.

```
(setf line1 '(all things in moderation))
```

```
(setf line2 '(moderation in the defense of liberty
              is no virtue))
```

```
> (set-difference line1 line2)
(ALL THINGS)
```

```
> (set-difference line2 line1)
(THE DEFENSE OF LIBERTY IS NO VIRTUE)
```

EXERCISES

6.19. What are the results of using NIL as an input to SET-DIFFERENCE?

6.20. Which of its two inputs does SET-DIFFERENCE need to copy? Which input never needs to be copied? Explain your reasoning.

6.6.5 SUBSETP

The SUBSETP predicate returns T if one set is contained in another, in other words, if every element of the first set is an element of the second set.

```
(subsetp '(a i) '(a e i o u))   ⇒   t
```

```
(subsetp '(a x) '(a e i o u))   ⇒   nil
```

EXERCISE

6.21. If set x is a subset of set y, then subtracting y from x should leave the empty set. Write MY-SUBSETP, a version of the SUBSETP predicate that returns T if its first input is a subset of its second input.

GENERAL SET EXERCISES

6.22. Suppose the global variable A is bound to the list (SOAP WATER). What will be the result of each of the following expressions?

```
(union a '(no soap radio))

(intersection a (reverse a))

(set-difference a '(stop for water))

(set-difference a a)

(member 'soap a)

(member 'water a)

(member 'washcloth a)
```

6.23. The **cardinality** of a set is the number of elements it contains. What Lisp primitive determines the cardinality of a set?

6.24. Sets are said to be equal if they contain exactly the same elements. Order does not matter in a set, so the sets (RED BLUE GREEN) and (GREEN BLUE RED) are considered equal. However, the EQUAL predicate does not consider them equal, because it treats them as lists, not as sets. Write a SET-EQUAL predicate that returns T if two things are equal as sets. (*Hint:* If two sets are equal, then each is a subset of the other.)

6.25. A set X is a **proper subset** of a set Y if X is a subset of Y but not equal to Y. Thus, (A C) is a proper subset of (C A B). (A B C) is a subset of (C A B), but not a proper subset of it. Write the PROPER-SUBSETP predicate, which returns T if its first input is a proper subset of its second input.

6.7 PROGRAMMING WITH SETS

Here is an example of how to solve a modest programming problem using sets. The problem is to write a function that adds a title to a name, turning "John Doe" into "Mr. John Doe" or "Jane Doe" into "Ms. Jane Doe." If a name already has a title, that title should be kept, but if it doesn't have one, we will try to determine the gender of the first name so that the appropriate title can be assigned.

To solve a problem like this, we must break it down into smaller pieces. Let's start with the question of whether a name has a title or not. Here's how we'd write a function to answer that question:

```
(defun titledp (name)
  (member (first name) '(mr ms miss mrs)))
```

> (titledp '(jane doe))   *"Jane" is not a title.*
NIL

> (titledp '(ms jane doe))   *"Ms." is in the set of titles.*
(MS MISS MRS)

The next step is to write functions to figure out whether a word is a male or female first name. We will use only a few instances of each type of name to keep the example brief.

```
(setf male-first-names
      '(john kim richard fred george))

(setf female-first-names
      '(jane mary wanda barbara kim))

(defun malep (name)
  (and (member name male-first-names)
       (not (member name female-first-names))))

(defun femalep (name)
  (and (member name female-first-names)
       (not (member name male-first-names))))
```

> (malep 'richard)   *"Richard" is in the set of males.*
T

> (malep 'barbara)   *"Barbara" is not a male name.*
NIL

> (femalep 'barbara)   *"Barbara" is a female name.*
T

> (malep 'kim)   *"Kim" can be either male or female,*
NIL   *so it's not exclusively male.*

Now we can write the GIVE-TITLE function that adds a title to a name. Of course, we will only add a title if the name doesn't already have one. If the first name isn't recognized as male or female, we'll play it safe and use "Mr. or Ms."

```
(defun give-title (name)
  "Returns a name with an appropriate title on
  the front."
  (cond ((titledp name) name)
        ((malep (first name)) (cons 'mr name))
        ((femalep (first name)) (cons 'ms name))
        (t (append '(mr or ms) name))))
```

> (give-title '(miss jane adams))   *Already has a title.*
(MISS JANE ADAMS)

> (give-title '(john q public))   *Untitled male name.*
(MR JOHN Q PUBLIC)

> (give-title '(barbara smith))   *Untitled female name.*
(MS BARBARA SMITH)

> (give-title '(kim johnson))   *Untitled, and gender*
(MR OR MS KIM JOHNSON)   *is ambiguous.*

The important features in this example are (1) breaking a problem down into simple little functions, and (2) writing and testing the functions one at a time. Once we had the TITLEDP, MALEP, and FEMALEP predicates written, GIVE-TITLE was easy to write.

Decomposing a problem into subproblems is an important skill. Experienced programmers can often see right away how a problem breaks down into logical subdivisions, but beginners must build up their intuition through practice.

Here are a few more things we can do with these lists of names. The functions below take no inputs, so their argument list is NIL.

```
(defun gender-ambiguous-names ()
  (intersection male-names female-names))

(gender-ambiguous-names)   ⇒   (kim)

(defun uniquely-male-names ()
  (set-difference male-names female-names))

(uniquely-male-names)
   ⇒   (john richard fred george)
```

So far, all the sets we've seen in this chapter contained only symbols and numbers. It is also quite easy to work with sets of lists, but a trick is required to use functions like MEMBER, UNION, INTERSECTION, and so on on sets of lists. See the discussion of the :TEST keyword in the Advanced Topics section.

MINI KEYBOARD EXERCISE

6.26. We are going to write a program that compares the descriptions of two objects and tells how many features they have in common. The descriptions will be represented as a list of features, with the symbol -VS- separating the first object from the second. Thus, when given a list like

```
(large red shiny cube -vs-
     small shiny red four-sided pyramid)
```

the program will respond with (2 COMMON FEATURES). We will compose this program from several small functions that you will write and test one at a time.

a. Write a function RIGHT-SIDE that returns all the features to the right of the -VS- symbol. RIGHT-SIDE of the list shown above should return (SMALL SHINY RED FOUR-SIDED PYRAMID). *Hint:* remember that the MEMBER function returns the entire sublist starting with the item for which you are searching. Test your function to make sure it works correctly.

b. Write a function LEFT-SIDE that returns all the features to the left of the -VS-. You can't use the MEMBER trick directly for this one, but you can use it if you do something to the list first.

c. Write a function COUNT-COMMON that returns the number of features the left and right sides of the input have in common.

d. Write the main function, COMPARE, that takes a list of features describing two objects, with a -VS- between them, and reports the number of features they have in common. COMPARE should return a list of form (*n* COMMON FEATURES).

e. Try the expression

```
(compare '(small red metal cube -vs-
               red plastic small cube))
```

You should get (3 COMMON FEATURES) as the result.

6.8 LISTS AS TABLES

Tables are another very useful structure we can build from lists. A table, or **association list** (**a-list** for short), is a list of lists. Each list is called an **entry**, and the car of each entry is its **key**. A table of five English words and their French equivalents is shown below. The table contains five entries; the keys are the English words.

```
(setf words
  '((one un)
    (two deux)
    (three trois)
    (four quatre)
    (five cinq)))
```

6.8.1 ASSOC

The ASSOC function looks up an entry in a table, given its key. Here are some examples.

```
(assoc 'three words)  ⇒  (three trois)

(assoc 'four words)  ⇒  (four quatre)

(assoc 'six words)  ⇒  nil
```

ASSOC goes through the table one entry at a time until it finds a key that matches the key for which it is searching; it returns that entry. If ASSOC can't find the key in the table, it returns NIL.

Notice that when ASSOC does find an entry with the given key, the value it returns is the entire entry. If we want only the French word and not the entire entry, we can take the second element of the result of ASSOC.

```
(defun translate (x)
  (second (assoc x words)))

(translate 'one)  ⇒  un

(translate 'five)  ⇒  cinq

(translate 'six)  ⇒  nil
```

EXERCISE

6.27. Should ASSOC be considered a predicate even though it never returns
T?

6.8.2 RASSOC

RASSOC is like ASSOC, except it looks at the cdr of each element of the
table instead of the car. (The name stands for "Reverse ASSOC.") To use
RASSOC with symbols as keys, the table must be a list of dotted pairs, like
this:

```
(setf sounds
  '((cow  . moo)
    (pig  . oink)
    (cat  . meow)
    (dog  . woof)
    (bird . tweet)))
```

```
(rassoc 'woof sounds)   ⇒   (dog . woof)
```

```
(assoc 'dog sounds)   ⇒   (dog . woof)
```

Both ASSOC and RASSOC return as soon as they find the first matching
table entry; the rest of the list is not searched.

EXERCISE

6.28. Set the global variable PRODUCE to this list:

```
((apple   . fruit)
 (celery  . veggie)
 (banana  . fruit)
 (lettuce . veggie))
```

Now write down the results of the following expressions:

```
(assoc 'banana produce)
```

```
(rassoc 'fruit produce)
```

```
(assoc 'lettuce produce)
```

```
(rassoc 'veggie produce)
```

6.9 PROGRAMMING WITH TABLES

Here is another example of the use of ASSOC. We will create a table of objects and their descriptions, where the descriptions are similar to those in the last mini keyboard exercise. We'll store the table of descriptions in the global variable THINGS. The table looks like this:

```
((object1 large green shiny cube)
 (object2 small red dull metal cube)
 (object3 red small dull plastic cube)
 (object4 small dull blue metal cube)
 (object5 small shiny red four-sided pyramid)
 (object6 large shiny green sphere))
```

Now we'll develop functions to tell us in which qualities two objects differ. We start by writing a function called DESCRIPTION to retrieve the description of an object:

```
(defun description (x)
   (rest (assoc x things)))

(description 'object3)  ⇒
   (red small dull plastic cube)
```

The differences between two objects are whatever properties appear in the description of the first but not the second, or the description of the second but not the first. The technical term for this is **set exclusive or**. There is a built-in Common Lisp function to compute it.

```
(defun differences (x y)
  (set-exclusive-or (description x)
                    (description y)))

(differences 'object2 'object3)  ⇒
  (metal plastic)
```

OBJECT2 is metal but OBJECT3 is plastic, so METAL and PLASTIC are properties not common to both. We can classify properties according to the type of quality to which they refer. Here is a table, represented as a list of dotted pairs:

```
(setf quality-table
   '((large     .  size)
     (small     .  size)
     (red       .  color)
     (green     .  color)
```

```
(blue        .   color)
(shiny       .   luster)
(dull        .   luster)
(metal       .   material)
(plastic     .   material)
(cube        .   shape)
(sphere      .   shape)
(pyramid     .   shape)
(four-sided  .   shape)))
```

We can use this table as part of a function that gives us the quality a given property refers to:

```
(defun quality (x)
  (cdr (assoc x quality-table)))
```

```
(quality 'red)   ⇒  color
```

```
(quality 'large)  ⇒  size
```

Using DIFFERENCES and QUALITY, we can write a function to tell us one quality that is different between a pair of objects.

```
(defun quality-difference (x y)
  (quality (first (differences x y))))
```

```
(quality-difference 'object2 'object3)
  ⇒  material
```

```
(quality-difference 'object1 'object6)
  ⇒  shape
```

```
(quality-difference 'object2 'object4)
  ⇒  color
```

What if we wanted a list of all the quality differences instead of just the first one? We would need some way to go from a list of differences like (RED BLUE METAL PLASTIC) to a list of corresponding qualities (COLOR COLOR MATERIAL MATERIAL), and then we'd have to eliminate duplicate elements. The first part can be accomplished with SUBLIS, which is discussed in the Advanced Topics section.

```
(differences 'object3 'object4)
  ⇒  (red blue metal plastic)
```

```
> (sublis quality-table
          (differences 'object3 'object4))
(COLOR COLOR MATERIAL MATERIAL)
```

Now all we have to do is eliminate duplicate entries in the result. Common Lisp provides a function called REMOVE-DUPLICATES for this purpose.

```
(defun contrast (x y)
  (remove-duplicates
    (sublis quality-table (differences x y))))
```

```
(contrast 'object3 'object4)   ⇒   (color material)
```

EXERCISES

6.29. What Lisp primitive returns the number of entries in a table?

6.30. Make a table called BOOKS of five books and their authors. The first entry might be (WAR-AND-PEACE LEO-TOLSTOY).

6.31. Write the function WHO-WROTE that takes the name of a book as input and returns the book's author.

6.32. Suppose we do (SETF BOOKS (REVERSE BOOKS)), which reverses the order in which the five books appear in the table. What will the WHO-WROTE function do now?

6.33. Suppose we wanted a WHAT-WROTE function that took an author's name as input and returned the title of one of his or her books. Could we create such a function using ASSOC and the current table? If not, how would the table have to be different?

6.34. Here is a table of states and some of their cities, stored in the global variable ATLAS:

```
((pennsylvania pittsburgh)
 (new-jersey newark)
 (pennsylvania johnstown)
 (ohio columbus)
 (new-jersey princeton)
 (new-jersey trenton))
```

Suppose we wanted to find all the cities a given state contains. ASSOC returns only the *first* entry with a matching key, not all such entries, so for this table ASSOC cannot solve our problem. Redesign the table so that ASSOC can be used successfully.

MINI KEYBOARD EXERCISE

6.35. In this problem we will simulate the behavior of a very simple-minded creature, *Nerdus Americanis* (also known as *Computerus Hackerus*). This creature has only five states: Sleeping, Eating, Waiting-for-a-Computer, Programming, and Debugging. Its behavior is cyclic: After it sleeps it always eats, after it eats it always waits for a computer, and so on, until after debugging it goes back to sleep for a while.

a. What type of data structure would be useful for representing the connection between a state and its successor? Write such a data structure for the five-state cycle given above, and store it in a global variable called NERD-STATES.

b. Write a function NERDUS that takes the name of a state as input and uses the data structure you designed to determine the next state the creature will be in. (NERDUS 'SLEEPING) should return EATING, for example. (NERDUS 'DEBUGGING) should return SLEEPING.

c. What is the result of (NERDUS 'PLAYING-GUITAR)?

d. When *Nerdus Americanis* ingests too many stimulants (caffeine overdose), it stops sleeping. After finishing Debugging, it immediately goes on to state Eating. Write a function SLEEPLESS-NERD that works just like NERDUS except it never sleeps. Your function should refer to the global variable NERD-STATES, as NERDUS does.

e. Exposing *Nerdus Americanis* to extreme amounts of chemical stimulants produces pathological behavior. Instead of an orderly advance to its next state, the creature advances two states. For example, it goes from Eating directly to Programming, and from there to Sleeping. Write a function NERD-ON-CAFFEINE that exhibits this unusual pathology. Your function should use the same table as NERDUS.

f. If a Nerd on caffeine is currently programming, how many states will it have to go through before it is debugging?

SUMMARY

Lists are an important data type in their own right, but in Lisp they are even more important because they are used to implement other data structures such as sets and tables.

As we saw in the mini keyboard exercises, the way to solve any nontrivial programming problem is to divide the problem into smaller, more manageable pieces. This is done by writing and testing several simple functions, then combining them to produce a solution to the main problem.

REVIEW EXERCISES

6.36. Write a function to swap the first and last elements of any list. (SWAP-FIRST-LAST '(YOU CANT BUY LOVE)) should return (LOVE CANT BUY YOU).

6.37. ROTATE-LEFT and ROTATE-RIGHT are functions that rotate the elements of a list. (ROTATE-LEFT '(A B C D E)) returns (B C D E A), whereas ROTATE-RIGHT returns (E A B C D). Write these functions.

6.38. Give an example of two sets X and Y such that (SET-DIFFERENCE X Y) equals (SET-DIFFERENCE Y X). Also give an example in which the set differences are *not* equal.

6.39. Recall the unary arithmetic system developed in the advanced topics section of Chapter 2. What list function performs unary addition?

6.40. Show how to transform the list (A B C D) into a table so that the ASSOC function using the table gives the same result as MEMBER using the list.

FUNCTIONS COVERED IN THIS CHAPTER

List functions: APPEND, REVERSE, NTH, NTHCDR, LAST, REMOVE.

Set functions: UNION, INTERSECTION, SET-DIFFERENCE, SET-EXCLUSIVE-OR, MEMBER, SUBSETP, REMOVE-DUPLICATES.

Table functions: ASSOC, RASSOC.

Lisp Toolkit: SDRAW

SDRAW is a tool for drawing cons cell representations of lists. It is not part of the Common Lisp standard; it is defined in Appendix A. There are several versions. The completely portable version will run in any Common Lisp implementation; it draws cons cell diagrams using ordinary characters, like so:

```
> (sdraw '(alpha (bravo) charlie))

[*|*]--->[*|*]---------->[*|*]--->NIL
 |        |               |
 v        v               v
ALPHA    [*|*]--->NIL    CHARLIE
          |
          v
         BRAVO
```

There are also a number of graphic versions, available on diskette from the publisher, which draw cons cells and arrows using graphics functions. They look much nicer that way. One graphic version uses CLX, the Common Lisp interface to the X Windows system. If your computer doesn't run X Windows, you won't be able to use this version, but if your Lisp provides some other graphics facility, it should be easy to adapt SDRAW to use it.

Another useful tool is the function SDRAW-LOOP, which acts like a read-eval-print loop except it draws the result as well as printing it. SDRAW-LOOP prompts for input with the string "S>." Here's an example.

```
> (sdraw-loop)

Type any Lisp expression, or (ABORT) to exit.

S> (cons '(birds dont have) 'noses)

[*|*]--->NOSES
 |
 v
[*|*]--->[*|*]--->[*|*]--->NIL
 |        |        |
 v        v        v
BIRDS    DONT     HAVE

Result:  ((BIRDS DONT HAVE) . NOSES)



S> (append '(they) '(have beaks))

[*|*]--->[*|*]--->[*|*]--->NIL
 |        |        |
 v        v        v
THEY     HAVE     BEAKS

Result:  (THEY HAVE BEAKS)



S> (abort)
```

A third function provided by the SDRAW program is called SCRAWL. It is an interactive version of SDRAW that allows you to "crawl around" a list by taking successive cars and cdrs, backing up, or returning to the start.

Keyboard Exercise

In this keyboard exercise we will write some routines for moving Robbie the robot around in a house. The map of the house appears in Figure 6-2. Robbie can move in any of four directions: north, south, east, or west.

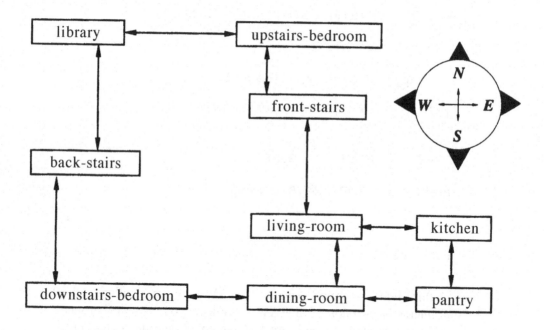

Figure 6-2 Map of the House.

The layout of the house is described in a table called ROOMS, with one element for each room:

```
((living-room ...)
 (upstairs-bedroom ...)
 (dining-room ...)
 (kitchen ...)
```

```
(pantry ...)
(downstairs-bedroom ...)
(back-stairs ...)
(front-stairs ...)
(library ...))
```

The entry for each room is in turn a table listing the directions that Robbie can travel from that room and where he ends up for each direction. The entire table is shown in Figure 6-3. The first element of the table is:

```
(living-room
  (north front-stairs)
  (south dining-room)
  (east kitchen))
```

If Robbie were in the living room, going north would take him to the front stairs, going south would take him to the dining room, and going east would take him to the kitchen. Since there is nothing listed for west, we assume that there is a wall there, so Robbie cannot travel west from the living room.

EXERCISE

6.41. If the table of rooms is already stored on the computer for you, load the file containing it. If not, you will have to type the table in as it appears in Figure 6-3. If you like, try (SDRAW ROOMS) or (SCRAWL ROOMS) to view the table as a cons cell structure.

a. Write a function CHOICES that takes the name of a room as input and returns the table of permissible directions Robbie may take from that room. For example, (CHOICES 'PANTRY) should return the list ((NORTH KITCHEN) (WEST DINING-ROOM)). Test your function to make sure it returns the correct result.

b. Write a function LOOK that takes two inputs, a direction and a room, and tells where Robbie would end up if he moved in that direction from that room. For example, (LOOK 'NORTH 'PANTRY) should return KITCHEN. (LOOK 'WEST 'PANTRY) should return DINING-ROOM. (LOOK 'SOUTH 'PANTRY) should return NIL. *Hint:* The CHOICES function will be a useful building block.

c. We will use the global variable LOC to hold Robbie's location. Type in an expression to set his location to be the pantry. The following function should be used whenever you want to change his location.

```
(defun set-robbie-location (place)
  "Moves Robbie to PLACE by setting
   the variable LOC."
  (setf loc place))
```

d. Write a function HOW-MANY-CHOICES that tells how many choices Robbie has for where to move to next. Your function should refer to the global variable LOC to find his current location. If he is in the pantry, (HOW-MANY-CHOICES) should return 2.

e. Write a predicate UPSTAIRSP that returns T if its input is an upstairs location. (The library and the upstairs bedroom are the only two locations upstairs.) Write a predicate ONSTAIRSP that returns T if its input is either FRONT-STAIRS or BACK-STAIRS.

f. Where's Robbie? Write a function of no inputs called WHERE that tells where Robbie is. If he is in the library, (WHERE) should say (ROBBIE IS UPSTAIRS IN THE LIBRARY). If he is in the kitchen, it should say (ROBBIE IS DOWNSTAIRS IN THE KITCHEN). If he is on the front stairs, it should say (ROBBIE IS ON THE FRONT-STAIRS).

g. Write a function MOVE that takes one input, a direction, and moves Robbie in that direction. MOVE should make use of the LOOK function you wrote previously, and should call SET-ROBBIE-LOCATION to move him. If Robbie can't move in the specified direction an appropriate message should be returned. For example, if Robbie is in the pantry, (MOVE 'SOUTH) should return something like (OUCH! ROBBIE HIT A WALL). (MOVE 'NORTH) should change Robbie's location and return (ROBBIE IS DOWNSTAIRS IN THE KITCHEN).

h. Starting from the pantry, take Robbie to the library via the back stairs. Then take him to the kitchen, but do not lead him through the downstairs bedroom on the way.

```
(setf rooms

    '((living-room        (north front-stairs)
                          (south dining-room)
                          (east kitchen))

     (upstairs-bedroom    (west library)
                          (south front-stairs))

     (dining-room         (north living-room)
                          (east pantry)
                          (west downstairs-bedroom))

     (kitchen             (west living-room)
                          (south pantry))

     (pantry              (north kitchen)
                          (west dining-room))

     (downstairs-bedroom  (north back-stairs)
                          (east dining-room))

     (back-stairs         (south downstairs-bedroom)
                          (north library))

     (front-stairs        (north upstairs-bedroom)
                          (south living-room))

     (library             (east upstairs-bedroom)
                          (south back-stairs))))
```

Figure 6-3 Table of Rooms.

6   Advanced Topics

6.10   TREES

Trees are nested lists. All the functions covered so far operate on the top level of a list; they do not look at any more of the structure than that. Lisp also includes a few functions that operate on the entire list structure. Two of these are SUBST and SUBLIS. In chapter 8 we will write many more functions that operate on trees.

6.10.1 SUBST

The SUBST function substitutes one item for another everywhere it appears in a list. It takes three inputs whose order is as in the phrase "substitute x for y in z." Here is an example of substituting FRED for BILL in a certain list:

```
> (subst 'fred 'bill
    '(bill jones sent me an itemized
          bill for the tires))
(FRED JONES SENT ME AN ITEMIZED
    FRED FOR THE TIRES)
```

If the symbol being sought doesn't appear at all in the list, SUBST returns the original list unchanged.

```
> (subst 'bill 'fred '(keep off the grass))
(KEEP OFF THE GRASS)

> (subst 'on 'off '(keep off the grass))
(KEEP ON THE GRASS)
```

SUBST looks at the entire structure of the list, not just the top-level elements.

```
> (subst 'the 'a
        '((a hatter) (a hare) and (a dormouse)))
((THE HATTER) (THE HARE) AND (THE DORMOUSE))
```

6.10.2 SUBLIS

SUBLIS is like SUBST, except it can make many substitutions simultaneously. The first input to SUBLIS is a table whose entries are dotted pairs. The second input is the list in which the substitutions are to be made.

```
> (sublis '((roses . violets)  (red . blue))
          '(roses are red))
(VIOLETS ARE BLUE)

(setf dotted-words
  '((one   . un)
    (two   . deux)
    (three . trois)
    (four  . quatre)
    (five  . cinq)))

> (sublis dotted-words '(three one four one five))
(TROIS UN QUATRE UN CINQ)
```

EXERCISE

6.42. Write a function called ROYAL-WE that changes every occurrence of the symbol I in a list to the symbol WE. Calling this function on the list (IF I LEARN LISP I WILL BE PLEASED) should return the list (IF WE LEARN LISP WE WILL BE PLEASED).

6.11 EFFICIENCY OF LIST OPERATIONS

At the beginning of the chapter we talked about how lists appear symmetric in parenthesis notation, but they really aren't. Another way this asymmetry shows up is in the relative speed or efficiency of certain operations. For example, it is trivial to extract the first element of a list, but expensive to extract the last. When extracting the first element, we start with a pointer to the first cons cell; the FIRST function merely has to get the pointer from the car of that cell and return it. Finding the last element of the list involves much more work, because the only way to get to it is to follow the chain of pointers from the first cell until we get to a cell whose cdr is an atom. Only then can we look in the car. If the original list is very long, it may take quite a while to find the last cell by "cdring down the list," as it is called.

Computers can follow chains of a hundred thousand cons cells or more in well under a second, so you won't normally notice the speed difference between FIRST and LAST if you are calling them from the top-level read-eval-print loop. But if you write a large program that involves many list operations, the difference will become noticeable.

Another factor affecting the speed of a function is how much consing it does. Creating new cons cells takes time, and it also fills up the computer's memory. Eventually some of these cells will be discarded, but they still take up space. In some Lisp implementations, memory can become completely full with useless cons cells, in which case the machine must stop temporarily and perform a **garbage collection**. The more consing a function does, the more frequent the garbage collections. Let's compare the efficiency of these two versions of ADD-TO-END:

```
(defun add-to-end-1 (x y)
  (append x (list y)))

(defun add-to-end-2 (x y)
  (reverse (cons y (reverse x))))
```

Suppose the first input to these functions is a list of n elements. ADD-TO-END-1 copies its first input using APPEND, which tacks the list containing the second input onto the end. It thus creates a total of $n+1$ cons cells. ADD-TO-END-2 begins by reversing its first input, which creates n new cons cells; it then conses the second input onto that, which makes one new cell; finally it reverses the result, which makes another $n+1$ new cells. So ADD-TO-END-2 creates a total of $n+1+(n+1)$ cons cells, of which the final $n+1$ form the result. The other $n+1$ are thrown away shortly after they are created; they become "garbage." Clearly, ADD-TO-END-1 is the more efficient function, since it creates fewer cons cells.

6.12 SHARED STRUCTURE

Two lists are said to share structure if they have cons cells in common. Lists that are typed in from the keyboard will never share structure, because READ builds every list it sees from fresh cons cells. But using CAR, CDR, and

CONS it is possible to create lists that do share structure. For example, we can make X and Y point to lists that share some structure by doing the following:

```
> (setf x '(a b c))
(A B C)

> (setf y (cons 'd (cdr x)))
(D B C)
```

The value of X is (A B C) and the value of Y is (D B C). The lists share the same cons cell structure for (B C), as the following indicates. The sharing comes about because we built Y from (CDR X). If we had simply said (SETF Y '(D B C)), no structure would be shared with X.

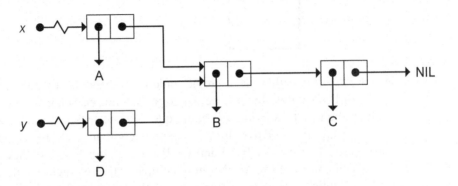

6.13 EQUALITY OF OBJECTS

In Lisp, symbols are unique, meaning there can be only one symbol in the computer's memory with a given name.[**] Every object in the memory has a numbered location, called its **address**. Since a symbol exists in only one place in memory, symbols have unique addresses. So in the list (TIME AFTER TIME), the two occurrences of the symbol TIME must refer to the same address. There cannot be two separate symbols named TIME.

[**]Note to instructors: We are assuming that only the standard packages are present, and there are no uninterned symbols. These details will not interest the beginning Lisper.

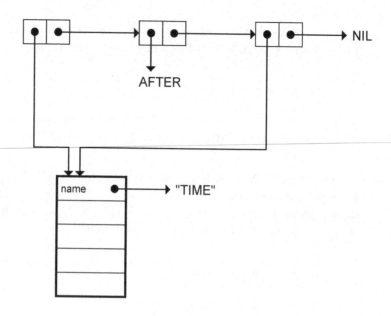

Lists, on the other hand, are not unique. We can easily have two different lists (A B C) simply by making two separate cons cell chains. The symbols to which the two lists point will be unique, but the lists themselves will not be. This means the EQUAL function cannot compare lists by comparing their addresses, because (A B C) and (A B C) are equal even if they are distinct cons cell chains. EQUAL therefore compares lists element by element. If the corresponding elements of two lists are equal, then the lists themselves are considered equal.

```
> (setf x1 (list 'a 'b 'c))        Make a fresh list (A B C).
(A B C)

> (setf x2 (list 'a 'b 'c))        Make another list (A B C).
(A B C)

> (equal x1 x2)                    The lists are EQUAL.
T
```

If we want to tell whether two pointers point to the same object, we must compare their addresses. The EQ predicate (pronounced ''eek'') does this. Lists are EQ to each other only if they have the same address; no element by element comparison is done.

```
> (eq x1 x2)                       The two lists are not EQ.
NIL
```

```
> (setf z x1)                 Now Z points to the same list as X1.
(A B C)

> (eq z x1)                   So Z and X1 are EQ.
T

> (eq z '(a b c))             These lists have different addresses.
NIL

> (equal z '(a b c))          But they have the same elements.
T
```

The EQ function is faster than the EQUAL function because EQ only has to compare an address against another address, whereas EQUAL has to first test if its inputs are lists, and if so it must compare each element of one against the corresponding element of the other. Due to its greater efficiency, programmers often use EQ instead of EQUAL when symbols are being compared. They don't usually use EQ on lists, unless they want to tell whether two cons cells are the same.

Numbers have different internal representations in different Lisp systems. In some implementations each number has a unique address, whereas in others this is not true. Therefore EQ should never be used to compare numbers.

The EQL predicate is a slightly more general variant of EQ. It compares the addresses of objects like EQ does, except that for two numbers of the same type (for example, both integers), it will compare their values instead. Numbers of different types are not EQL, even if their values are the same.

```
(eql 'foo 'foo)  ⇒  t

(eql 3 3)  ⇒  t

(eql 3 3.0)  ⇒  nil      Different types.
```

EQL is the "standard" comparison predicate in Common Lisp. Functions such as MEMBER and ASSOC that contain implicit equality tests do them using EQL unless told to use some other predicate.

For comparing numbers of disparate types, there is yet another equality predicate called =. This predicate is the most efficient way to compare two numbers. It is an error to give it any other kind of input.

```
(= 3 3.0)  ⇒  t

(= 'foo 'foo)  ⇒  Error! FOO is not a number.
```

Finally, the EQUALP predicate is similar to EQUAL, but in a few ways more liberal. One example is ignoring case distinctions in strings.

```
(equal "foo bar" "Foo BAR")   ⇒   nil

(equalp "foo bar" "Foo BAR")   ⇒   t
```

Beginners are frequently confused by the profusion of equality tests in Common Lisp. I recommend forgetting about all of these specialized functions; just remember two bits of advice. First, use EQUAL: It does what you want. Second, remember that built-in functions like MEMBER and ASSOC, which involve implicit equality tests, use EQL by default, for efficiency reasons. That means they will not compare lists correctly unless you tell them to use a different equality predicate. The next section explains how to do that. To summarize:

- EQ is the fastest equality test: It compares addresses. Experts use it to compare symbols quickly, and to test whether two cons cells are physically the same object. It should not be used to compare numbers.

- EQL is like EQ except it can safely compare numbers of the same type, such as two integers or two floating point numbers. It is the default equality test in Common Lisp.

- EQUAL is the predicate beginners should use. It compares lists element by element; otherwise it works like EQL.

- EQUALP is more liberal than EQUAL: It ignores case distinctions in strings, among other things.

- = is the most efficient way to compare numbers, and the only way to compare numbers of disparate types, such as 3 and 3.0. It only accepts numbers.

6.14 KEYWORD ARGUMENTS

Many Common Lisp functions that work on lists can take extra, optional arguments called **keyword arguments**. For example, the REMOVE function takes an optional argument called :COUNT that tells it how many instances of the item to remove.

```
(setf text '(b a n a n a - p a n d a))
```

```
> (remove 'a text)                    Remove all As.
(B N N - P N D)

> (remove 'a text :count 3)           Remove 3 As.
(B N N - P A N D A)
```

Remove also accepts a :FROM-END keyword. If its value is non-NIL, then REMOVE starts from the end of the list instead of from the beginning. So, to remove the last two As in the list, we could write:

```
> (remove 'a text :count 2 :from-end t)
(B A N A N A - P N D)
```

A keyword is a special type of symbol whose name is always preceded by a colon. The symbols COUNT and :COUNT are not the same; they are different objects and not EQ to each other.[***]   Keywords always evaluate to themselves, so they do not need to be quoted. It is an error to try to change the value of a keyword. The KEYWORDP predicate returns T if its input is a keyword.

```
:count  ⇒  :count

(symbolp :count)  ⇒  t

(equal :count 'count)  ⇒  nil

(keywordp 'count)  ⇒  nil

(keywordp :count)  ⇒  t
```

Another function that takes keyword arguments is MEMBER. Normally, MEMBER uses EQL to test whether an item appears in a set. EQL will work correctly for both symbols and numbers. But suppose our set contains lists? In that case we must use EQUAL for the equality test, or else MEMBER won't find the item we're looking for:

```
(setf cards
   '((3 clubs) (5 diamonds) (ace spades)))

(member '(5 diamonds) cards)  ⇒  nil

(second cards)  ⇒  (5 diamonds)
```

[***]Even though these symbols have the same name, they exist in different "packages" and so are distinct.

```
(eql (second cards) '(5 diamonds))   ⇒   nil

(equal (second cards) '(5 diamonds))   ⇒   t
```

The :TEST keyword can be used with MEMBER to specify a different function for the equality test. We write #'EQUAL to specially quote the function for use as an input to MEMBER.

```
> (member '(5 diamonds) cards :test #'equal)
((5 DIAMONDS) (ACE SPADES))
```

All list functions that include equality tests accept a :TEST keyword argument. REMOVE is another example. We can't remove (5 DIAMONDS) from CARDS unless we tell REMOVE to use EQUAL for its equality test.

```
> (remove '(5 diamonds) cards)
((3 CLUBS) (5 DIAMONDS) (ACE SPADES))

> (remove '(5 diamonds) cards :test #'equal)
((3 CLUBS) (ACE SPADES))
```

Other functions that accept a :TEST keyword are UNION, INTERSECTION, SET-DIFFERENCE, ASSOC, RASSOC, SUBST, and SUBLIS. To find out which keywords a function accepts, use the online documentation. It is an error to supply a keyword to a function that isn't expecting that keyword.

```
> (remove '(ace spades) cards :reason 'bad-luck)
Error! :REASON is an invalid keyword argument
to REMOVE.
```

FUNCTIONS COVERED IN ADVANCED TOPICS

Tree functions: SUBST, SUBLIS.

Additional equality functions: EQ, EQL, EQUALP, =.

Keyword predicate: KEYWORDP.

7

Applicative Programming

7.1 INTRODUCTION

The three programming styles we will cover in this book are **applicative** programming, **recursion**, and **iteration**. Many instructors prefer to teach recursion first, but I believe applicative programming is the easiest for beginners to learn. To accomodate everyone's taste, Chapters 7 and 8 have been made independent; they can be covered in either order.

Applicative programming is based on the idea that functions are data, just like symbols and lists are data, so one should be able to pass functions as inputs to other functions, and also return functions as values. The **applicative operators** we will study in this chapter are functions that take another function as input and apply it to the elements of a list in various ways. These operators are all built from a primitive function known as FUNCALL. In the Advanced Topics section we will write our own applicative operator, and also write a function that constructs and returns new functions.

7.2 FUNCALL

FUNCALL calls a function on some inputs. We can use FUNCALL to call the CONS function on the inputs A and B like this:

```
(funcall #'cons 'a 'b)  ⇒  (a . b)
```

The #' (or ''sharp quote'') notation is the correct way to quote a function in Common Lisp. If you want to see what the function CONS looks like in your implementation, try the following example in your Lisp:

```
> (setf fn #'cons)
#<Compiled-function CONS {6041410}>

> fn
#<Compiled-function CONS {6041410}>

> (type-of fn)
COMPILED-FUNCTION

> (funcall fn 'c 'd)
(C . D)
```

The value of the variable FN is a function object. TYPE-OF shows that the object is of type COMPILED-FUNCTION. So you see that functions and symbols are not the same. The symbol CONS serves as the name of the CONS function, but it is not the actual function. The relationship between functions and the symbols that name them is explained in Advanced Topics section 3.18.

Note that only ordinary functions can be quoted with #'. It is an error to quote a macro function or special function this way, or to quote a symbol with #' if that symbol does not name a function.

```
> #'if
Error: IF is not an ordinary function.

> #'turnips
Error: TURNIPS is an undefined function.
```

7.3 THE MAPCAR OPERATOR

MAPCAR is the most frequently used applicative operator. It applies a function to each element of a list, one at a time, and returns a list of the results. Suppose we have written a function to square a single number. By itself, this function cannot square a list of numbers, because * doesn't work on lists.

```
(defun square (n) (* n n))

(square 3)  ⇒  9
```

```
(square '(1 2 3 4 5))   ⇒   Error! Wrong type input to *.
```

With MAPCAR we can apply SQUARE to each element of the list individually. To pass the SQUARE function as an input to MAPCAR, we quote it by writing #'SQUARE.

```
>   (mapcar #'square '(1 2 3 4 5))
(1 4 9 16 25)

>   (mapcar #'square '(3 8 -3 5 2 10))
(9 64 9 25 4 100)
```

Here is a graphical description of the MAPCAR operator. As you can see, each element of the input list is mapped independently to a corresponding element in the output.

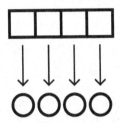

When MAPCAR is used on a list of length *n*, the resulting list also has exactly *n* elements. So if MAPCAR is used on the empty list, the result is the empty list.

```
(mapcar #'square '())   ⇒   nil
```

7.4 MANIPULATING TABLES WITH MAPCAR

Suppose we set the global variable WORDS to a table of English and French words:

```
(setf words
  '((one un)
    (two deux)
    (three trois)
    (four quatre)
    (five cinq)))
```

We can perform several useful manipulations on this table with MAPCAR. We can extract the English words by taking the first component of each table entry:

```
> (mapcar #'first words)
(ONE TWO THREE FOUR FIVE)
```

We can extract the French words by taking the second component of each entry:

```
> (mapcar #'second words)
(UN DEUX TROIS QUATRE CINQ)
```

We can create a French–English dictionary from the English–French one by reversing each table element:

```
> (mapcar #'reverse words)
((UN ONE)
 (DEUX TWO)
 (TROIS THREE)
 (QUATRE FOUR)
 (CINQ FIVE))
```

Given a function TRANSLATE, defined using ASSOC, we can translate a string of English digits into a string of French ones:

```
(defun translate (x)
  (second (assoc x words)))

> (mapcar #'translate '(three one four one five))
(TROIS UN QUATRE UN CINQ)
```

Besides MAPCAR, there are several other applicative operators built in to Common Lisp. Many more are defined by programmers as they are needed, using FUNCALL.

EXERCISES

7.1. Write an ADD1 function that adds one to its input. Then write an expression to add one to each element of the list (1 3 5 7 9).

7.2. Let the global variable DAILY-PLANET contain the following table:

```
((olsen jimmy 123-76-4535 cub-reporter)
 (kent  clark 089-52-6787 reporter)
 (lane  lois  951-26-1438 reporter)
 (white perry 355-16-7439 editor))
```

Each table entry consists of a last name, a first name, a social security number, and a job title. Use MAPCAR on this table to extract a list of social security numbers.

7.3. Write an expression to apply the ZEROP predicate to each element of the list (2 0 3 4 0 -5 -6). The answer you get should be a list of Ts and NILs.

7.4. Suppose we want to solve a problem similar to the preceding one, but instead of testing whether an element is zero, we want to test whether it is greater than five. We can't use > directly for this because > is a function of two inputs; MAPCAR will only give it one input. Show how first writing a one-input function called GREATER-THAN-FIVE-P would help.

7.5 LAMBDA EXPRESSIONS

There are two ways to specify the function to be used by an applicative operator. The first way is to define the function with DEFUN and then specify it by #'*name*, as we have been doing. The second way is to pass the function definition directly. This is done by writing a list called a **lambda expression**. For example, the following lambda expression computes the square of its input:

```
(lambda (n) (* n n))
```

Since lambda expressions are functions, they can be passed directly to MAPCAR by quoting them with #'. This saves you the trouble of writing a separate DEFUN before calling MAPCAR.

```
> (mapcar #'(lambda (n) (* n n)) '(1 2 3 4 5))
(1 4 9 16 25)
```

Lambda expressions look similar to DEFUNs, except that the function name is missing and the word LAMBDA appears in place of DEFUN. But lambda expressions are actually unnamed functions. LAMBDA is not a macro or special function that has to be evaluated, like DEFUN. Rather it is a marker that says "this list represents a function."

Lambda expressions are especially useful for synthesizing one-input functions from related functions of two inputs. For example, suppose we wanted to multiply every element of a list by 10. We might be tempted to write something like:

```
(mapcar #'* '(1 2 3 4 5))
```

but where is the 10 supposed to go? The * function needs two inputs, but MAPCAR is only going to give it one. The correct way to solve this problem is to write a lambda expression of *one* input that multiplies its input by 10. Then we can feed the lambda expression to MAPCAR.

```
> (mapcar #'(lambda (n) (* n 10)) '(1 2 3 4 5))
(10 20 30 40 50)
```

Here is another example of the use of MAPCAR along with a lambda expression. We will turn each element of a list of names into a list (HI THERE *name*).

```
> (mapcar #'(lambda (x) (list 'hi 'there x))
          '(joe fred wanda))
((HI THERE JOE) (HI THERE FRED) (HI THERE WANDA))
```

If you type in a quoted lambda expression at top level, the result you get back depends on the particular Lisp implementation you're using. It might look like any of the following:

```
> (lambda (n) (* n 10))              Don't forget to quote it!
Error: Undefined function LAMBDA.

> #'(lambda (n) (* n 10))
(LAMBDA (N) (* N 10))

> #'(lambda (n) (* n 10))
#<Interpreted-function 3515162>

> #'(lambda (n) (* n 10))
#<Lexical-closure {7142156}>
```

Throughout this book we will refer to the objects you get back from a #'(LAMBDA...) expression as **lexical closures**. They will be discussed in more detail in the Advanced Topics section.

EXERCISES

7.5. Write a lambda expression to subtract seven from a number.

7.6. Write a lambda expression that returns T if its input is T or NIL, but NIL for any other input.

7.7. Write a function that takes a list such as (UP DOWN UP UP) and "flips" each element, returning (DOWN UP DOWN DOWN). Your function should include a lambda expression that knows how to flip an individual element, plus an applicative operator to do this to every element of the list.

7.6 THE FIND-IF OPERATOR

FIND-IF is another applicative operator. If you give FIND-IF a predicate and a list as input, it will find the first element of the list for which the predicate returns *true* (any non-NIL value). FIND-IF returns that element.

```
> (find-if #'oddp '(2 4 6 7 8 9))
7

> (find-if #'(lambda (x) (> x 3))
           '(2 4 6 7 8 9))
4
```

Here is a graphical description of what FIND-IF does:

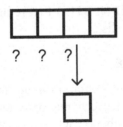

If no elements satisfy the predicate, FIND-IF returns NIL.

```
(find-if #'oddp '(2 4 6 8))   ⇒   nil
```

7.7 WRITING ASSOC WITH FIND-IF

ASSOC searches for a table entry with a specified key. We can write a simple version of ASSOC that uses FIND-IF to search the table.

```
(defun my-assoc (key table)
    (find-if #'(lambda (entry)
                (equal key (first entry)))
             table))
```

```
(my-assoc 'two words)  ⇒  (TWO DEUX)
```

The lambda expression (actually a lexical closure) that MY-ASSOC passes to FIND-IF takes a table entry such as (ONE UN) as input. It returns T if the first element of the entry matches the key that is the first input to MY-ASSOC. FIND-IF calls the closure on each entry in the table, until it finds one that makes the closure return T.

Notice that the expression (EQUAL KEY (FIRST ENTRY)) that appears in the body of the lambda expression refers to two variables. ENTRY is local to the lambda expression, but KEY is not. KEY is local to MY-ASSOC. This illustrates an important point about lambda expressions: Inside the body of a lambda expression we can not only reference its local variables, we can also reference any local variables of the function containing the lambda expression.

EXERCISES

7.8. Write a function that takes two inputs, X and K, and returns the first number in the list X that is roughly equal to K. Let's say that "roughly equal" means no less than K − 10 and no more than K + 10.

7.9. Write a function FIND-NESTED that returns the first element of a list that is itself a non-NIL list.

MINI KEYBOARD EXERCISE

7.10. In this exercise we will write a program to transpose a song from one key to another. In order to manipulate notes more efficiently, we will translate them into numbers. Here is the correspondence between notes and numbers for a one-octave scale:

C	=	1	F-SHARP	=	7
C-SHARP	=	2	G	=	8
D	=	3	G-SHARP	=	9
D-SHARP	=	4	A	=	10
E	=	5	A-SHARP	=	11
F	=	6	B	=	12

a. Write a table to represent this information. Store it in a global variable called NOTE-TABLE.

b. Write a function called NUMBERS that takes a list of notes as input and returns the corresponding list of numbers. (NUMBERS '(E D C D E E)) should return (5 3 1 3 5 5). This list represents the first seven notes of "Mary Had a Little Lamb."

c. Write a function called NOTES that takes a list of numbers as input and returns the corresponding list of notes. (NOTES '(5 3 1 3 5 5 5)) should return (E D C D E E E). *Hint:* Since NOTE-TABLE is keyed by note, ASSOC can't look up numbers in it; neither can RASSOC, since the elements are lists, not dotted pairs. Write your own table-searching function to search NOTE-TABLE by number instead of by note.

d. Notice that NOTES and NUMBERS are mutual inverses:

> *For X a list of notes:*
> X   =   (NOTES (NUMBERS X))

> *For X a list of numbers:*
> X   =   (NUMBERS (NOTES X))

What can be said about (NOTES (NOTES X)) and (NUMBERS (NUMBERS X))?

e. To transpose a piece of music up by *n* half steps, we begin by adding the value *n* to each note in the piece. Write a function called RAISE that takes a number *n* and a list of numbers as input and raises each number in the list by the value *n*. (RAISE 5 '(5 3 1 3 5 5 5)) should return (10 8 6 8 10 10 10), which is "Mary Had a Little Lamb" transposed five half steps from the key of C to the key of F.

f. Sometimes when we raise the value of a note, we may raise it right into the next octave. For instance, if we raise the triad C-E-G represented by the list (1 5 8) into the key of F by adding five to each note, we get (6 10 13), or F-A-C. Here the C note, represented by the number 13, is an octave above the regular C, represented by 1. Write a function called NORMALIZE that takes a list of numbers as input and "normalizes" them to make them be between 1 and 12. A number greater than 12 should have 12 subtracted from it; a number less than 1 should have 12 added to it. (NORMALIZE '(6 10 13)) should return (6 10 1).

g. Write a function TRANSPOSE that takes a number *n* and a song as input, and returns the song transposed by *n* half steps. (TRANSPOSE 5 '(E D C D E E E)) should return (A G F G A A A). Your solution should assume the availability of the NUMBERS, NOTES, RAISE, and NORMALIZE functions. Try transposing "Mary Had a Little Lamb" up by 11 half steps. What happens if you transpose it by 12 half steps? How about −1 half steps?

7.8 REMOVE-IF AND REMOVE-IF-NOT

REMOVE-IF is another applicative operator that takes a predicate as input. REMOVE-IF removes all the items from a list that satisfy the predicate, and returns a list of what's left.

```
> (remove-if #'numberp '(2 for 1 sale))
(FOR SALE)
```

```
> (remove-if #'oddp '(1 2 3 4 5 6 7))
(2 4 6)
```

Here is a graphical description of REMOVE-IF:

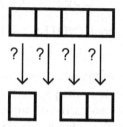

Suppose we want to find all the positive elements in a list of numbers. The PLUSP predicate tests if a number is greater than zero. To invert the sense of this predicate we wrap a NOT around it using a lambda expression, as shown in the following. After removing all the elements that satisfy (NOT (PLUSP *x*)), what we have left are the positive elements.

```
> (remove-if #'(lambda (x) (not (plusp x)))
             '(2 0 -4 6 -8 10))
(2 6 10)
```

The REMOVE-IF-NOT operator is used more frequently than REMOVE-IF. It works just like REMOVE-IF except it automatically inverts the sense of the predicate. This means the only items that will be removed are those for which the predicate returns NIL. So REMOVE-IF-NOT returns a list of all the items that *satisfy* the predicate. Thus, if we choose PLUSP as the predicate, REMOVE-IF-NOT will find all the positive numbers in a list.

```
> (remove-if-not #'plusp '(2 0 -4 6 -8 10))
(2 6 10)
   > (remove-if-not #'oddp '(2 0 -4 6 -8 10))
NIL
```

Here are some additional examples of REMOVE-IF-NOT:

```
> (remove-if-not #'(lambda (x) (> x 3))
                 '(2 4 6 8 4 2 1))
(4 6 8 4)

> (remove-if-not #'numberp
       '(3 apples 4 pears and 2 little plums))
(3 4 2)

> (remove-if-not #'symbolp
       '(3 apples 4 pears and 2 little plums))
(APPLES PEARS AND LITTLE PLUMS)
```

Here is a function, COUNT-ZEROS, that counts how many zeros appear in a list of numbers. It does this by taking the subset of the list elements that are zero, and then taking the length of the result.

```
(remove-if-not #'zerop '(34 0 0 95 0))  ⟹  (0 0 0)

(defun count-zeros (x)
    (length (remove-if-not #'zerop x)))

(count-zeros '(34 0 0 95 0))  ⟹  3

(count-zeros '(1 0 63 0 38))  ⟹  2

(count-zeros '(0 0 0 0 0))  ⟹  5

(count-zeros '(1 2 3 4 5))  ⟹  0
```

EXERCISES

7.11. Write a function to pick out those numbers in a list that are greater than one and less than five.

7.12. Write a function that counts how many times the word ''the'' appears in a sentence.

7.13. Write a function that picks from a list of lists those of exactly length two.

7.14. Here is a version of SET-DIFFERENCE written with REMOVE-IF:

```
(defun my-setdiff (x y)
  (remove-if #'(lambda (e) (member e y))
             x))
```

Show how the INTERSECTION and UNION functions can be written using REMOVE-IF or REMOVE-IF-NOT.

MINI KEYBOARD EXERCISE

7.15. In this keyboard exercise we will manipulate playing cards with applicative operators. A card will be represented by a list of form (*rank suit*), for example, (ACE SPADES) or (2 CLUBS). A hand will be represented by a list of cards.

a. Write the functions RANK and SUIT that return the rank and suit of a card, respectively. (RANK '(2 CLUBS)) should return 2, and (SUIT '(2 CLUBS)) should return CLUBS.

b. Set the global variable MY-HAND to the following hand of cards:

```
((3 hearts)
 (5 clubs)
 (2 diamonds)
 (4 diamonds)
 (ace spades))
```

Now write a function COUNT-SUIT that takes two inputs, a suit and a hand of cards, and returns the number of cards belonging to that suit. (COUNT-SUIT 'DIAMONDS MY-HAND) should return 2.

c. Set the global variable COLORS to the following table:

```
((clubs black)
 (diamonds red)
 (hearts red)
 (spades black))
```

Now write a function COLOR-OF that uses the table COLORS to retrieve the color of a card. (COLOR-OF '(2 CLUBS)) should return BLACK. (COLOR-OF '(6 HEARTS)) should return RED.

d. Write a function FIRST-RED that returns the first card of a hand that is of a red suit, or NIL if none are.

e. Write a function BLACK-CARDS that returns a list of all the black cards in a hand.

f. Write a function WHAT-RANKS that takes two inputs, a suit and a hand, and returns the ranks of all cards belonging to that suit. (WHAT-RANKS 'DIAMONDS MY-HAND) should return the list (2 4). (WHAT-RANKS 'SPADES MY-HAND) should return the list (ACE). *Hint:* First extract all the cards of the specified suit, then use another operator to get the ranks of those cards.

g. Set the global variable ALL-RANKS to the list

```
(2 3 4 5 6 7 8 9 10 jack queen king ace)
```

Then write a predicate HIGHER-RANK-P that takes two cards as input and returns true if the first card has a higher rank than the second. *Hint:* look at the BEFOREP predicate on page 171 of Chapter 6.

h. Write a function HIGH-CARD that returns the highest ranked card in a hand. *Hint:* One way to solve this is to use FIND-IF to search a list of ranks (ordered from high to low) to find the highest rank that appears in the hand. Then use ASSOC on the hand to pick the card with that rank. Another solution would be to use REDUCE (defined in the next section) to repeatedly pick the highest card of each pair.

7.9 THE REDUCE OPERATOR

REDUCE is an applicative operator that reduces the elements of a list into a single result. REDUCE takes a function and a list as input, but unlike the other operators we've seen, REDUCE must be given a function that accepts *two* inputs. Example: To add up a list of numbers with REDUCE, we use + as the reducing function.

```
(reduce #'+ '(1 2 3))  ⇒  6

(reduce #'+ '(10 9 8 7 6))  ⇒  40

(reduce #'+ '(5))  ⇒  5

(reduce #'+ nil)  ⇒  0
```

Similarly, to multiply a bunch of numbers together, we use * as the reducing function:

```
(reduce #'* '(2 4 5))  ⇒  40

(reduce #'* '(3 4 0 7))  ⇒  0

(reduce #'* '(8))  ⇒  8
```

We can also apply reduction to lists of lists. To turn a table into a one-level list, we use APPEND as the reducing function:

```
> (reduce #'append
           '((one un) (two deux) (three trois)))
(ONE UN TWO DEUX THREE TROIS)
```

Here is A graphical description of REDUCE:

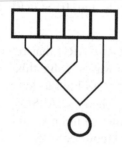

EXERCISES

7.16. Suppose we had a list of sets ((A B C) (C D A) (F B D) (G)) that we wanted to collapse into one big set. If we use APPEND for our reducing function, the result won't be a true set, because some elements will appear more than once. What reducing function should be used instead?

7.17. Write a function that, given a list of lists, returns the total length of all the lists. This problem can be solved two different ways.

7.18. (REDUCE #'+ NIL) returns 0, but (REDUCE #'* NIL) returns 1. Why do you think this is?

7.10 EVERY

EVERY takes a predicate and a list as input. It returns T if there is no element that causes the predicate to return false. Examples:

```
> (every #'numberp '(1 2 3 4 5))
T

> (every #'numberp '(1 2 A B C 5))
NIL
```

```
> (every #'(lambda (x) (> x 0)) '(1 2 3 4 5))
T

> (every #'(lambda (x) (> x 0)) '(1 2 3 -4 5))
NIL
```

If EVERY is called with NIL as its second argument, it simply returns T, since the empty list has no elements that could fail to satisfy the predicate.

```
> (every #'oddp nil)
T

> (every #'evenp nil)
T
```

EVERY can also operate on multiple lists, given a predicate that accepts multiple inputs.

```
> (every #'> '(10 20 30 40) '(1 5 11 23))
T
```

Since 10 is greater than 1, 20 greater than 5, 30 greater than 11, and 40 greater than 23, EVERY returns T.

EXERCISES

7.19. Write a function ALL-ODD that returns T if every element of a list of numbers is odd.

7.20. Write a function NONE-ODD that returns T if every element of a list of numbers is not odd.

7.21. Write a function NOT-ALL-ODD that returns T if not every element of a list of numbers is odd.

7.22. Write a function NOT-NONE-ODD that returns T if it is not the case that a list of numbers contains no odd elements.

7.23. Are all four of the above functions distinct from one another, or are some of them the same? Can you think of better names for the last two?

SUMMARY

Applicative operators are functions that apply other functions to data structures. There are many possible applicative operators, only a few of which are built in to Lisp. Advanced Lisp programmers make up their own operators whenever they need new ones.

MAPCAR applies a function to every element of a list and returns a list of the results. FIND-IF searches a list and returns the first element that satisfies a predicate. REMOVE-IF removes all the elements of a list that satisfy a predicate, so the list it returns contains only those elements that fail to satisfy it. REMOVE-IF-NOT is used more frequently than REMOVE-IF. It returns all the elements that *do* satisfy the predicate, having removed those that don't satisfy it. EVERY returns T only if every element of a list satisfies a predicate. REDUCE uses a reducing function to reduce a list to a single value.

REVIEW EXERCISES

7.24. What is an applicative operator?

7.25. Why are lambda expressions useful? Is it possible to do without them?

7.26. Show how to write FIND-IF given REMOVE-IF-NOT.

7.27. Show how to write EVERY given REMOVE-IF.

7.28. Devise a graphical description for the EVERY operator.

FUNCTIONS COVERED IN THIS CHAPTER

Applicative operators: MAPCAR, FIND-IF, REMOVE-IF, REMOVE-IF-NOT, REDUCE, EVERY.

Lisp Toolkit: TRACE and DTRACE

The TRACE macro is used to watch particular functions as they are called and as they return. With each call you will see the arguments to the function; when the function returns you will see the return values. Each Lisp implementation has its own style for displaying trace information. The example below is typical:

```
(defun half (n) (* n 0.5))

(defun average (x y)
  (+ (half x) (half y)))
```

```
> (trace half average)
(HALF AVERAGE)

> (average 3 7)
 0: (AVERAGE 3 7)
   1: (HALF 3)
   1: returned 1.5
   1: (HALF 7)
   1: returned 3.5
 0: returned 5.0
5.0
```

If you call TRACE with no arguments, it returns the list of currently traced functions.

```
> (trace)
(HALF AVERAGE)
```

The UNTRACE macro turns off tracing for one or more functions. Since UNTRACE is a macro function like TRACE, its arguments should not be quoted.

```
> (untrace half)
(HALF)
```

Calling UNTRACE with no arguments untraces all currently traced functions.

```
> (untrace)
(AVERAGE)
```

In the remainder of this book we will use a more detailed tracing format that shows each variable in the argument list along with the value to which it is bound. For example:

```
> (average 3 7)
----Enter AVERAGE
|      X = 3
|      Y = 7
|    ----Enter HALF
|    |      N = 3
|     \--HALF returned 1.5
|    ----Enter HALF
|    |      N = 7
|     \--HALF returned 3.5
 \--AVERAGE returned 5.0
5.0
```

If your Lisp's TRACE isn't this detailed, don't panic, you can use mine. It's called DTRACE, and the full program listing is given in an appendix at the end of the book. This style of trace is especially helpful when tracing functions with several inputs, and even more so when the inputs are long, possibly nested, lists.

```
(defun add-to-end (x y)
   (append x (list y)))

(defun repeat-first (phrase)
   (add-to-end phrase (first phrase)))

> (dtrace add-to-end repeat-first)
(ADD-TO-END REPEAT-FIRST)

> (repeat-first '(for whom the bell tolls))
----Enter REPEAT-FIRST
|       PHRASE = (FOR WHOM THE BELL TOLLS)
|    ----Enter ADD-TO-END
|    |       X = (FOR WHOM THE BELL TOLLS)
|    |       Y = FOR
|    \--ADD-TO-END returned
|            (FOR WHOM THE BELL TOLLS FOR)
\--REPEAT-FIRST returned
         (FOR WHOM THE BELL TOLLS FOR)
(FOR WHOM THE BELL TOLLS FOR)
```

DUNTRACE undoes the effect of DTRACE. Don't try to trace a function with both TRACE and DTRACE at the same time: You may get very strange results.

We can use DTRACE to observe the behavior of applicative operators like FIND-IF. We will trace the ODDP function and then use ODDP as an input to FIND-IF.

```
(defun find-first-odd (x)
   (find-if #'oddp x))

> (dtrace find-first-odd oddp)
(FIND-FIRST-ODD ODDP)
```

```
> (find-first-odd '(2 4 6 7 8))
----Enter FIND-FIRST-ODD
|     X = (2 4 6 7 8)
|    ----Enter ODDP
|    |     NUMBER = 2
|     \--ODDP returned NIL
|    ----Enter ODDP
|    |     NUMBER = 4
|     \--ODDP returned NIL
|    ----Enter ODDP
|    |     NUMBER = 6
|     \--ODDP returned NIL
|    ----Enter ODDP
|    |     NUMBER = 7
|     \--ODDP returned T
 \--FIND-FIRST-ODD returned 7
7
```

This brings up one last point about the use of TRACE and DTRACE. Although they may be used to trace built-in functions such as ODDP, this sometimes turns out to be dangerous. Avoid tracing the most fundamental built-in functions such as EVAL, CONS, and +. Otherwise your Lisp might end up in an infinite loop, and you will have to abandon it and start over.

Keyboard Exercise

In this keyboard exercise we will develop a system for representing knowledge about "blocks world" scenes such as Figure 7-1. Assertions about the objects in a scene are represented as triples of form (*block attribute value*). Here are some assertions about block B2's attributes:

```
(b2 shape brick)
(b2 color red)
(b2 size small)
(b2 supports b1)
(b2 left-of b3)
```

A collection (in other words, a list) of assertions is called a **database**.

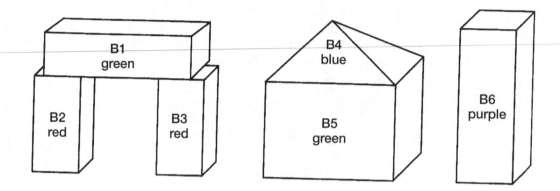

Figure 7-1 A typical blocks world scene.

Given a database describing the blocks in the figure, we can write functions to answer questions such as, "What color is block B2?" or "What blocks support block B1?" To answer these questions, we will use a function called a **pattern matcher** to search the database for us. For example, to find out the color of block B2, we use the pattern (B2 COLOR ?).

```
> (fetch '(b2 color ?))
((B2 COLOR RED))
```

To find which blocks support B1, we use the pattern (? SUPPORTS B1):

```
> (fetch '(? supports b1))
((B2 SUPPORTS B1) (B3 SUPPORTS B1))
```

FETCH returns those assertions from the database that match a given pattern. It should be apparent from the preceding examples that a pattern is a triple, like an assertion, with some of its elements replaced by question marks. Figure 7-2 shows some patterns and their English interpretations.

A question mark in a pattern means any value can match in that position. Thus, the pattern (B2 COLOR ?) can match assertions like (B2 COLOR RED), (B2 COLOR GREEN), (B2 COLOR BLUE), and so on. It cannot match the assertion (B1 COLOR RED), because the first element of the pattern is the symbol B2, whereas the first element of the assertion is B1.

Pattern	English Interpretation
(b1 color ?)	*What color is b1?*
(? color red)	*Which blocks are red?*
(b1 color red)	*Is b1 known to be red?*
(b1 ? b2)	*What relation is b1 to b2?*
(b1 ? ?)	*What is known about b1?*
(? supports ?)	*What support relationships exist?*
(? ? b1)	*What blocks are related to b1?*
(? ? ?)	*What's in the database?*

Figure 7-2 Some patterns and their interpretations.

EXERCISE

7.29. If the blocks database is already stored on the computer for you, load the file containing it. If not, you will have to type it in as it appears in Figure 7-3. Save the database in the global variable DATABASE.

a. Write a function MATCH-ELEMENT that takes two symbols as inputs. If the two are equal, or if the second is a question mark, MATCH-ELEMENT should return T. Otherwise it should return NIL. Thus (MATCH-ELEMENT 'RED 'RED) and (MATCH-ELEMENT 'RED '?) should return T, but (MATCH-ELEMENT 'RED 'BLUE) should return NIL. Make sure your function works correctly before proceeding further.

b. Write a function MATCH-TRIPLE that takes an assertion and a pattern as input, and returns T if the assertion matches the pattern. Both inputs will be three-element lists. (MATCH-TRIPLE '(B2 COLOR RED) '(B2 COLOR ?)) should return T. (MATCH-TRIPLE '(B2 COLOR RED) '(B1 COLOR GREEN)) should return NIL. *Hint:* Use MATCH-ELEMENT as a building block.

c. Write the function FETCH that takes a pattern as input and returns all assertions in the database that match the pattern. Remember that

DATABASE is a global variable. (FETCH '(B2 COLOR ?)) should return ((B2 COLOR RED)), and (FETCH '(? SUPPORTS B1)) should return ((B2 SUPPORTS B1) (B3 SUPPORTS B1)).

d. Use FETCH with patterns you construct yourself to answer the following questions. What shape is block B4? Which blocks are bricks? What relation is block B2 to block B3? List the color of every block. What facts are known about block B4?

e. Write a function that takes a block name as input and returns a *pattern* asking the color of the block. For example, given the input B3, your function should return the list (B3 COLOR ?).

f. Write a function SUPPORTERS that takes one input, a block, and returns a list of the blocks that support it. (SUPPORTERS 'B1) should return the list (B2 B3). Your function should work by constructing a pattern containing the block's name, using that pattern as input to FETCH, and then extracting the block names from the resulting list of assertions.

g. Write a predicate SUPP-CUBE that takes a block as input and returns true if that block is supported by a cube. (SUPP-CUBE 'B4) should return a true value; (SUPP-CUBE 'B1) should not because B1 is supported by bricks but not cubes. *Hint:* Use the result of the SUPPORTERS function as a starting point.

h. We are going to write a DESCRIPTION function that returns the description of a block. (DESCRIPTION 'B2) will return (SHAPE BRICK COLOR RED SIZE SMALL SUPPORTS B1 LEFT-OF B3). We will do this in steps. First, write a function DESC1 that takes a block as input and returns all assertions dealing with that block. (DESC1 'B6) should return ((B6 SHAPE BRICK) (B6 COLOR PURPLE) (B6 SIZE LARGE)).

i. Write a function DESC2 of one input that calls DESC1 and strips the block name off each element of the result. (DESC2 'B6) should return the list ((SHAPE BRICK) (COLOR PURPLE) (SIZE LARGE)).

j. Write the DESCRIPTION function. It should take one input, call DESC2, and merge the resulting list of lists into a single list. (DESCRIPTION 'B6) should return (SHAPE BRICK COLOR PURPLE SIZE LARGE).

k. What is the description of block B1? Of block B4?

l. Block B1 is made of wood, but block B2 is made of plastic. How would you add this information to the database?

```
(setf database
      '((b1 shape brick)
        (b1 color green)
        (b1 size small)
        (b1 supported-by b2)
        (b1 supported-by b3)
        (b2 shape brick)
        (b2 color red)
        (b2 size small)
        (b2 supports b1)
        (b2 left-of b3)
        (b3 shape brick)
        (b3 color red)
        (b3 size small)
        (b3 supports b1)
        (b3 right-of b2)
        (b4 shape pyramid)
        (b4 color blue)
        (b4 size large)
        (b4 supported-by b5)
        (b5 shape cube)
        (b5 color green)
        (b5 size large)
        (b5 supports b4)
        (b6 shape brick)
        (b6 color purple)
        (b6 size large)))
```

Figure 7-3 The blocks database.

7   Advanced Topics

7.11   OPERATING ON MULTIPLE LISTS

In the beginning of this chapter we used MAPCAR to apply a one-input function to the elements of a list. MAPCAR is not restricted to one-input functions, however. Given a function of *n* inputs, MAPCAR will map it over *n* lists. For example, given a list of people and a list of jobs, we can use MAPCAR with a two-input function to pair each person with a job:

```
> (mapcar #'(lambda (x y) (list x 'gets y))
           '(fred wilma george diane)
           '(job1 job2 job3 job4))
((FRED GETS JOB1)
 (WILMA GETS JOB2)
 (GEORGE GETS JOB3)
 (DIANE GETS JOB4))
```

MAPCAR goes through the two lists in parallel, taking one element from each at each step. If one list is shorter than the other, MAPCAR stops when it reaches the end of the shortest list.

Another example of operating on multiple lists is the problem of adding two lists of numbers pairwise:

```
> (mapcar #'+ '(1 2 3 4 5) '(60 70 80 90 100))
(61 72 83 94 105)

> (mapcar #'+ '(1 2 3) '(10 20 30 40 50))
(11 22 33)
```

EXERCISE

7.30. Recall the English–French dictionary we stored in the global variable WORDS earlier in the chapter. Given this dictionary plus the list or corresponding Spanish words (UNO DOS TRES QUATRO CINCO), write an expression to return a trilingual dictionary. The first entry of the dictionary should be (ONE UN UNO).

7.12 THE FUNCTION SPECIAL FUNCTION

Just as ' is shorthand for the QUOTE special function, #' is shorthand for the FUNCTION special function. Writing #'CONS is therefore equivalent to writing (FUNCTION CONS).

QUOTE always returns its unevaluated argument, but FUNCTION works a little differently. It returns the *functional interpretation* of its unevaluated argument. If the argument is a symbol, it generally returns the contents of the symbol's function cell. Often this is a compiled code object.

```
> 'cons
CONS

> #'cons
#<Compiled-function CONS 6041410>
```

On the other hand, if the argument to FUNCTION is a lambda expression, the result is usually a lexical closure.

```
> #'(lambda (x) (+ x 2))
#<Lexical-closure 3471524>
```

The result returned by FUNCTION is always some kind of function object. These objects are a form of data, just like symbols and lists. For example, we can store them in variables. We can also call them, using FUNCALL or APPLY. (APPLY was discussed in Advanced Topics section 3.21.)

```
> (setf g #'(lambda (x) (* x 10)))
#<Lexical-closure 41653824>

> (funcall g 12)
120
```

The value of the variable G is a lexical closure, which is a function. But G itself is not the name of any function; if we wrote (G 12) we would get an undefined function error.

7.13   KEYWORD ARGUMENTS TO APPLICATIVE OPERATORS

Some applicative operators, such as FIND-IF, REMOVE-IF, REMOVE-IF-NOT, and REDUCE, accept optional keyword arguments. For example, the :FROM-END keyword, if given a non-NIL value, causes the list to be processed from right to left.

```
> (find-if #'oddp '(2 3 4 5 6))      Find the first odd number.
3
```

```
> (find-if #'oddp '(2 3 4 5 6)       Find the last odd number.
          :from-end t)
5
```

The :FROM-END keyword is particularly interesting with REDUCE; it causes elements to be reduced from right to left instead of the usual left to right.

```
> (reduce #'cons '(a b c d e))
((((A . B) . C) . D) . E)
```

```
> (reduce #'cons '(a b c d e) :from-end t)
(A B C D . E)
```

REMOVE-IF and REMOVE-IF-NOT also accept a :COUNT keyword that specifies the maximum number of elements to be removed. See the online documentation or your Common Lisp reference manual for the complete list of keyword arguments accepted by a particular function. MAPCAR and EVERY do not accept any keyword arguments; they accept a variable number of lists instead.

7.14   SCOPING AND LEXICAL CLOSURES

Recall the MY-ASSOC example from section 7.7. Since the lambda expression is passed to FIND-IF and called from inside the body of FIND-IF, how is it possible for it to refer to the local variables of MY-ASSOC? Why is it unable to see the local variables, if any, of FIND-IF itself?

```
(defun my-assoc (key table)
   (find-if #'(lambda (entry)
                (equal key (first entry)))
            table))
```

```
(my-assoc 'two words)   ⇒   (TWO DEUX)
```

First, it is important to remember that what is passed to FIND-IF is not the raw lambda expression, but rather a lexical closure created by FUNCTION (abbreviated as #'). The closure remembers its lexical environment. In the following evaltrace diagram, a hollow arrow shows the scope boundary of the body of the closure. An arc links this arrow to the scope boundary for its parent context, the body of MY-ASSOC.

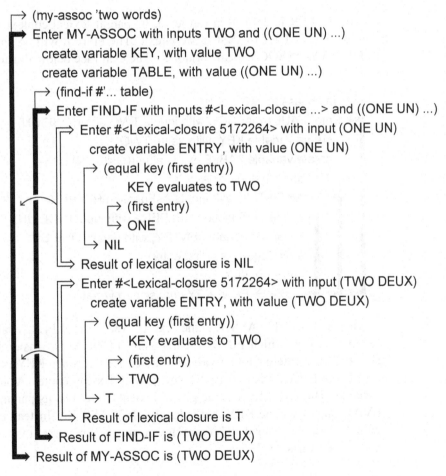

The scope rule for closures is that any variable not local to the closure is looked up in the closure's parent context. Every lexical context has a parent context. The thick solid lines we've been using for the bodies of functions like MY-ASSOC and FIND-IF denote lexical contexts whose parent is the global context. That's why when EVAL hits one of these thick lines while looking up a variable, it immediately looks for a global variable with that name.

Suppose we wrote a function FAULTY-ASSOC that replaced the lambda expression with an independent function called HELPER:

```
(defun helper (entry)
  (equal key (first entry)))
```

```
(defun faulty-assoc (key table)
  (find-if #'helper table))
```

Since HELPER is defined at top level, its parent lexical context is the global context, not FAULTY-ASSOC's context. Therefore it will be unable to refer to FAULTY-ASSOC's local variables. The evaltrace below illustrates this.

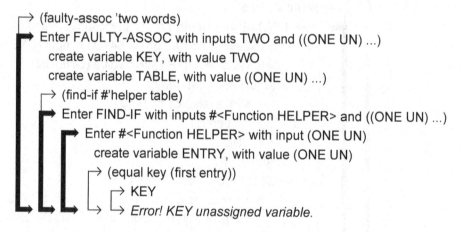

Inside FAULTY-ASSOC, the expression #'HELPER evaluates to a function object, which FIND-IF calls using FUNCALL. Inside the body of HELPER is a reference to a variable named KEY. Since KEY is not local to HELPER, EVAL tries to find some parent lexical context containing this variable. But HELPER has the global lexical context as its parent context, so EVAL cannot see the KEY that is local to MY-ASSOC. Instead it looks for a global variable named KEY. The result is an error message: ''KEY unassigned variable.''

7.15 WRITING AN APPLICATIVE OPERATOR

Using FUNCALL, we can write our own applicative operator that takes a function as input. Our operator will be called INALIENABLE-RIGHTS. It applies its input to a particular list, drawn from the American Declaration of Independence.

```
(defun inalienable-rights (fn)
  (funcall fn
    '(life liberty and the pursuit of happiness)))

> (inalienable-rights #'length)
7

> (inalienable-rights #'reverse)
(HAPPINESS OF PURSUIT THE AND LIBERTY LIFE)

> (inalienable-rights #'first)
LIFE

> (inalienable-rights #'rest)
(LIBERTY AND THE PURSUIT OF HAPPINESS)
```

It is an error to call INALIENABLE-RIGHTS on something that is not a function, because FUNCALL requires a function as its first input.

```
> (inalienable-rights 5)
Error! 5 is not a function.
```

The input to INALIENABLE-RIGHTS must be a function that can take a single list as its argument. We can't use the CONS function as an input because CONS requires two arguments.

```
> (inalienable-rights #'cons)
Error! CONS requires two inputs, but only got one.
```

However, we can use CONS inside a lambda expression that takes one argument, like so:

```
> (inalienable-rights
    #'(lambda (x) (cons 'high x)))
(HIGH LIFE LIBERTY AND THE PURSUIT OF HAPPINESS)
```

7.16 FUNCTIONS THAT MAKE FUNCTIONS

It is possible to write a function whose value is another function. Suppose we want to make a function that returns true if its input is greater than a certain number N. We can make this function by constructing a lambda expression that refers to N, and returning that lambda expression:

```
(defun make-greater-than-predicate (n)
  #'(lambda (x) (> x n)))
```

The value returned by MAKE-GREATER-THAN-PREDICATE will be a lexical closure. We can store this value away somewhere, or pass it as an argument to FUNCALL or any applicative operator.

```
> (setf pred (make-greater-than-predicate 3))
#<Lexical-closure 7315225>

(funcall pred 2)  ⇒  nil

(funcall pred 5)  ⇒  t

(find-if pred '(2 3 4 5 6 7 8 9))  ⇒  4
```

FUNCTIONS COVERED IN ADVANCED TOPICS

Special function for quoting functions: FUNCTION.

8

Recursion

8.1 INTRODUCTION

Because some instructors prefer to teach recursion as the first major control structure, this chapter and the preceding one may be taught in either order. They are independent.

Recursion is one of the most fundamental and beautiful ideas in computer science. A function is said to be "recursive" if it calls itself. Recursive control structure is the main topic of this chapter, but we will also take a look at recursive data structures in the Advanced Topics section. The insight necessary to recognize the recursive nature of many problems takes a bit of practice to develop, but once you "get it," you'll be amazed at the interesting things you can do with just a three- or four-line recursive function.

We will use a combination of three techniques to illustrate what recursion is all about: dragon stories, program traces, and recursion templates. Dragon stories are the most controversial technique: Students enjoy them and find them helpful, but computer science professors aren't always as appreciative. If you don't like dragons, you may skip Sections 8.2, 8.4, 8.6, and 8.9. The intervening sections will still make sense; they just won't be as much fun.

8.2 MARTIN AND THE DRAGON

In ancient times, before computers were invented, alchemists studied the mystical properties of numbers. Lacking computers, they had to rely on dragons to do their work for them. The dragons were clever beasts, but also lazy and bad-tempered. The worst ones would sometimes burn their keeper to a crisp with a single fiery belch. But most dragons were merely uncooperative, as violence required too much energy. This is the story of how Martin, an alchemist's apprentice, discovered recursion by outsmarting a lazy dragon.

One day the alchemist gave Martin a list of numbers and sent him down to the dungeon to ask the dragon if any were odd. Martin had never been to the dungeon before. He took a candle down with him, and in the furthest, darkest corner found an old dragon, none too friendly looking. Timidly, he stepped forward. He did not want to be burnt to a crisp.

"What do *you* want?" grumped the dragon as it eyed Martin suspiciously.

"Please, dragon, I have a list of numbers, and I need to know if any of them are odd" Martin began. "Here it is." He wrote the list in the dirt with his finger:

```
(3142 5798 6550 8914)
```

The dragon was in a disagreeable mood that day. Being a dragon, it always was. "Sorry, boy" the dragon said. "I might be willing to tell you if the *first* number in that list is odd, but that's the best I could possibly do. Anything else would be too complicated; probably not worth my trouble."

"But I need to know if *any* number in the list is odd, not just the first number" Martin explained.

"Too bad for you!" the dragon said. "I'm only going to look at the first number of the list. But I'll look at as many lists as you like if you give them to me one at a time."

Martin thought for a while. There had to be a way around the dragon's orneriness. "How about this first list then?" he asked, pointing to the one he had drawn on the ground:

```
(3142 5798 6550 8914)
```

"The first number in that list is not odd," said the dragon.

Martin then covered the first part of the list with his hand and drew a new left parenthesis, leaving

```
(5798 6550 8914)
```

and said ''How about this list?''

''The first number in that list is not odd,'' the dragon replied.

Martin covered some more of the list. ''How about this list then?''

```
(6550 8914)
```

''The first number in that list isn't odd either,'' said the dragon. It sounded bored, but at least it was cooperating.

''And this one?'' asked Martin.

```
(8914)
```

''Not odd.''

''And this one?''

```
()
```

''That's the empty list!'' the dragon snorted. ''There can't be an odd number in there, because there's *nothing* in there.''

''Well,'' said Martin, ''I now know that not one of the numbers in the list the alchemist gave me is odd. They're *all* even.''

''I NEVER said that!!!'' bellowed the dragon. Martin smelled smoke. ''I only told you about the *first* number in each list you showed me.''

''That's true, Dragon. Shall I write down all of the lists you looked at?''

''If you wish,'' the dragon replied. Martin wrote in the dirt:

```
(3142 5798 6550 8914)
     (5798 6550 8914)
          (6550 8914)
               (8914)
                   ()
```

''Don't you see?'' Martin asked. ''By telling me that the first element of each of those lists wasn't odd, you told me that *none* of the elements in my original list was odd.''

''That's pretty tricky,'' the dragon said testily. ''It looks liked you've discovered recursion. But don't ask me what that means—you'll have to figure it out for yourself.'' And with that it closed its eyes and refused to utter another word.

8.3   A FUNCTION TO SEARCH FOR ODD NUMBERS

Here is a recursive function ANYODDP that returns T if any element of a list of numbers is odd. It returns NIL if none of them are.

```
(defun anyoddp (x)
   (cond ((null x) nil)
         ((oddp (first x)) t)
         (t (anyoddp (rest x)))))
```

If the list of numbers is empty, ANYODDP should return NIL, since as the dragon noted, there can't be an odd number in a list that contains nothing. If the list is not empty, we go to the second COND clause and test the first element. If the first element is odd, there is no need to look any further; ANYODDP can stop and return T. When the first element is even, ANYODDP must call itself on the rest of the list to keep looking for odd elements. That is the recursive part of the definition.

To see better how ANYODDP works, we can use DTRACE to announce every call to the function and every return value. (The DTRACE tool used here was introduced in the Lisp Toolkit section of Chapter 7. If your Lisp doesn't have DTRACE, use TRACE instead.)

```
(defun anyoddp (x)
  (cond ((null x) nil)
        ((oddp (first x)) t)
        (t (anyoddp (rest x)))))
```

```
(dtrace anyoddp)
```

We'll start with the simplest cases: an empty list, and a list with one odd number.

```
> (anyoddp nil)
----Enter ANYODDP
|     X = NIL
 \--ANYODDP returned NIL          First COND clause returns NIL.
NIL

> (anyoddp '(7))
----Enter ANYODDP
|     X = (7)
 \--ANYODDP returned T            Second COND clause returns T.
T
```

Now let's consider the case where the list contains one even number. The tests in the first two COND clauses will be false, so the function will end up at the third clause, where it calls itself recursively on the REST of the list. Since the REST is NIL, this reduces to a previously solved problem: (ANYODDP NIL) is NIL due to the first COND clause.

```
> (anyoddp '(6))
----Enter ANYODDP
|      X = (6)
|      ----Enter ANYODDP          Third clause: recursive call.
|      |     X = NIL
|      \--ANYODDP returned NIL    First clause returns NIL.
 \--ANYODDP returned NIL
NIL
```

If the list contains two elements, an even number followed by an odd number, the recursive call will trigger the second COND clause instead of the first:

```
> (anyoddp '(6 7))
----Enter ANYODDP
|      X = (6 7)
|      ----Enter ANYODDP          Third clause:  recursive call.
|      |     X = (7)
|      \--ANYODDP returned T      Second COND clause returns T.
 \--ANYODDP returned T
T
```

Finally, let's consider the general case where there are multiple even and odd numbers:

```
> (anyoddp '(2 4 6 7 8 9))
----Enter ANYODDP
|      X = (2 4 6 7 8 9)
|      ----Enter ANYODDP
|      |      X = (4 6 7 8 9)
|      |      ----Enter ANYODDP
|      |      |      X = (6 7 8 9)
|      |      |      ----Enter ANYODDP
|      |      |      |     X = (7 8 9)
|      |      |      \--ANYODDP returned T
|      |      \--ANYODDP returned T
|      \--ANYODDP returned T
 \--ANYODDP returned T
T
```

Note that in this example the function did not have to recurse all the way down to NIL. Since the FIRST of (7 8 9) is odd, ANYODDP could stop and return T at that point.

EXERCISES

8.1. Use a trace to show how ANYODDP would handle the list (3142 5798 6550 8914). Which COND clause is never true in this case?

8.2. Show how to write ANYODDP using IF instead of COND.

8.4 MARTIN VISITS THE DRAGON AGAIN

"Hello Dragon!" Martin called as he made his way down the rickety dungeon staircase.

"Hmmmph! You again. I'm on to your recursive tricks." The dragon did not sound glad to see him.

"I'm supposed to find out what five factorial is," Martin said. "What's *factorial* mean, anyway?"

At this the dragon put on a most offended air and said, "I'm not going to tell you. Look it up in a book."

"All right," said Martin. "Just tell me what five factorial is and I'll leave you alone."

"You don't know what factorial means, but you want *me* to tell you what factorial of five is??? All right buster, I'll tell you, not that it will do you any good. Factorial of five is five times factorial of four. I hope you're satisfied. Don't forget to bolt the door on your way out."

"But what's factorial of four?" asked Martin, not at all pleased with the dragon's evasiveness.

"Factorial of four? Why, it's four times factorial of three, of course."

"And I suppose you're going to tell me that factorial of three is three times factorial of two," Martin said.

"What a clever boy you are!" said the dragon. "Now go away."

"Not yet," Martin replied. "Factorial of two is two times factorial of one. Factorial of one is one times factorial of zero. Now what?"

"Factorial of zero is one," said the dragon. "That's really all you ever need to remember about factorials."

"Hmmm," said Martin. "There's a pattern to this factorial function. Perhaps I should write down the steps I've gone through." Here is what he wrote:

```
Factorial(5) = 5 × Factorial(4)
             = 5 × 4 × Factorial(3)
             = 5 × 4 × 3 × Factorial(2)
             = 5 × 4 × 3 × 2 × Factorial(1)
             = 5 × 4 × 3 × 2 × 1 × Factorial(0)
             = 5 × 4 × 3 × 2 × 1 × 1
```

"Well," said the dragon, "you've recursed all the way down to factorial of zero, which you know is one. Now why don't you try working your way back up to...." When it realized what it was doing, the dragon stopped in mid-sentence. Dragons aren't supposed to be helpful.

Martin started to write again:

```
            1 × 1= 1
        2 × 1 × 1= 2
    3 × 2 × 1 × 1= 6
  4 × 3 × 2 × 1 × 1= 24
5 × 4 × 3 × 2 × 1 × 1= 120
```

"Hey!" Martin yelped. "Factorial of 5 is 120. That's the answer! Thanks!!"

"*I* didn't tell you the answer," the dragon said testily. "*I* only told you that factorial of zero is one, and factorial of *n* is *n* times factorial of *n*−1. You did the rest yourself. Recursively, I might add."

"That's true," said Martin. "Now if I only knew what 'recursively' really meant."

8.5 A LISP VERSION OF THE FACTORIAL FUNCTION

The dragon's words gave a very precise definition of factorial: *n* factorial is *n* times *n*−1 factorial, and zero factorial is one. Here is a function called FACT that computes factorials recursively:

```
(defun fact (n)
  (cond ((zerop n) 1)
        (t (* n (fact (- n 1)))))))
```

And here is how Lisp would solve Martin's problem:

```
(dtrace fact)

> (fact 5)
----Enter FACT
|      N = 5
|    ----Enter FACT
|    |    N = 4
|    |  ----Enter FACT
|    |  |    N = 3
|    |  |  ----Enter FACT
|    |  |  |    N = 2
|    |  |  |  ----Enter FACT
|    |  |  |  |    N = 1
|    |  |  |  |  ----Enter FACT
|    |  |  |  |  |    N = 0
|    |  |  |  |  \--FACT returned 1
|    |  |  |  \--FACT returned 1
|    |  |  \--FACT returned 2
|    |  \--FACT returned 6
|    \--FACT returned 24
\--FACT returned 120
120
```

EXERCISE

8.3. Why does (FACT 20.0) produce a different result than (FACT 20)? Why do (FACT 0.0) and (FACT 0) both produce the same result?

8.6 THE DRAGON'S DREAM

The next time Martin returned to the dungeon, he found the dragon rubbing its eyes, as if it had just awakened from a long sleep.

"I had a most curious dream," the dragon said. "It was a recursive dream, in fact. Would you like to hear about it?"

Martin was stunned to find the dragon in something resembling a friendly mood. He forgot all about the alchemist's latest problem. "Yes, please do tell me about your dream," he said.

"Very well," began the dragon. "Last night I was looking at a long loaf of bread, and I wondered how many slices it would make. To answer my question I actually went and cut one slice from the loaf. I had one slice, and one slightly shorter loaf of bread, but no answer. I puzzled over the problem until I fell asleep.''

''And that's when you had the dream?'' Martin asked.

''Yes, a very curious one. I dreamt about another dragon who had a loaf of bread just like mine, except his was a slice shorter. And he too wanted to know how many slices his loaf would make, but he had the same problem I did. He cut off a slice, like me, and stared at the remaining loaf, like me, and then he fell asleep like me as well.''

''So neither one of you found the answer,'' Martin said disappointedly. ''You don't know how long your loaf is, and you don't know how long his is either, except that it's one slice shorter than yours.''

''But I'm not done yet,'' the dragon said. ''When the dragon in *my* dream fell asleep, *he* had a dream as well. He dreamt about—if you can imagine this—a dragon whose loaf of bread was one slice shorter than *his own* loaf. And this dragon also wanted to find out how many slices his loaf would make, and he tried to find out by cutting a slice, but that didn't tell him the answer, so he fell asleep thinking about it.''

''Dreams within dreams!!'' Martin exclaimed. ''You're making my head swim. Did that last dragon have a dream as well?''

''Yes, and he wasn't the last either. Each dragon dreamt of a dragon with a loaf one slice shorter than his own. I was piling up a pretty deep stack of dreams there.''

''How did you manage to wake up then?'' Martin asked.

''Well,'' the dragon said, ''eventually one of the dragons dreamt of a dragon whose loaf was so small it wasn't there at all. You might call it 'the empty loaf.' That dragon could see his loaf contained no slices, so he knew the answer to his question was zero; he didn't fall asleep.

''When the dragon who dreamt of that dragon woke up, he knew that since his own loaf was one slice longer, it must be exactly one slice long. So he awoke knowing the answer to his question.

''And, when the dragon who dreamt of *that* dragon woke up, *he* knew that his loaf had to be two slices long, since it was one slice longer than that of the dragon he dreamt about. And when the dragon who dreamt of *him* woke up....''

''I get it!'' Martin said. ''He added one to the length of the loaf of the dragon he dreamed about, and that answered his own question. And when *you* finally woke up, you had the answer to yours. How many slices did your loaf make?''

"Twenty-seven,'' said the dragon. ''It was a very long dream.''

8.7 A RECURSIVE FUNCTION FOR COUNTING SLICES OF BREAD

If we represent a slice of bread by a symbol, then a loaf can be represented as a list of symbols. The problem of finding how many slices a loaf contains is thus the problem of finding how many elements a list contains. This is of course what LENGTH does, but if we didn't have LENGTH, we could still count the slices recursively.

```
(defun count-slices (loaf)
  (cond ((null loaf) 0)
        (t (+ 1 (count-slices (rest loaf))))))
```

```
(dtrace count-slices)
```

If the input is the empty list, then its length is zero, so COUNT-SLICES simply returns zero.

```
> (count-slices nil)
----Enter COUNT-SLICES
|      LOAF = NIL
 \--COUNT-SLICES returned 0
0
```

If the input is the list (X), COUNT-SLICES calls itself recursively on the REST of the list, which is NIL, and then adds one to the result.

```
> (count-slices '(x))
----Enter COUNT-SLICES
|      LOAF = (X)
|    ----Enter COUNT-SLICES
|    |      LOAF = NIL
|     \--COUNT-SLICES returned 0
 \--COUNT-SLICES returned 1
1
```

When the input is a longer list, COUNT-SLICES has to recurse more deeply to get to the empty list so it can return zero. Then as each recursive call returns, one is added to the result.

```
> (count-slices '(x x x x x))
----Enter COUNT-SLICES
|      LOAF = (X X X X X)
|    ----Enter COUNT-SLICES
|    |      LOAF = (X X X X)
|    |    ----Enter COUNT-SLICES
|    |    |      LOAF = (X X X)
|    |    |    ----Enter COUNT-SLICES
|    |    |    |      LOAF = (X X)
|    |    |    |    ----Enter COUNT-SLICES
|    |    |    |    |      LOAF = (X)
|    |    |    |    |    ----Enter COUNT-SLICES
|    |    |    |    |    |      LOAF = NIL
|    |    |    |    |    \--COUNT-SLICES returned 0
|    |    |    |    \--COUNT-SLICES returned 1
|    |    |    \--COUNT-SLICES returned 2
|    |    \--COUNT-SLICES returned 3
|    \--COUNT-SLICES returned 4
\--COUNT-SLICES returned 5
5
```

8.8 THE THREE RULES OF RECURSION

The dragon, beneath its feigned distaste for Martin's questions, actually enjoyed teaching him about recursion. One day it decided to formally explain what recursion means. The dragon told Martin to approach every recursive problem as if it were a journey. If he followed three rules for solving problems recursively, he would always complete the journey successfully. The dragon explained the rules this way:

1. Know when to stop.

2. Decide how to take one step.

3. Break the journey down into that step plus a smaller journey.

Let's see how each of these rules applies to the Lisp functions we wrote. The first rule, "know when to stop," warns us that any recursive function must check to see if the journey has been completed before recursing further. Usually this is done in the first COND clause. In ANYODDP the first clause checks if the input is the empty list, and if so the function stops and returns NIL, since the empty list doesn't contain any numbers. The factorial function, FACT, stops when the input gets down to zero. Zero factorial is one, and, as the dragon said, that's all you ever need to remember about factorial. The rest is computed recursively. In COUNT-SLICES the first COND clause checks for NIL, "the empty loaf." COUNT-SLICES returns zero if NIL is the input. Again, this is based on the realization that the empty loaf contains no slices, so we do not have to recurse any further.

The second rule, "decide how to take one step," asks us to break off from the problem one tiny piece that we instantly know how to solve. In ANYODDP we check whether the FIRST of a list is an odd number; if so we return T. In the factorial function we perform a single multiplication, multiplying the input N by factorial of N−1. In COUNT-SLICES the step is the + function: For each slice we cut off the loaf, we add one to whatever the length of the resulting loaf turned out to be.

The third rule, "break the journey down into that step plus a smaller journey," means find a way for the function to call itself recursively on the slightly smaller problem that results from breaking a tiny piece off. The ANYODDP function calls itself on the REST of the list, a shorter list than the original, to see if there are any odd numbers there. The factorial function recursively computes factorial of N-1, a slightly simpler problem than factorial of N, and then uses the result to get factorial of N. In COUNT-SLICES we use a recursive call to count the number of slices in the REST of a loaf, and then add one to the result to get the size of the whole loaf.

The Dragon's Three Recursive Functions				
Function	Stop When Input Is	Return	Step to Take	Rest of Problem
ANYODDP	NIL	NIL	(ODDP (FIRST X))	(ANYODDP (REST X))
FACT	0	1	N × . . .	(FACT (- N 1))
COUNT-SLICES	NIL	0	1 + . . .	(COUNT-SLICES (REST LOAF))

Table 8-1 Applying the three rules of recursion.

Table 8-1 sums up our understanding of how the three rules apply to ANYODDP, FACT, and COUNT-SLICES. Now that you know the rules, you can write your own recursive functions.

FIRST RECURSION EXERCISE

8.4. We are going to write a function called LAUGH that takes a number as input and returns a list of that many HAs. (LAUGH 3) should return the list (HA HA HA). (LAUGH 0) should return a list with no HAs in it, or, as the dragon might put it, "the empty laugh."

Here is a skeleton for the LAUGH function:

```
(defun laugh (n)
  (cond (α   β)
        (t (cons 'ha γ))))
```

Under what condition should the LAUGH function stop recursing? Replace the symbol α in the skeleton with that condition. What value should LAUGH return for that case? Replace symbol β in the skeleton with that value. Given that a single step for this problem is to add a HA onto the result of a subproblem, fill in that subproblem by replacing the symbol γ.

Type your LAUGH function into the computer. Then type (DTRACE LAUGH) to trace it, and (LAUGH 5) to test it. Do you get the result you want? What happens for (LAUGH 0)? What happens for (LAUGH -1)?

Note: If the function looks like it's in an infinite loop, break out of it and get back to the read-eval-print loop. (Exactly how this is done depends on the particular version of Lisp you use. Ask your local Lisp expert if you need help.) Then use DTRACE to help you understand what's going on.

EXERCISES

8.5. In this exercise we are going to write a function ADD-UP to add up all the numbers in a list. (ADD-UP '(2 3 7)) should return 12. You already know how to solve this problem applicatively with REDUCE; now you'll learn to solve it recursively. Before writing ADD-UP we must answer three questions posed by our three rules of recursion.

  a. When do we stop? Is there any list for which we immediately *know* what the sum of all its elements is? What is that list? What value should the function return if it gets that list as input?

  b. Do we know how to take a single step? Look at the second COND clause in the definition of COUNT-SLICES or FACT.

Does this give you any ideas about what the single step should be for ADD-UP?

c. How should ADD-UP call itself recursively to solve the rest of the problem? Look at COUNT-SLICES or FACT again if you need inspiration.

Write down the complete definition of ADD-UP. Type it into the computer. Trace it, and then try adding up a list of numbers.

8.6. Write ALLODDP, a recursive function that returns T if all the numbers in a list are odd.

8.7. Write a recursive version of MEMBER. Call it REC-MEMBER so you don't redefine the built-in MEMBER function.

8.8. Write a recursive version of ASSOC. Call it REC-ASSOC.

8.9. Write a recursive version of NTH. Call it REC-NTH.

8.10. For x a nonnegative integer and y a positive integer, $x+y$ equals $x+1+(y-1)$. If y is zero then $x+y$ equals x. Use these equations to build a recursive version of + called REC-PLUS out of ADD1, SUB1, COND and ZEROP. You'll have to write ADD1 and SUB1 too.

8.9 MARTIN DISCOVERS INFINITE RECURSION

On his next trip down to the dungeon Martin brought with him a parchment scroll. "Look dragon," he called, "someone else must know about recursion. I found this scroll in the alchemist's library."

The dragon peered suspiciously as Martin unrolled the scroll, placing a candlestick at each end to hold it flat. "This scroll makes no sense," the dragon said. "For one thing, it's got far too many parentheses."

"The writing *is* a little strange," Martin agreed, "but I think I've figured out the message. It's an algorithm for computing Fibonacci numbers."

"I already know how to compute Fibonacci numbers," said the dragon.

"Oh? How?"

"Why, I wouldn't *dream* of spoiling the fun by telling you," the dragon replied.

"I didn't think you would," Martin shot back. "But the scroll says that Fib of n equals Fib of $n-1$ plus Fib of $n-2$. That's a *recursive* definition, and I already know how to work with recursion.''

''What else does the scroll say?'' the dragon asked.

''Nothing else. Should it say more?''

Suddenly the dragon assumed a most ingratiating tone. Martin found the change startling. ''Dearest boy! Would you do a poor old dragon one tiny little favor? Compute a Fibonacci number for me. I promise to only ask you for a small one.''

''Well, I'm supposed to be upstairs now, cleaning the cauldrons,'' Martin began, but seeing the hurt look on the dragon's face he added, ''but I guess I have time for a *small* one.''

''You won't regret it,'' promised the dragon. ''Tell me: What is Fib of four?''

Martin traced his translation of the Fibonacci algorithm in the dust:

```
Fib(n)   =   Fib(n-1) + Fib(n-2)
```

Then he began to compute Fib of four:

```
Fib(4)    =   Fib(3) + Fib(2)
Fib(3)    =   Fib(2) + Fib(1)
Fib(2)    =   Fib(1) + Fib(0)
Fib(1)    =   Fib(0) + Fib(-1)
Fib(0)    =   Fib(-1) + Fib(-2)
Fib(-1)   =   Fib(-2) + Fib(-3)
Fib(-2)   =   Fib(-3) + Fib(-4)
Fib(-3)   =   Fib(-4) + Fib(-5)
```

''Finished?'' the dragon asked innocently.

''No,'' Martin replied. ''Something is wrong. The numbers are becoming increasingly negative.''

''Well, will you be finished soon?''

''It looks like I won't ever be finished,'' Martin said. ''This recursion keeps going on forever.''

''Aha! You see? You're stuck in an *infinite* recursion!'' the dragon gloated. ''I noticed it at once.''

''Then why didn't you say something?'' Martin demanded.

The dragon grimaced and gave a little snort; blue flame appeared briefly in its nostrils. ''How will you *ever* come to master recursion if you rely on a dragon to do your thinking for you?''

Martin wasn't afraid, but he stepped back a bit anyway to let the smoke clear. "Well, how did you spot the problem so *quickly*, dragon?"

"Elementary, boy. The scroll told how to take a single step, and how to break the journey down to a smaller one. It said nothing at all about when you get to stop. Ergo," the dragon grinned, "you don't."

8.10 INFINITE RECURSION IN LISP

Lisp functions can be made to recurse infinitely by ignoring the dragon's first rule of recursion, which is to know when to stop. Here is the Lisp implementation of Martin's algorithm:

```
(defun fib (n)
  (+ (fib (- n 1))
     (fib (- n 2))))

(dtrace fib)

> (fib 4)
----Enter FIB
|     N = 4
|   ----Enter FIB
|   |   N = 3
|   |   ----Enter FIB
|   |   |   N = 2
|   |   |   ----Enter FIB
|   |   |   |   N = 1
|   |   |   |   ----Enter FIB
|   |   |   |   |   N = 0
|   |   |   |   |   ----Enter FIB
|   |   |   |   |   |   N = -1
|   |   |   |   |   |   ----Enter FIB
|   |   |   |   |   |   |   N = -2
|   |   |   |   |   |   |   ----Enter FIB
|   |   |   |   |   |   |   |   N = -3
```

                    ad infinitum

Usually a good programmer can tell just by looking at a function whether it will exhibit infinite recursion, but in some cases this can be quite difficult to determine. Try tracing the following function C, giving it inputs that are small positive integers:

```
(defun c (n)
  (cond ((equal n 1) t)
        ((evenp n) (c (/ n 2)))
        (t (c (+ (* 3 n) 1))))))
```

```
> (c 3)
----Enter C
|      N = 3
|    ----Enter C
|    |     N = 10
|    |   ----Enter C
|    |   |     N = 5
|    |   |   ----Enter C
|    |   |   |     N = 16
|    |   |   |   ----Enter C
|    |   |   |   |     N = 8
|    |   |   |   |   ----Enter C
|    |   |   |   |   |     N = 4
|    |   |   |   |   |   ----Enter C
|    |   |   |   |   |   |     N = 2
|    |   |   |   |   |   |   ----Enter C
|    |   |   |   |   |   |   |     N = 1
|    |   |   |   |   |   |   \--C returned T
|    |   |   |   |   |   \--C returned T
|    |   |   |   |   \--C returned T
|    |   |   |   \--C returned T
|    |   |   \--C returned T
|    |   \--C returned T
|    \--C returned T
\--C returned T
T
```

Try calling C on other values between one and ten. Notice that there is no obvious relationship between the size of the input and the number of recursive calls that result. Number theorists believe the function returns T for every positive integer, in other words, there are no inputs which cause it to recurse infinitely. This is known as Collatz's conjecture. But until the conjecture is proved, we can't say for certain whether or not C always returns.

EXERCISES

8.11. The missing part of Martin's Fibonacci algorithm is the rule for Fib(1) and Fib(0). Both of these are defined to be one. Using this information, write a correct version of the FIB function. (FIB 4) should return five. (FIB 5) should return eight.

8.12. Consider the following version of ANY-7-P, a recursive function that searches a list for the number seven:

```
(defun any-7-p (x)
   (cond ((equal (first x) 7) t)
         (t (any-7-p (rest x)))))
```

Give a sample input for which this function will work correctly. Give one for which the function will recurse infinitely.

8.13. Review the definition of the factorial function, FACT, given previously. What sort of input could you give it to cause an infinite recursion?

8.14. Write the very shortest infinite recursion function you can.

8.15. Consider the circular list shown below. What is the car of this list? What is the cdr? What will the COUNT-SLICES function do when given this list as input?

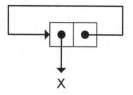

X

8.11 RECURSION TEMPLATES

Most recursive Lisp functions fall into a few standard forms. These are described by **recursion templates**, which capture the essence of the form in a fill-in-the-blanks pattern. You can create new functions by choosing a template and filling in the blanks. Also, once you've mastered them, you can use the templates to analyze existing functions to see which pattern they fit.

8.11.1 Double-Test Tail Recursion

The first template we'll study is double-test tail recursion, which is shown in Figure 8-1. "Double-test" indicates that the recursive function has two end tests; if either is true, the corresponding end value is returned instead of proceeding with the recursion. When both end tests are false, we end up at the

Double-Test Tail Recursion

Template:

```
(DEFUN func (X)
   (COND (end-test-1 end-value-1)
         (end-test-2 end-value-2)
         (T (func reduced-x)))))
```

Example:

Func:	ANYODDP
End-test-1:	(NULL X)
End-value-1:	NIL
End-test-2:	(ODDP (FIRST X))
End-value-2:	T
Reduced-x:	(REST X)

```
(defun anyoddp (x)
   (cond ((null x) nil)
         ((oddp (first x)) t)
         (t (anyoddp (rest x)))))
```

Figure 8-1 Template for double-test tail recursion.

last COND clause, where the function reduces the input somehow and then calls itself recursively. This template is said to be **tail-recursive** because the action part of the last COND clause does not do any work after the recursive call. Whatever result the recursive call produces, that is what the COND returns, so that is what each parent call returns. ANYODDP is an example of a tail-recursive function.

EXERCISES

8.16. What would happen if we switched the first and second COND clauses in ANYODDP?

8.17. Use double-test tail recursion to write FIND-FIRST-ODD, a function that returns the first odd number in a list, or NIL if there are none. Start by copying the recursion template values for ANYODDP; only a small change is necessary to derive FIND-FIRST-ODD.

8.11.2 Single-Test Tail Recursion

A simpler but less frequently used template is single-test tail recursion, which is shown in Figure 8-2. Suppose we want to find the first atom in a list, where the list may be nested arbitrarily deeply. We can do this by taking successive FIRSTs of the list until we reach an atom. The function FIND-FIRST-ATOM does this:

```
(find-first-atom '(ooh ah eee))  ⇒  ooh

(find-first-atom '((((a f)) i) r))  ⇒  a

(find-first-atom 'fred)  ⇒  fred
```

In general, single-test recursion is used when we know the function will always find what it's looking for eventually; FIND-FIRST-ATOM is guaranteed to find an atom if it keeps taking successive FIRSTs of its input. We use double-test recursion when there is the possibility the function might not find what it's looking for. In ANYODDP, for example, the second test checked if it had found an odd number, but first a test was needed to see if the function had run off the end of the list, in which case it should return NIL.

EXERCISES

8.18. Use single-test tail recursion to write LAST-ELEMENT, a function that returns the last element of a list. LAST-ELEMENT should recursively

Single-Test Tail Recursion

Template:

```
(DEFUN func (X)
  (COND (end-test end-value)
        (T (func reduced-x))))
```

Example:

Func:	FIND-FIRST-ATOM
End-test:	(ATOM X)
End-value:	X
Reduced-x:	(FIRST X)

```
(defun find-first-atom (x)
  (cond ((atom x) x)
        (t (find-first-atom (first x)))))
```

Figure 8-2 Template for single-test tail recursion.

travel down the list until it reaches the last cons cell (a cell whose cdr is an atom); then it should return the car of this cell.

8.19. Suppose we decided to convert ANYODDP to single-test tail recursion by simply eliminating the COND clause with the NULL test. For which inputs would it still work correctly? What would happen in those cases where it failed to work correctly?

8.11.3 Augmenting Recursion

Augmenting recursive functions like COUNT-SLICES build up their result bit-by-bit. We call this process **augmentation**. Instead of dividing the problem into an initial step plus a smaller journey, they divide it into a smaller journey plus a final step. The final step consists of choosing an augmentation value and applying it to the result of the previous recursive call. In COUNT-SLICES, for example, we built up the result by first making a recursive call and then adding one to the result. A template for single-test augmenting recursion is shown in Figure 8-3.

No augmentation of the result is permitted in tail-recursive functions. Therefore, the value returned by a tail-recursive function is always equal to one of the end-values in the function definition; it isn't built up bit-by-bit as each recursive call returns. Compare ANYODDP, which always returns T or NIL; it never augments its result.

EXERCISES

8.20. Of the three templates we've seen so far, which one describes FACT, the factorial function? Write down the values of the various template components for FACT.

8.21. Write a recursive function ADD-NUMS that adds up the numbers N, N−1, N−2, and so on, down to 0, and returns the result. For example, (ADD-NUMS 5) should compute 5+4+3+2+1+0, which is 15.

8.22. Write a recursive function ALL-EQUAL that returns T if the first element of a list is equal to the second, the second is equal to the third, the third is equal to the fourth, and so on. (ALL-EQUAL '(I I I I)) should return T. (ALL-EQUAL '(I I E I)) should return NIL. ALL-EQUAL should return T for lists with less than two elements. Does this problem require augmentation? Which template will you use to solve it?

Single-Test Augmenting Recursion

Template:

```
(DEFUN func (X)
  (COND (end-test end-value)
        (T (aug-fun aug-val
                    (func reduced-x))))))
```

Example:

Func:	COUNT-SLICES
End-test:	(NULL X)
End-value:	0
Aug-fun:	+
Aug-val:	1
Reduced-x:	(REST X)

```
(defun count-slices (x)
  (cond ((null x) 0)
        (t (+ 1 (count-slices (rest x))))))
```

Figure 8-3 Template for single-test augmenting recursion.

8.12 VARIATIONS ON THE BASIC TEMPLATES

The templates we've learned so far have many uses. Certain ways of using them are especially common in Lisp programming, and deserve special mention. In this section we'll cover four variations on the basic templates.

8.12.1 List-Consing Recursion

List-consing recursion is used very frequently in Lisp. It is a special case of augmenting recursion where the augmentation function is CONS. As each recursive call returns, we create one new cons cell. Thus, the depth of the recursion is equal to the length of the resulting cons cell chain, plus one (because the last call returns NIL instead of a cons). The LAUGH function you wrote in the first recursion exercise is an example of list-consing recursion. See Figure 8-4 for the template.

EXERCISES

8.23. Suppose we evaluate (LAUGH 5). Make a table showing, for each call to LAUGH, the value of N (from five down to zero), the value of the first input to CONS, the value of the second input to CONS, and the result returned by LAUGH.

8.24. Write COUNT-DOWN, a function that counts down from *n* using list-consing recursion. (COUNT-DOWN 5) should produce the list (5 4 3 2 1).

8.25. How could COUNT-DOWN be used to write an applicative version of FACT? (You may skip this problem if you haven't read Chapter 7 yet.)

8.26. Suppose we wanted to modify COUNT-DOWN so that the list it constructs ends in zero. For example, (COUNT-DOWN 5) would produce (5 4 3 2 1 0). Show two ways this can be done.

8.27. Write SQUARE-LIST, a recursive function that takes a list of numbers as input and returns a list of their squares. (SQUARE-LIST '(3 4 5 6)) should return (9 16 25 36).

List-Consing Recursion
(A Special Case of Augmenting Recursion)

Template:

```
(DEFUN func (N)
  (COND (end-test NIL)
        (T (CONS new-element
                 (func reduced-n))))))
```

Example:

Func:	LAUGH
End-test:	(ZEROP N)
New-element:	'HA
Reduced-n:	(- N 1)

```
(defun laugh (n)
  (cond ((zerop n) nil)
        (t (cons 'ha (laugh (- n 1))))))
```

Figure 8-4 Template for list-consing recursion.

8.12.2 Simultaneous Recursion on Several Variables

Simultaneous recursion on multiple variables is a straightforward extension to any recursion template. Instead of having only one input, the function has several, and one or more of them is "reduced" with each recursive call. For example, suppose we want to write a recursive version of NTH, called MY-NTH. Recall that (NTH 0 *x*) is (FIRST *x*); this tells us which end test to use. With each recursive call we reduce *n* by one and take successive RESTs of the list *x*. The resulting function demonstrates single-test tail recursion with simultaneous recursion on two variables. The template is shown in Figure 8-5. Here is a trace in which you can see the two variables being reduced simultaneously.

```
(defun my-nth (n x)
  (cond ((zerop n) (first x))
        (t (my-nth (- n 1) (rest x))))))
```

```
> (my-nth 2 '(a b c d e))
----Enter MY-NTH
|     N = 2
|     X = (A B C D E)
|    ----Enter MY-NTH
|    |     N = 1
|    |     X = (B C D E)
|    |    ----Enter MY-NTH
|    |    |     N = 0
|    |    |     X = (C D E)
|    |    \--MY-NTH returned C
|    \--MY-NTH returned C
\--MY-NTH returned C
C
```

EXERCISES

8.28. The expressions (MY-NTH 5 '(A B C)) and (MY-NTH 1000 '(A B C)) both run off the end of the list. and hence produce a NIL result. Yet the second expression takes quite a bit longer to execute than the first. Modify MY-NTH so that the recursion stops as soon the function runs off the end of the list.

8.29. Write MY-MEMBER, a recursive version of MEMBER. This function will take two inputs, but you will only want to reduce one of them with each successive call. The other should remain unchanged.

Simultaneous Recursion on Several Variables
(Using the Single-Test Tail Recursion Template)

Template:

```
(DEFUN func (N X)
   (COND (end-test end-value)
         (T (func reduced-n reduced-x))))
```

Example:

Func:	MY-NTH
End-test:	(ZEROP N)
End-value:	(FIRST X)
Reduced-n:	(- N 1)
Reduced-x:	(REST X)

```
(defun my-nth (n x)
   (cond ((zerop n) (first x))
         (t (my-nth (- n 1) (rest x)))))
```

Figure 8-5 Template for simultaneous recursion on several variables, using single-test tail recursion.

8.30. Write MY-ASSOC, a recursive version of ASSOC.

8.31. Suppose we want to tell as quickly as possible whether one list is shorter than another. If one list has five elements and the other has a million, we don't want to have to go through all one million cons cells before deciding that the second list is longer. So we must not call LENGTH on the two lists. Write a recursive function COMPARE-LENGTHS that takes two lists as input and returns one of the following symbols: SAME-LENGTH, FIRST-IS-LONGER, or SECOND-IS-LONGER. Use triple-test simultaneous recursion. *Hint:* If x is shorter than y and both are nonempty, then (REST x) is shorter than (REST y).

8.12.3 Conditional Augmentation

In some list-processing problems we want to skip certain elements of the list and use only the remaining ones to build up the result. This is known as **conditional augmentation**. For example, in EXTRACT-SYMBOLS, defined on the facing page, only elements that are symbols will be included in the result.

```
> (extract-symbols '(3 bears and 1 girl))
----Enter EXTRACT-SYMBOLS
|      X = (3 BEARS AND 1 GIRL)
|    ----Enter EXTRACT-SYMBOLS
|    |      X = (BEARS AND 1 GIRL)
|    |    ----Enter EXTRACT-SYMBOLS
|    |    |      X = (AND 1 GIRL)
|    |    |    ----Enter EXTRACT-SYMBOLS
|    |    |    |      X = (1 GIRL)
|    |    |    |    ----Enter EXTRACT-SYMBOLS
|    |    |    |    |      X = (GIRL)
|    |    |    |    |    ----Enter EXTRACT-SYMBOLS
|    |    |    |    |    |      X = NIL
|    |    |    |    |    \--EXTRACT-SYMBOLS returned NIL
|    |    |    |    \--EXTRACT-SYMBOLS returned (GIRL)
|    |    |    \--EXTRACT-SYMBOLS returned (GIRL)
|    |    \--EXTRACT-SYMBOLS returned (AND GIRL)
|    \--EXTRACT-SYMBOLS returned (BEARS AND GIRL)
\--EXTRACT-SYMBOLS returned (BEARS AND GIRL)
(BEARS AND GIRL)
```

The body of EXTRACT-SYMBOLS contains two recursive calls. One call is nested inside an augmentation expression, which in this case conses a new

Conditional Augmentation

Template:

```
(DEFUN func (X)
   (COND (end-test end-value)
         (aug-test (aug-fun aug-val
                              (func reduced-x))
         (T (func reduced-x)))))
```

Example:

Func:	EXTRACT-SYMBOLS
End-test:	(NULL X)
End-value:	NIL
Aug-test:	(SYMBOLP (FIRST X))
Aug-fun:	CONS
Aug-val:	(FIRST X)
Reduced-x:	(REST X)

```
(defun extract-symbols (x)
  (cond ((null x) nil)
        ((symbolp (first x))
         (cons (first x)
               (extract-symbols (rest x))))
        (t (extract-symbols (rest x)))))
```

Figure 8-6 Template for conditional augmentation.

element onto the result list. The other call is unaugmented; instead its result is simply returned. In the preceding trace output you'll note that sometimes two successive calls return the same value, such as two lists (GIRL) and two lists (BEARS AND GIRL); that's because one of each pair of calls chose the unaugmented COND clause. When the augmented clause was chosen, the result got longer, as when we went from NIL to (GIRL), from there to (AND GIRL), and from there to (BEARS AND GIRL). See Figure 8-6 for the general template for conditional augmentation.

EXERCISES

8.32. Write the function SUM-NUMERIC-ELEMENTS, which adds up all the numbers in a list and ignores the non-numbers. (SUM-NUMERIC-ELEMENTS '(3 BEARS 3 BOWLS AND 1 GIRL)) should return seven.

8.33. Write MY-REMOVE, a recursive version of the REMOVE function.

8.34. Write MY-INTERSECTION, a recursive version of the INTERSECTION function.

8.35. Write MY-SET-DIFFERENCE, a recursive version of the SET-DIFFERENCE function.

8.36. The function COUNT-ODD counts the number of odd elements in a list of numbers; for example, (COUNT-ODD '(4 5 6 7 8)) should return two. Show how to write COUNT-ODD using conditional augmentation. Then write another version of COUNT-ODD using the regular augmenting recursion template. (To do this you will need to write a conditional expression for the augmentation value.)

8.12.4 Multiple Recursion

A function is **multiple recursive** if it makes more than one recursive call with each invocation. (Don't confuse simultaneous with multiple recursion. The former technique just reduces several variables simultaneously; it does not involve multiple recursive calls with each invocation.) The Fibonacci function is a classic example of multiple recursion. Fib(N) calls itself twice: once for Fib(N−1) and again for Fib(N−2). The results of the two calls are combined using +. A general template for multiple recursion is shown in Figure 8-7.

A good way to visualize the process of multiple recursion is to look at the shape of the nested calls in the trace output. Let's define a **terminal call** as a

Multiple Recursion

Template:

```
(DEFUN func (N)
   (COND (end-test-1 end-value-1)
         (end-test-2 end-value-2)
         (T (combiner (func first-reduced-n)
                      (func second-reduced-n))))))
```

Example:

Func:	FIB
End-test-1:	(EQUAL N 0)
End-value-1:	1
End-test-2:	(EQUAL N 1)
End-value-2:	1
Combiner:	+
First-reduced-n:	(− N 1)
Second-reduced-n:	(− N 2)

```
(defun fib (n)
   (cond ((equal n 0) 1)
         ((equal n 1) 1)
         (t (+ (fib (- n 1))
               (fib (- n 2)))))))
```

Figure 8-7 Template for multiple recursion.

call that does not recurse any further. In all previous functions, successive
calls were nested strictly one inside the other, and the innermost call was the
only terminal call. Then, the return values flowed in a straight line from the
innermost call back to the outermost. But with a multiple-recursive function
such as FIB, each call produces *two* new calls. The two are nested inside the
parent call, but they cannot nest inside each other. Instead they appear side by
side within the parent. Multiple recursive functions therefore have many
terminal calls. In the following trace output, there are three terminal calls and
two nonterminal calls.

```
> (fib 3)
----Enter FIB
|     N = 3
|   ----Enter FIB
|   |     N = 2
|   |   ----Enter FIB
|   |   |     N = 1
|   |   \--FIB returned 1
|   |   ----Enter FIB
|   |   |     N = 0
|   |   \--FIB returned 1
|   \--FIB returned 2
|   ----Enter FIB
|   |     N = 1
|   \--FIB returned 1
\--FIB returned 3
3
```

EXERCISE

8.37. Define a simple function COMBINE that takes two numbers as input
and returns their sum. Now replace the occurence of + in FIB with
COMBINE. Trace FIB and COMBINE, and try evaluating (FIB 3) or
(FIB 4). What can you say about the relationship between COMBINE,
terminal calls, and nonterminal calls?

8.13 TREES AND CAR/CDR RECURSION

Sometimes we want to process all the elements of a nested list, not just the
top-level elements. If the list is irregularly shaped, such as (((GOLDILOCKS
. AND)) (THE . 3) BEARS), this might appear difficult. When we write our
function, we won't know how long or how deeply nested its inputs will be.

CAR/CDR Recursion
(A Special Case of Multiple Recursion)

Template:

```
(DEFUN func (X)
   (COND (end-test-1  end-value-1)
         (end-test-2  end-value-2)
         (T (combiner (func (CAR X))
                      (func (CDR X)))))))
```

Example:

Func:	FIND-NUMBER
End-test-1:	(NUMBERP X)
End-value-1:	X
End-test-2:	(ATOM X)
End-value-2:	NIL
Combiner:	OR

```
(defun find-number (x)
   (cond ((numberp x) x)
         ((atom x) nil)
         (t (or (find-number (car x))
                (find-number (cdr x))))))
```

Figure 8-8 Template for CAR/CDR recursion.

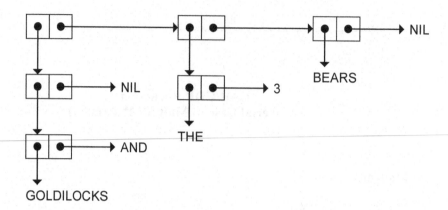

The trick to solving this problem is not to think of the input as an irregularly shaped nested list, but rather as a binary tree (see the following illustration.) Binary trees are very regular: Each node is either an atom or a cons with two branches, the car and the cdr. Therefore all our function has to do is process the atoms, and call itself recursively on the car and cdr of each cons. This technique is called CAR/CDR recursion; it is a special case of multiple recursion.

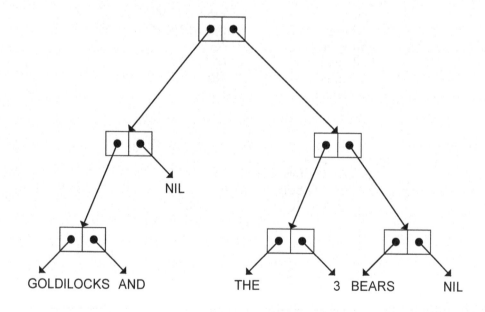

For example, suppose we want a function FIND-NUMBER to search a tree and return the first number that appears in it, or NIL if there are none. Then we should use NUMBERP and ATOM as our end tests and OR as the combiner. (See the template in Figure 8-8.) Note that since OR is a conditional, as soon as one clause of the OR evaluates to true, the OR stops and returns that value. Thus we don't have to search the whole tree; the function will stop recursing as soon as any call results in a non-NIL value.

Besides tree searching, another common use for CAR/CDR recursion is to build trees by using CONS as the combiner. For example, here is a function that takes a tree as input and returns a new tree in which every non-NIL atom has been replaced by the symbol Q.

```
(defun atoms-to-q (x)
  (cond ((null x) nil)
        ((atom x) 'q)
        (t (cons (atoms-to-q (car x))
                 (atoms-to-q (cdr x))))))
```

```
> (atoms-to-q '(a . b))
(Q . Q)
```

```
> (atoms-to-q '(hark (harold the angel) sings))
(Q (Q Q Q) Q)
```

EXERCISES

8.38. What would be the effect of deleting the first COND clause in ATOMS-TO-Q?

8.39. Write a function COUNT-ATOMS that returns the number of atoms in a tree. (COUNT-ATOMS '(A (B) C)) should return five, since in addition to A, B, and C there are two NILs in the tree.

8.40. Write COUNT-CONS, a function that returns the number of cons cells in a tree. (COUNT-CONS '(FOO)) should return one. (COUNT-CONS '(FOO BAR)) should return two. (COUNT-CONS '((FOO))) should also return two, since the list ((FOO)) requires two cons cells. (COUNT-CONS 'FRED) should return zero.

8.41. Write a function SUM-TREE that returns the sum of all the numbers appearing in a tree. Nonnumbers should be ignored. (SUM-TREE '((3 BEARS) (3 BOWLS) (1 GIRL))) should return seven.

8.42. Write MY-SUBST, a recursive version of the SUBST function.

8.43. Write FLATTEN, a function that returns all the elements of an arbitrarily nested list in a single-level list. (FLATTEN '((A B (R)) A C (A D ((A (B)) R) A))) should return (A B R A C A D A B R A).

8.44. Write a function TREE-DEPTH that returns the maximum depth of a binary tree. (TREE-DEPTH '(A . B)) should return one. (TREE-DEPTH '((A B C D))) should return five, and (TREE-DEPTH '((A . B) . (C . D))) should return two.

8.45. Write a function PAREN-DEPTH that returns the maximum depth of nested parentheses in a list. (PAREN-DEPTH '(A B C)) should return one, whereas TREE-DEPTH would return three. (PAREN-DEPTH '(A B ((C) D) E)) should return three, since there is an element C that is nested in three levels of parentheses. *Hint:* This problem can be solved by CAR/CDR recursion, but the CAR and CDR cases will not be exactly symmetric.

8.14 USING HELPING FUNCTIONS

For some problems it is useful to structure the solution as a helping function plus a recursive function. The recursive function does most of the work. The helping function is the one that you call from top level; it performs some special service either at the beginning or the end of the recursion. For example, suppose we want to write a function COUNT-UP that counts from one up to n:

```
(count-up 5)  ⇒  (1 2 3 4 5)

(count-up 0)  ⇒  nil
```

This problem is harder than COUNT-DOWN because the innermost recursive call must terminate the recursion when the input reaches five (in the preceding example), not zero. In general, how will the function know when to stop? The easiest way is to supply the original value of N to the recursive function so it can decide when to stop. We must also supply an extra argument: a counter that tells the function how far along it is in the recursion. The job of the helping function is to provide the initial value for the counter.

```
(defun count-up (n)
  (count-up-recursively 1 n))
```

```
(defun count-up-recursively (cnt n)
  (cond ((> cnt n) nil)
        (t (cons cnt
                 (count-up-recursively
                   (+ cnt 1)
                   n)))))

(dtrace count-up count-up-recursively)
```

```
> (count-up 3)
----Enter COUNT-UP
|     N = 3
|   ----Enter COUNT-UP-RECURSIVELY
|   |     CNT = 1
|   |     N = 3
|   |   ----Enter COUNT-UP-RECURSIVELY
|   |   |     CNT = 2
|   |   |     N = 3
|   |   |   ----Enter COUNT-UP-RECURSIVELY
|   |   |   |     CNT = 3
|   |   |   |     N = 3
|   |   |   |   ----Enter COUNT-UP-RECURSIVELY
|   |   |   |   |     CNT = 4
|   |   |   |   |     N = 3
|   |   |   |   \--COUNT-UP-RECURSIVELY returned NIL
|   |   |   \--COUNT-UP-RECURSIVELY returned (3)
|   |   \--COUNT-UP-RECURSIVELY returned (2 3)
|   \--COUNT-UP-RECURSIVELY returned (1 2 3)
\--COUNT-UP returned (1 2 3)
(1 2 3)
```

EXERCISES

8.46. Another way to solve the problem of counting upward is to to add an element to the end of the list with each recursive call instead of adding elements to the beginning. This approach doesn't require a helping function. Write this version of COUNT-UP.

8.47. Write MAKE-LOAF, a function that returns a loaf of size N. (MAKE-LOAF 4) should return (X X X X). Use IF instead of COND.

8.48. Write a recursive function BURY that buries an item under *n* levels of parentheses. (BURY 'FRED 2) should return ((FRED)), while (BURY 'FRED 5) should return (((((FRED))))). Which recursion template did you use?

8.49. Write PAIRINGS, a function that pairs the elements of two lists. (PAIRINGS '(A B C) '(1 2 3)) should return ((A 1) (B 2) (C 3)). You may assume that the two lists will be of equal length.

8.50. Write SUBLISTS, a function that returns the successive sublists of a list. (SUBLISTS '(FEE FIE FOE)) should return ((FEE FIE FOE) (FIE FOE) (FOE)).

8.51. The simplest way to write MY-REVERSE, a recursive version of REVERSE, is with a helping function plus a recursive function of two inputs. Write this version of MY-REVERSE.

8.52. Write MY-UNION, a recursive version of UNION.

8.53. Write LARGEST-EVEN, a recursive function that returns the largest even number in a list of nonnegative integers. (LARGEST-EVEN '(5 2 4 3)) should return four. (LARGEST-EVEN NIL) should return zero. Use the built-in MAX function, which returns the largest of its inputs.

8.54. Write a recursive function HUGE that raises a number to its own power. (HUGE 2) should return 2^2, (HUGE 3) should return $3^3 = 27$, (HUGE 4) should return $4^4 = 256$, and so on. Do not use REDUCE.

8.15 RECURSION IN ART AND LITERATURE

Recursion can be found not only in computer programs, but also in stories and in paintings. The classic *One Thousand and One Arabian Nights* contains stories within stories within stories, giving it a recursive flavor. A similar effect is expressed visually in some of Dr. Seuss's drawings in *The Cat in the Hat Comes Back*. One of these is shown in Figure 8-9. The nesting of cats within hats is like the nesting of contexts when a recursive function calls itself. In the story, each cat's taking off his hat plays the role of a recursive function call. Little cat B has his hat on at this point, but the recursion eventually gets all the way to Z, and terminates with an explosion. (If this story has any moral, it would appear to be, "Know when to stop!")

Some of the most imaginative representations of recursion and self-referentiality in art are the works of the Dutch artist M. C. Escher, whose lithograph "Drawing Hands" appears in Figure 8-10. Douglas Hofstadter discusses the role of recursion in music, art, and mathematics in his book *Godel, Escher, Bach: An Eternal Golden Braid.* The dragon stories in this

Figure 8-9 Recursively nested cats, from *The Cat in the Hat Comes Back*, by Dr. Seuss. Copyright (c) 1958 by Dr. Seuss. Reprinted by permission of Random House, Inc.

chapter were inspired by characters in Hofstadter's book.

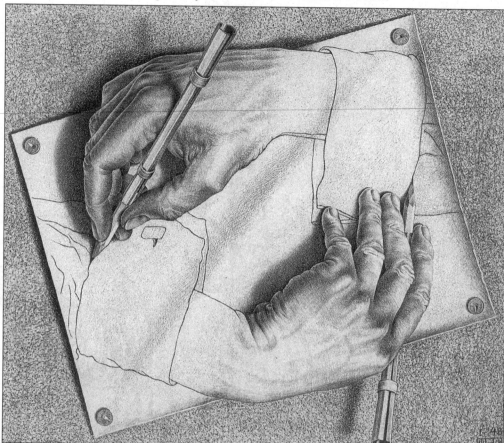

Figure 8-10 ''Drawing Hands'' by M. C. Escher. Copyright (c) 1989 M. C. Escher heirs/Cordon Art–Baarn–Holland.

SUMMARY

Recursion is a very powerful control structure, and one of the most important ideas in computer science. A function is said to be ''recursive'' if it calls itself. To write a recursive function, we must solve three problems posed by the Dragon's three rules of recursion:

1. Know when to stop.

2. Decide how to take one step.

3. Break the journey down into that step plus a smaller journey.

We've seen a number of recursion templates in this chapter. Recursion templates capture the essence of certain stereotypical recursive solutions. They can be used for writing new functions, or for analyzing existing functions. The templates we've seen so far are:

1. Double-test tail recursion.
2. Single-test tail recursion.
3. Single-test augmenting recursion.
4. List-consing recursion.
5. Simultaneous recursion on several variables.
6. Conditional augmentation.
7. Multiple recursive calls.
8. CAR/CDR recursion.

REVIEW EXERCISES

8.55. What distinguishes a recursive function from a nonrecursive one?

8.56. Write EVERY-OTHER, a recursive function that returns every other element of a list—the first, third, fifth, and so on. (EVERY-OTHER '(A B C D E F G)) should return (A C E G). (EVERY-OTHER '(I CAME I SAW I CONQUERED)) should return (I I I).

8.57. Write LEFT-HALF, a recursive function in two parts that returns the first $n/2$ elements of a list of length n. Write your function so that the list does not have to be of even length. (LEFT-HALF '(A B C D E)) should return (A B C). (LEFT-HALF '(1 2 3 4 5 6 7 8)) should return (1 2 3 4). You may use LENGTH but not REVERSE in your definition.

8.58. Write MERGE-LISTS, a function that takes two lists of numbers, each in increasing order, as input. The function should return a list that is a merger of the elements in its inputs, in order. (MERGE-LISTS '(1 2 6 8 10 12) '(2 3 5 9 13)) should return (1 2 2 3 5 6 8 9 10 12 13).

8.59. Here is another definition of the factorial function:

```
Factorial(0) = 1
Factorial(N) = Factorial(N+1) / (N+1)
```

Verify that these equations are true. Is the definition recursive? Write a Lisp function that implements it. For which inputs will the function return the correct answer? For which inputs will it fail to return the correct answer? Which of the three rules of recursion does the definition violate?

Lisp Toolkit: The Debugger

All beginning Lispers quickly learn one debugger command, because as soon as they type something wrong, that's where they end up: in the debugger. They have to learn how to get out! Lisp implementations differ substantially when it comes to debuggers, so there will be no standard way to recover from an error. Some of you have probably been typing Q for Quit or :A for Abort, while others may be typing Control-C or Control-G. In any case, now that you're confident you can exit the debugger whenever you like, why not stay around a while?

The debugger does not actually remove bugs from programs. What it does is let you examine the state of the computation when an error has occurred. This also makes it a good tool for learning about recursion. We can use the BREAK function to enter the debugger at a strategic point in the computation. The argument to BREAK is a message, in string quotes, to be printed when the debugger is entered. Here is a modified version of FACT that demonstrates the use of BREAK:

```
(defun fact (n)
  (cond ((zerop n) (break "N is zero."))
        (t (* n (fact (- n 1))))))
```

```
> (fact 5)
N is zero.
Entering the debugger:

Debug>
```

We are now sitting in the debugger; ''Debug>'' is the debugger's prompt. (Your debugger may use a different prompt.) One of the things we can do at this point is display a **backtrace** of the control stack, which shows all the recursive calls that are currently stacked up. If you're not familiar with terms like ''control stack'' and ''stack frame,'' just play around with the debugger for a while and you'll get the hang of what's going on. (The control stack is Lisp's way of keeping track of a collection of nested function calls. A stack frame is an entry on the stack that describes one of these function calls.) In my debugger the command for displaying a backtrace is BK.

```
Debug> bk
(BREAK "N is zero.")
(FACT (- N 1))
(FACT (- N 1))
(FACT (- N 1))
(FACT (- N 1))
(FACT (- N 1))
(FACT 5)
<Bottom of Stack>
```

Variants of the BK command allow different sorts of control stack information to be displayed. In my debugger, BKFV gives a display of function names and their local variables.

```
Debug> bkfv
BREAK
  N = 0
FACT
  N = 1
FACT
  N = 2
FACT
  N = 3
FACT
  N = 4
FACT
  N = 5
FACT
<Bottom of Stack>
```

While inside the debugger we can look at the values of variables, and type arbitrary Lisp expressions using them.

```
Debug> n
0

Debug> (cons 'foo n)
(FOO . 0)
```

When we enter the debugger, we are sitting at the top of the stack. We can move around the stack using the commands called (in my debugger) UP and DOWN. If we move down the stack, we can see other local variables named N.

```
Debug> down
(FACT (- N 1))
```

```
Debug> down
(FACT (- N 1))

Debug> down
(FACT (- N 1))

Debug> bkv
(BREAK "N is zero.")
  N = 0
(FACT (- N 1))
  N = 1
(FACT (- N 1))
  N = 2
(FACT (- N 1))      <-- Current stack frame
  N = 3
(FACT (- N 1))
  N = 4
(FACT (- N 1))
  N = 5
(FACT 5)
<Bottom of Stack>

Debug> n
3
```

Finally, we can use the debugger to return from any one of the function calls currently on the stack. This causes the computation to resume as if the function had returned normally:

```
Debug> return 10
600
```

When we returned 10 from the current stack frame, the computation resumed at that point, and the value produced was $5 \times 4 \times 3 \times 10 = 600$.

Your debugger won't look exactly like mine, and it may provide somewhat different capabilities, but the basic idea of examining the control stack is common to all Lisp debuggers. Look in the user's manual for your Lisp implementation to see which debugger commands are offered. Typing HELP or :H or "?" to your debugger may also produce a list of commands.

Keyboard Exercise

In this exercise we will extract different sorts of information from a genealogical database. The database gives information for five generations of a family, as shown in Figure 8-11. Such diagrams are usually called family trees, but this family's genealogical history is not a simple tree structure. Marie has married her first cousin Nigel. Wanda has had one child with Vincent and another with Ivan. Zelda and Robert, the parents of Yvette, have two great grandparents in common. (This might explain why Yvette turned out so weird.) And only Tamara knows who Frederick's father is; she's not telling.

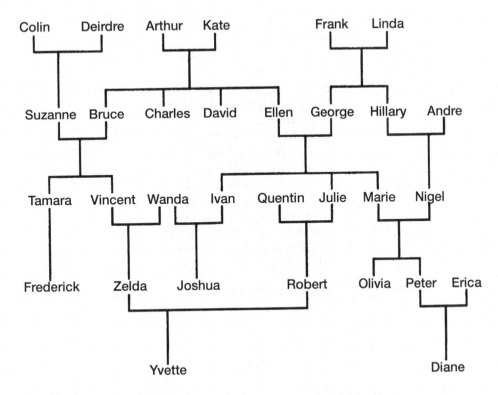

Figure 8-11 Genealogy information for five generations of a family.

```
(setf family
      '((colin nil nil)
        (deirdre nil nil)
        (arthur nil nil)
        (kate nil nil)
        (frank nil nil)
        (linda nil nil)
        (suzanne colin deirdre)
        (bruce arthur kate)
        (charles arthur kate)
        (david arthur kate)
        (ellen arthur kate)
        (george frank linda)
        (hillary frank linda)
        (andre nil nil)
        (tamara bruce suzanne)
        (vincent bruce suzanne)
        (wanda nil nil)
        (ivan george ellen)
        (julie george ellen)
        (marie george ellen)
        (nigel andre hillary)
        (frederick nil tamara)
        (zelda vincent wanda)
        (joshua ivan wanda)
        (quentin nil nil)
        (robert quentin julie)
        (olivia nigel marie)
        (peter nigel marie)
        (erica nil nil)
        (yvette robert zelda)
        (diane peter erica)))
```

Figure 8-12 The genealogy database.

Each person in the database is represented by an entry of form

(*name   father   mother*)

When someone's father or mother is unknown, a value of NIL is used.

The functions you write in this keyboard exercise need not be recursive, except where indicated. For functions that return lists of names, the exact order in which these names appear is unimportant, but there should be no duplicates.

EXERCISE

8.60. If the genealogy database is already stored on the computer for you, load the file containing it. If not, you will have to type it in as it appears in Figure 8-12. Store the database in the global variable FAMILY.

a. Write the functions FATHER, MOTHER, PARENTS, and CHILDREN that return a person's father, mother, a list of his or her known parents, and a list of his or her children, respectively. (FATHER 'SUZANNE) should return COLIN. (PARENTS 'SUZANNE) should return (COLIN DEIRDRE). (PARENTS 'FREDERICK) should return (TAMARA), since Frederick's father is unknown. (CHILDREN 'ARTHUR) should return the set (BRUCE CHARLES DAVID ELLEN). If any of these functions is given NIL as input, it should return NIL. This feature will be useful later when we write some recursive functions.

b. Write SIBLINGS, a function that returns a list of a person's siblings, including genetic half-siblings. (SIBLINGS 'BRUCE) should return (CHARLES DAVID ELLEN). (SIBLINGS 'ZELDA) should return (JOSHUA).

c. Write MAPUNION, an applicative operator that takes a function and a list as input, applies the function to every element of the list, and computes the union of all the results. An example is (MAPUNION #'REST '((1 A B C) (2 E C J) (3 F A B C D))), which should return the set (A B C E J F D). *Hint:* MAPUNION can be defined as a combination of two applicative operators you already know.

d. Write GRANDPARENTS, a function that returns the set of a person's grandparents. Use MAPUNION in your solution.

e. Write COUSINS, a function that returns the set of a person's genetically related first cousins, in other words, the children of any of their parents' siblings. (COUSINS 'JULIE) should return the set (TAMARA VINCENT NIGEL). Use MAPUNION in your solution.

f. Write the two-input *recursive* predicate DESCENDED-FROM that returns a true value if the first person is descended from the second. (DESCENDED-FROM 'TAMARA 'ARTHUR) should return T. (DESCENDED-FROM 'TAMARA 'LINDA) should return NIL. (*Hint:* You are descended from someone if he is one of your parents, or if either your father or mother is descended from him. This is a recursive definition.)

g. Write the *recursive* function ANCESTORS that returns a person's set of ancestors. (ANCESTORS 'MARIE) should return the set (ELLEN ARTHUR KATE GEORGE FRANK LINDA). (*Hint:* A person's ancestors are his parents plus his parents' ancestors. This is a recursive definition.)

h. Write the *recursive* function GENERATION-GAP that returns the number of generations separating a person and one of his or her ancestors. (GENERATION-GAP 'SUZANNE 'COLIN) should return one. (GENERATION-GAP 'FREDERICK 'COLIN) should return three. (GENERATION-GAP 'FREDERICK 'LINDA) should return NIL, because Linda is not an ancestor of Frederick.

i. Use the functions you have written to answer the following questions:

1. Is Robert descended from Deirdre?
2. Who are Yvette's ancestors?
3. What is the generation gap between Olivia and Frank?
4. Who are Peter's cousins?
5. Who are Olivia's grandparents?

8 Advanced Topics

8.16 ADVANTAGES OF TAIL RECURSION

Remember that tail-recursive functions do no work after the recursive call; the function returns whatever the recursive call returns. ANYODDP is a tail-recursive function, but COUNT-SLICES is not. If we look at the trace of COUNT-SLICES again, we see that each call produces a different return value (owing to augmentation). In a tail-recursive function, all calls return the same value as the terminal call.

```
> (count-slices '(x x x x))
----Enter COUNT-SLICES
|    LOAF = (X X X X)
|    ----Enter COUNT-SLICES
|    |    LOAF = (X X X)
|    |    ----Enter COUNT-SLICES
|    |    |    LOAF = (X X)
|    |    |    ----Enter COUNT-SLICES
|    |    |    |    LOAF = (X)
|    |    |    |    ----Enter COUNT-SLICES
|    |    |    |    |    LOAF = NIL
|    |    |    |    \--COUNT-SLICES returned 0
|    |    |    \--COUNT-SLICES returned 1
|    |    \--COUNT-SLICES returned 2
|    \--COUNT-SLICES returned 3
\--COUNT-SLICES returned 4
```

In general, it is better to write recursive functions in tail-recursive form whenever possible, because Lisp systems can execute tail-recursive functions more efficiently than ordinary recursive functions. They do this by replacing the recursive call with a jump. Many Lisp compilers perform this optimization automatically; some interpreters do as well.

A common technique for producing a tail-recursive version of an ordinary recursive function is to introduce an extra variable for accumulating augmentation values. For example, here is a tail-recursive function called TR-COUNT-SLICES that sets up the initial call to TR-CS1. TR-CS1 uses an extra variable N to hold the count of the number of slices seen so far.

```
(defun tr-count-slices (loaf)
  (tr-cs1 loaf 0))
```

```
(defun tr-cs1 (loaf n)
  (cond ((null loaf) n)
        (t (tr-cs1 (rest loaf) (+ n 1)))))
```

In the trace of TR-COUNT-SLICES you will note that the value of N increases with each call. The terminal call computes the return value, four; this value is then passed back unchanged by each level.

Another example of how augmentation can be eliminated by introducing an extra variable is the REVERSE function. To reverse a list of length *n*, we can reverse the REST of the list recursively, then tack the FIRST element onto the end, like so:

```
(defun my-reverse (x)
  (cond ((null x) nil)
        (t (append (reverse (rest x))
                   (list (first x))))))
```

But this definition isn't tail recursive. After the recursive call returns, the result is augmented by APPEND. Here is a two-part, tail-recursive definition of REVERSE that uses an extra variable to build up the result with (rather than after) each recursive call.

```
(defun tr-reverse (x)
  (tr-rev1 x nil))
```

```
(defun tr-rev1 (x result)
  (cond ((null x) result)
        (t (tr-rev1
            (rest x)
            (cons (first x) result)))))
```

```
(dtrace tr-reverse tr-rev1)
```

```
> (tr-reverse '(a b c d))
----Enter TR-REVERSE
|      X = (A B C D)
|    ----Enter TR-REV1
|    |      X = (A B C D)
|    |      RESULT = NIL
|    |    ----Enter TR-REV1
|    |    |      X = (B C D)
|    |    |      RESULT = (A)
|    |    |    ----Enter TR-REV1
|    |    |    |      X = (C D)
|    |    |    |      RESULT = (B A)
|    |    |    |    ----Enter TR-REV1
|    |    |    |    |      X = (D)
|    |    |    |    |      RESULT = (C B A)
|    |    |    |    |    ----Enter TR-REV1
|    |    |    |    |    |      X = NIL
|    |    |    |    |    |      RESULT = (D C B A)
|    |    |    |    |    \--TR-REV1 returned (D C B A)
|    |    |    |    \--TR-REV1 returned (D C B A)
|    |    |    \--TR-REV1 returned (D C B A)
|    |    \--TR-REV1 returned (D C B A)
|    \--TR-REV1 returned (D C B A)
 \--TR-REVERSE returned (D C B A)
(D C B A)
```

Not all recursive functions have tail-recursive versions. Any function that is multiple recursive, such as FIB, cannot be made tail recursive simply by introducing an extra variable, since after the first recursive call returns there is another one waiting to be done.

EXERCISES

8.61. Write a tail-recursive version of COUNT-UP.

8.62. Write a tail-recursive version of FACT.

8.63. Write tail-recursive versions of UNION, INTERSECTION, and SET-DIFFERENCE. Your functions need not return results in the same order as the built-in functions.

8.17   WRITING NEW APPLICATIVE OPERATORS

We can use FUNCALL to invoke a function that the user supplies. This allows us to write our own applicative operators. For example, here is a simplified version of MAPCAR that only maps over a single list.

```
(defun my-mapcar (fn x)
  (cond ((null x) nil)
        (t (cons (funcall fn (first x))
                 (my-mapcar fn (rest x)))))))
```

The function we supply to MY-MAPCAR must be a function of one input, since that's how many inputs it will be FUNCALLed with.

EXERCISE

8.64. Write a TREE-FIND-IF operator that returns the first non-NIL atom of a tree that satisfies a predicate. (TREE-FIND-IF #'ODDP '((2 4) (5 6) 7)) should return 5.

8.18   THE LABELS SPECIAL FUNCTION

Up to now we've been writing helping functions as separate DEFUNs. This is a little bit sloppy, since if the helping function is defined at top level, someone might call it accidentally. A second, more serious difficulty is that helping functions defined with DEFUN cannot access any of the main function's local variables. Both these problems can be solved with LABELS.

The LABELS special function allows us to establish local function definitions inside the body of the main function, just as LET allows us to establish local variables. The syntax of these two forms is similar. For LABELS, it looks like this:

```
(LABELS  ( (fn-1  args-1  body-1)
              . . .
           (fn-n  args-2  body-2) )
   body)
```

The body can call any of the local functions. The local functions can call each other, and can also reference their parent's variables.

In the following example, notice that COUNT-UP-RECURSIVELY references N, the input to COUNT-UP.

```
(defun count-up (n)
  (labels ((count-up-recursively (cnt)
              (if (> cnt n) nil
                (cons cnt
                    (count-up-recursively
                        (+ cnt 1)))))))
    (count-up-recursively 1)))
```

One disadvantage of using LABELS is that in most Lisp implementations, there is no way to trace functions that are defined inside a LABELS expression. But you can still use STEP to step through the evaluation manually, if necessary.

EXERCISE

8.65. Use LABELS to write versions of TR-COUNT-SLICES and TR-REVERSE.

8.19 RECURSIVE DATA STRUCTURES

This chapter has been devoted to writing functions with recursive definitions. Data structures may also have recursive definitions. Consider the following definition of an **S-expression** (''symbolic expression''):

An S-expression is either an atom, or a cons cell whose CAR and CDR parts are S-expressions.

The term "S-expression" is used inside its own definition. That is what makes the definition recursive. S-expressions are instances of a very common recursive data structure, with important applications in all areas of computer science, called a **tree**. Here is another example of a tree, this time representing an arithmetic expression:

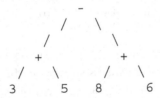

The bottom nodes of the tree are called **terminal nodes** because they have no branches descending from them. The remaining nodes are called **nonterminal nodes**. A tree can be defined recursively just as S-expressions were:

> *A tree is either a single terminal node, or a nonterminal node whose branches are trees.*

Trees are naturally represented by lists. The tree above corresponds to the list ((3 + 5) - (8 + 6)). Let's look at another arithmetic expression tree:

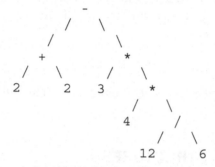

This tree illustrates the fact that the branches of a nonterminal node need not be of the same length. The list representation of this tree is ((2 + 2) - (3 * (4 * (12 / 6)))). We can define arithmetic expressions recursively as:

> *An arithmetic expression is either a number, or a three-element list whose first and third elements are arithmetic expressions and whose middle element is one of +, -, *, or /.*

EXERCISES

8.66. Write ARITH-EVAL, a function that evaluates arithmetic expressions. (ARITH-EVAL '(2 + (3 * 4))) should return 14.

8.67. Write a predicate LEGALP that returns T if its input is a legal arithmetic expression. For example, (LEGALP 4) and (LEGALP '((2 * 2) - 3)) should return T. (LEGALP NIL) and (LEGALP '(A B C D)) should return NIL.

8.68. A ''proper list'' is a cons cell chain ending in NIL. Lists that aren't proper lists are called dotted lists, because they must be written with a dot. If we wanted to define the concept of proper list recursively, we could say "NIL is a proper list, and so is any cons cell whose...." Fill in the rest of the definition.

8.69. Of the positive integers greater than one, some are **primes** while others are not. Primes are numbers that are divisible only by themselves and by 1. A nonprime, which is known as a **composite** number, can always be factored into primes. Here is a factorization tree for the number 60 that was obtained by successive divisions by primes:

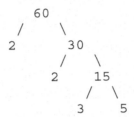

The number 60 has factors 2, 2, 3, and 5, which means $60 = 2 \times 2 \times 3 \times 5$. Write a recursive definition for positive integers greater than one in terms of prime numbers.

8.70. Following is a function FACTORS that returns the list of prime factors of a number. It uses the built-in REM function to compute the remainder of dividing one number by another. (FACTORS 60) returns (2 2 3 5). Try tracing the helping function to see how it works. Write a similar function, FACTOR-TREE, that returns a factorization tree instead. (FACTOR-TREE 60) should return the list (60 2 (30 2 (15 3 5))).

```
(defun factors (n)
  (factors-help n 2))

(defun factors-help (n p)
  (cond ((equal n 1) nil)
        ((zerop (rem n p))
         (cons p (factors-help (/ n p) p)))
        (t (factors-help n (+ p 1)))))
```

8.71. The trees for arithmetic expressions are called **binary trees**, because each nonterminal node has exactly two branches. Any list can be viewed as a binary tree. Draw a binary tree representing the cons cell structure of the list (A B (C D) E). What are the terminal nodes of the tree? What are the nonterminal nodes?

8.72. More general types of trees are possible, in which different nodes have different numbers of branches. Pick some concept or object and describe it in terms of a general tree structure.

9

Input/Output

9.1 INTRODUCTION

Input/output, or "i/o," is the way a computer communicates with the world. Lisp's read-eval-print loop provides a simple kind of i/o, since it reads expressions from the keyboard and prints the results on the display. Sometimes we want to do more. Using the i/o functions described in this chapter, you can make your program print any message you like. You can even make it print out questions and wait for the user to type responses on the keyboard.

Another use for i/o functions is to read data from a disk file, or write data into a file so you can read it back some other day. It's easier to do this in Common Lisp than in most other languages.

Historically, input/output has been one of the areas of greatest disagreement among Lisp systems. Even today there is no standard window system interface, for example, and no standard way to control a mouse or produce graphic designs. Each Lisp vendor provides his own tools for doing these things. Fortunately, the most basic i/o routines *have* finally been standardized. We will stick to the basics in this book.

9.2 CHARACTER STRINGS

In order to get the computer to print informative messages on the display, we must first learn about **character strings**. Character strings (strings for short) are a type of sequence; they are similar in some ways to lists, and are a subtype of vectors (discussed in Chapter 13), but they have a different set of primitive operations.

Strings evaluate to themselves, as numbers do. Notice in the following examples that strings do not get converted to all uppercase the way symbols do. Strings are not symbols. The STRINGP predicate returns T if its input is a string.

```
> "strings are things"
"strings are things"

> (setf a "This object is a string.")
"This object is a string."

> (stringp a)
T

> a
"This object is a string."

> (setf b 'this-object-is-a-symbol)
THIS-OBJECT-IS-A-SYMBOL

> (stringp b)
NIL
```

As you can see, character strings must be enclosed in double quote characters (",) which are not the same as the apostrophe (') we use to quote symbols and lists. Two apostrophes will not work here; you must use the double quote key on your keyboard in order to type a string.

9.3 THE FORMAT FUNCTION

The FORMAT function normally returns NIL, but as a side effect it causes things to be written on the display or to a file. The first argument to FORMAT should be the symbol T when we want to write to the display. (Different values are used when writing to a disk file.) The second argument

must be a string, called the **format control string.** FORMAT writes the
string, without the quotes, and then returns NIL.

```
> (format t "Hi, mom!")
Hi, mom!
NIL
```

The format control string can also contain special formatting directives,
which begin with a tilde, ''~,'' character. For example, the ~% directive
causes FORMAT to move to a new line. Two ~% directives right next to each
other result in a blank line in the output.

```
> (format t "Time flies~%like an arrow.")
Time flies
like an arrow.
NIL


> (format t "Fruit flies~%~%like bananas.")
Fruit flies

like bananas.
NIL
```

The ~& directive tells FORMAT to move to a new line unless it knows it is
already at the beginning of a new line. So two or three successive ~&
directives have the same effect as a single one. The ~& directive is useful
because we don't always know where the cursor will have been left when our
function gets called. For instance, some Common Lisp implementations
require the user to press carriage return after the final right parenthesis when
typing an expression to the read-eval-print loop, while other implementations
do not, so at the time that FORMAT is called, the cursor will be in a different
place depending on whether the user had to hit the return key or not.

In programs that produce several lines of output, it is good practice to
begin each format control string with ~& so that the cursor is guaranteed to be
on a fresh line before printing each message.

```
(defun mary ()
  (format t "~&Mary had a little bat.")
  (format t "~&Its wings were long and brown.")
  (format t "~&And everywhere that Mary went")
  (format t "~&The bat went, upside-down."))
```

```
> (mary)
Mary had a little bat.
Its wings were long and brown.
And everywhere that Mary went
The bat went, upside-down.
NIL
```

Another important formatting directive is ~S, which inserts the printed representation of a Lisp object into the message that FORMAT prints. (The S stands for "S-expression," or "symbolic expression," a somewhat archaic term for a Lisp object.) For each occurrence of ~S in the format control string, FORMAT requires one extra argument. In the following example, the first ~S is replaced by the symbol BOSTON, the second ~S is replaced by the list (NEW YORK), and the third ~S is replaced by the number 55.

```
> (format t "From ~S to ~S in ~S minutes!"
           'boston '(new york) 55)
From BOSTON to (NEW YORK) in 55 minutes!
NIL
```

Here is another example. The function SQUARE-TALK takes a number as input and tells you the square of that number. It does not return the square; it returns NIL because that is the result returned by FORMAT.

```
(defun square-talk (n)
   (format t  "~&~S squared is ~S" n (* n n)))


> (square-talk 10)
10 squared is 100
NIL


> (mapcar #'square-talk '(1 2 3 4 5))
1 squared is 1
2 squared is 4
3 squared is 9
4 squared is 16
5 squared is 25
(NIL NIL NIL NIL NIL)
```

The result returned by the MAPCAR is a list of NILs because each call to SQUARE-TALK returns NIL.

The ~A directive prints an object without using **escape characters**. The easiest way to explain this is to compare how ~A and ~S print strings. ~S includes the quotation marks, whereas ~A does not. A quotation mark is one kind of escape character.

```
(defun test (x)
  (format t "~&With escape characters:   ~S" x)
  (format t "~&Without escape characters:   ~A" x))

> (test "Hi, mom")
With escape characters:   "Hi, mom"
Without escape characters:   Hi, mom
NIL
```

EXERCISES

9.1. Write a function to print the following saying on the display: ''There are old pilots, and there are bold pilots, but there are no old bold pilots.'' Your function should break up the quotation into several lines.

9.2. Write a recursive function DRAW-LINE that draws a line of a specified length by doing (FORMAT T "*") the correct number of times. (DRAW-LINE 10) should produce

```
* * * * * * * * * *
```

9.3. Write a recursive function DRAW-BOX that calls DRAW-LINE repeatedly to draw a box of specified dimensions. (DRAW-BOX 10 4) should produce

```
* * * * * * * * * *
* * * * * * * * * *
* * * * * * * * * *
* * * * * * * * * *
```

9.4. Write a recursive function NINETY-NINE-BOTTLES that sings the well-known song ''Ninety-nine Bottles of Beer on the Wall.'' The first verse of this song is

```
99 bottles of beer on the wall,
99 bottles of beer!
Take one down,
Pass it around,
98 bottles of beer on the wall.
```

NINETY-NINE-BOTTLES should take a number N as input and start counting from N down to zero. (This is so you can run it on three bottles instead of all ninety nine.) Your function should also leave a blank line between each verse, and say something appropriate when it runs out of beer.

9.5. Part of any tic-tac-toe playing program is a function to display the board. Write a function PRINT-BOARD that takes a list of nine elements as input. Each element will be an X, an O, or NIL. PRINT-BOARD should display the corresponding board. For example, (PRINT-BOARD '(X O O NIL X NIL O NIL X)) should print:

9.4 THE READ FUNCTION

READ is a function that reads one Lisp object (a number, symbol, list, or whatever) from the keyboard and returns that object as its value. The object does not have to be quoted because it will not be evaluated. By placing calls to READ inside a function, we can make the computer read data from the keyboard under program control. Here are some examples. User type-in in response to READ is underlined.

```
(defun my-square ()
  (format t "Please type in a number: ")
  (let ((x (read)))
    (format t "The number ~S squared is ~S.~%"
      x (* x x))))
```

```
> (my-square)
Please type in a number: 7
The number 7 squared is 49.
NIL
```

```
> (my-square)
Please type in a number: -4
The number -4 squared is 16.
NIL
```

EXERCISES

9.6. Write a function to compute an hourly worker's gross pay given an hourly wage in dollars and the number of hours he or she worked. Your function should prompt for each input it needs by printing a message in English. It should display its answer in English as well.

9.7. The COOKIE-MONSTER function keeps reading data from the terminal until it reads the symbol COOKIE. Write COOKIE-MONSTER. Here is a sample interaction:

```
> (cookie-monster)
Give me cookie!!!
Cookie? rock
No want ROCK...

Give me cookie!!!
Cookie? hairbrush
No want HAIRBRUSH...

Give me cookie!!!
Cookie? cookie
Thank you!...Munch munch munch...BURP
NIL
```

9.5 THE YES-OR-NO-P FUNCTION

The YES-OR-NO-P function takes a format control string as input and asks the user a yes or no question. The user must respond by typing ''yes,'' in which case the function returns T, or ''no,'' in which case it returns NIL.

```
(defun riddle ()
  (if (yes-or-no-p
         "Do you seek Zen enlightenment? ")
      (format t "Then do not ask for it!")
      (format t "You have found it.")))
```

```
> (riddle)
Do you seek Zen enlightenment? yes
Then do not ask for it!
NIL
```

```
> (riddle)
Do you seek Zen enlightenment? no
You have found it.
NIL
```

There is also a shorter form of this function, called Y-OR-N-P, that only requires the user to type ''y'' or ''n'' in response.

9.6 READING FILES WITH WITH-OPEN-FILE

The WITH-OPEN-FILE macro provides a convenient way to read data from a file. Its syntax is:

```
(WITH-OPEN-FILE (var pathname)
    body)
```

WITH-OPEN-FILE creates a local variable (just like LET) and sets it to a **stream object** representing a connection to that file. Stream objects are a special Lisp datatype for describing connections to files. If you want to see one, take a look at the value of the global variable *TERMINAL-IO*. It holds the stream object Lisp uses to read from the keyboard and write to the display. Here's what it looks like on my Lisp system:

```
> *terminal-io*
#<Typescript stream, TS=18>
```

Within the body of WITH-OPEN-FILE the stream object can be passed as an optional argument to READ to read data from the file instead of from the keyboard. On leaving the WITH-OPEN-FILE form, the connection to the file is closed automatically.

Let's try an example of reading data from a file. Suppose the file "timber.dat" in the directory /usr/dst contains these lines:[*]

```
"The North Slope"
((45 redwood) (12 oak) (43 maple))
100
```

We can read this data with the following program:

```
(defun get-tree-data ()
  (with-open-file (stream "/usr/dst/timber.dat")
    (let* ((tree-loc (read stream))
           (tree-table (read stream))
           (num-trees (read stream)))
      (format t "~&There are ~S trees on ~S."
              num-trees tree-loc)
      (format t "~&They are:   ~S" tree-table))))
```

---

[*]Common Lisp understands file names using whatever syntax is appropriate to the machine on which the Lisp is running. On Unix machines, the pathname /usr/dst/timber.dat is interpreted as file timber.dat in directory /usr/dst.

```
> (get-tree-data)
There are 100 trees on "The North Slope".
They are:  ((45 REDWOOD) (12 OAK) (43 MAPLE))
NIL
```

9.7 WRITING FILES WITH WITH-OPEN-FILE

We can also use WITH-OPEN-FILE to open files for output by passing it the special keyword argument :DIRECTION :OUTPUT. The stream that WITH-OPEN-FILE creates can then be used in place of the usual T as a first argument to FORMAT.

```
(defun save-tree-data (tree-loc tree-table
                         num-trees)
  (with-open-file (stream "/usr/dst/timber.newdat"
                     :direction :output)
    (format stream "~S~%" tree-loc)
    (format stream "~S~%" tree-table)
    (format stream "~S~%" num-trees)))

> (save-tree-data
    "The West Ridge"
    '((45 redwood) (22 oak) (43 maple))
    110)
NIL
```

If we write data to a file using just the ~S directive, we are assured of being able to read it back in again. It is of course possible to write arbitrary messages to a file, containing strange punctuation, unbalanced parentheses, or what have you, but we would not be able to read the file back into Lisp using READ. Such a file might still be useful, though, because it could be read by people. If necessary it could be read by Lisp a character at a time, using techniques not covered here.

SUMMARY

The FORMAT function takes two or more arguments. The first argument should be T to print on the display; the second must be a format control string. The remaining arguments are used to fill in information required by ~S directives in the format control string. The ~% directive causes FORMAT to begin a new line; the ~& directive begins a new line only if not already at the beginning of a new line.

The READ function reads one Lisp object from the terminal and returns that object. The object does not have to be quoted because it will not be evaluated. YES-OR-NO-P and Y-OR-N-P print questions (using a format control string) and then return T or NIL depending on the answer the user gives.

WITH-OPEN-FILE opens a file for either input or output, and binds a local variable to a stream object that represents the connection to that file. This stream object can be passed to READ or FORMAT to do file i/o.

REVIEW EXERCISES

9.8. How are strings different from symbols?

9.9. What is printed by each of the following?

```
(format t "a~S" 'b)

(format t "always~%broke")

(format t "~S~S" 'alpha 'bet)
```

FUNCTIONS COVERED IN THIS CHAPTER

String predicate: STRINGP.

Input/output functions: FORMAT, READ, YES-OR-NO-P, Y-OR-N-P.

Macro for simple file i/o: WITH-OPEN-FILE.

Keyboard Exercise

In this exercise we will write a program for producing a graph of an arbitrary function. The program will prompt for a function name F and then plot $y = F(x)$ for a specified range of x values. Here is an example of how the program works:

```
> (make-graph)
Function to graph?  square
Starting x value?  -7
Ending x value?  7
Plotting string?  "****"
```

```
                                                            ****
                                                    ****
                                            ****
                                    ****
                            ****
                    ****
              ****
          ****
      ****
        ****
            ****
                    ****
                            ****
                                    ****
                                            ****
                                                    ****
                                                            ****
```

T

EXERCISE

9.10. As you write each of the following functions, test it by calling it from
top level with appropriate inputs before proceeding on to the next
function.

a. Write a recursive function SPACE-OVER that takes a number N as
input and moves the cursor to the right by printing N spaces, one at a
time. SPACE should print ''Error!'' if N is negative. Test it by
using the function TEST. Try (TEST 5) and (TEST −5).

```
(defun test (n)
  (format t "~%>>>")
  (space-over n)
  (format t "<<<"))
```

b. Write a function PLOT-ONE-POINT that takes two inputs,
PLOTTING-STRING and Y-VAL, prints PLOTTING-STRING
(without the quotes) in column Y-VAL, and then moves to a new
line. The leftmost column is numbered zero.

c. Write a function PLOT-POINTS that takes a string and a list of y values as input and plots them. (PLOT-POINTS "<>" '(4 6 8 10 8 6 4)) should print

d. Write a function GENERATE that takes two numbers M and N as input and returns a list of the integers from M to N. (GENERATE -3 3) should return (-3 -2 -1 0 1 2 3).

e. Write the MAKE-GRAPH function. MAKE-GRAPH should prompt for the values of FUNC, START, END, and PLOTTING-STRING, and then graph the function. *Note:* You can pass FUNC as an input to MAPCAR to generate the list of y values for the function. What will the second input to MAPCAR be?

f. Define the SQUARE function and graph it over the range -7 to 7. Use your first name as the plotting symbol.

Lisp Toolkit: DRIBBLE

The DRIBBLE function records part of a Lisp session in a file. This is useful if you want to make a printout of an interactive session to show to someone else. Given a file name as an argument, DRIBBLE opens that file for output and starts recording. If called with no arguments, it closes the file in which it was recording. Here is an example:

```
> (dribble "session1.log")
Now recording in file /usr/dst/session1.log

> (cons t 3)
(T . 3)
```

```
> (list 'do 'not 'feed 'the 'weeds)
(DO NOT FEED THE WEEDS)

> (dribble)
Finished recording in file /usr/dst/session1.log
```

The file ''session1.log'' now contains the following:

```
;Recording in /usr/dst/session1.log
;Recording started at 1:45pm 1-Mar-89:

> (cons t 3)
(T . 3)

> (list 'do 'not 'feed 'the 'weeds)
(DO NOT FEED THE WEEDS)

> (dribble)
```

9   Advanced Topics

9.8 PARAMETERS TO FORMAT DIRECTIVES

Some format directives accept prefix parameters that further specify their behavior. Prefix parameters appear between the ~ and the directive. For example, the ~S directive accepts a *width* parameter. By using an explicit width, like ~10S, we can produce columnar output.

```
(setf glee-club
  '((john smith) (barbara wilson) (mustapha ali)))

(defun print-one-name (name)
  (format t "~&~10S ~S"
    (second name)
    (first name)))
```

```
(defun print-all-names (x)
  (mapcar #'print-one-name x)
  'done)

> (print-all-names glee-club)
SMITH       JOHN
WILSON      BARBARA
ALI         MUSTAPHA
DONE
```

9.9 ADDITIONAL FORMAT DIRECTIVES

The ~D directive prints an integer in decimal notation (that is, base 10). It is also possible to print numbers in other bases, and even in roman numerals, but we won't get into that here. The ~F directive prints floating point numbers in a fixed-format notation that always includes a decimal point. All of these directives take prefix parameters. The first prefix parameter is used to specify a fixed width for the output: how many characters it should take up. (Lisp will pad the output with blanks if necessary.) We will consider just one other prefix parameter. With the ~F directive, the second prefix parameter specifies how many digits are to appear after the decimal point. For example, ~7,5F specifies a seven character field, with five digits appearing after the decimal point:

```
(defun sevenths (x)
  (mapcar #'(lambda (numerator)
              (format t "~&~4,2F / 7 is ~7,5F"
                numerator
                (/ numerator 7.0)))
          x)
  'done)

> (sevenths '(1 3/2 2 2.5 3))
1.00 / 7 is  0.14286
1.50 / 7 is  0.21429
2.00 / 7 is  0.28571
2.50 / 7 is  0.35714
3.00 / 7 is  0.42857
DONE
```

9.10 THE LISP 1.5 OUTPUT PRIMITIVES

The primitive i/o functions TERPRI, PRIN1, PRINC, and PRINT were
defined in Lisp 1.5 (the ancestor of all modern Lisp systems) and are still
found in Common Lisp today. They are included in the Advanced Topics
section as a historical note; you can get the same effect with FORMAT.
TERPRI stands for *ter*minate *pri*nt. It moves the cursor to a new line. PRIN1
and PRINC take a Lisp object as input and print it on the terminal. PRIN1
prints the object with whatever escape characters are necessary to assure that it
can be read back in with READ; PRINC prints the object without escape
characters. Basically, the ~S format directive works like PRIN1, and the ~A
directive works like PRINC. Both PRIN1 and PRINC return their first
argument.

```
> (setf a "Wherefore art thou, Romeo?")
"Wherefore art thou, Romeo?"

> (prin1 a)
"Wherefore art thou, Romeo?"
"Wherefore art thou, Romeo?"

> (princ a)
Wherefore art thou, Romeo?
"Wherefore art thou, Romeo?"
```

The PRINT function is a combination of the preceding three functions. It
goes to a newline with TERPRI, prints its argument with PRIN1, and then
prints a space with PRINC. A simple version of PRINT could be defined as
follows:

```
(defun my-print (x)
  (terpri)
  (prin1 x)
  (princ " ")
  x)

> (mapcar #'my-print '(0 1 2 3 4))

0
1
2
3
4
(0 1 2 3 4)
```

TERPRI, PRIN1, and PRINC accept an optional stream argument just like READ; this allows them to be used for file i/o.

9.11 HANDLING END-OF-FILE CONDITIONS

Sometimes it's necessary to read a file without knowing in advance how many objects it contains. When your program gets to the end of the file, the next READ will generate an **end-of-file error**, and you'll end up in the debugger. It is possible to tell READ to return a special value, called an eof indicator, instead of generating an error on end of file. We do this by supplying two extra arguments to READ: a NIL (meaning don't generate an error), and the value we want to use as the eof indicator. We must be careful what value we choose for this. If we used something common, like the symbol FOO, then if the file actually contains a FOO, our program will be fooled into thinking it has reached the end. Therefore, a good choice for an eof indicator is a freshly generated cons cell. We will use EQ rather than EQUAL to make sure that exactly that cons cell is returned.

Here is an example of a program that reads an arbitrary file of Lisp objects, tells how many objects were read, and returns a list of them. It uses the cons cell (EOF) as its special end-of-file value, but any freshly generated cons cell will do, since only the cell's address is important, not its contents.

```
(defun read-my-file ()
  (with-open-file (stream "/usr/dst/sample-file")
    (let ((contents
            (read-all-objects stream (list '$eof$))))
      (format t "~&Read ~S objects from the file."
        (length contents))
      contents)))

(defun read-all-objects (stream eof-indicator)
  (let ((result (read stream nil eof-indicator)))
    (if (eq result eof-indicator)
        nil
        (cons result
              (read-all-objects stream
                                eof-indicator)))))
```

Suppose our sample file contains the following lines:

```
35 cat (moose
meat) 98.6 "Frozen yogurt"
```

```
3.14159
```

The program would produce the following result:

```
> (read-my-file)
Read 6 objects from the file.
(35 CAT (MOOSE MEAT) 98.6 "Frozen yogurt" 3.14159)
```

9.12 PRINTING IN DOT NOTATION

Dot notation is a variant of cons cell notation. In dot notation each cons cell is displayed as a left parenthesis, the car part, a dot, the cdr part, and a right parenthesis. The car and cdr parts, if lists, are themselves displayed in dot notation, making this a recursive definition. For example, the list (A) is represented by a single cons cell whose car is the symbol A and whose cdr is NIL. In dot notation this list is written (A . NIL). Here are some more examples of dot notation:

List Notation	Dot Notation
NIL	NIL
A	A
(A)	(A . NIL)
(A B)	(A . (B . NIL))
(A B C)	(A . (B . (C . NIL)))
(A (B) C)	(A . ((B . NIL) . (C . NIL)))

EXERCISES

9.11. Write a function DOT-PRIN1 that takes a list as input and prints it in dot notation. DOT-PRIN1 will print parentheses by (FORMAT T "(") and (FORMAT T ")"), and dots by (FORMAT T " . "), and will call itself recursively to print lists within lists. DOT-PRIN1 should return NIL as its result. Try (DOT-PRIN1 '(A (B) C)) and see if your output matches the result in the table above. Then try (DOT-PRIN1 '((((A)))).

9.12. Lisp can also *read* lists in dot notation. Try (DOT-PRIN1 '(A . (B . C))). Be sure to type a space before and after each dot.

9.13. If you type in the quoted list '(A . NIL), Lisp types back (A). What happens when you type '(A . B)?

9.14. Consider the following two circular list structures, each composed of a single cons cell. What will be the behavior of DOT-PRIN1 if it is given the first structure as input? What will it do if given the second structure as input?

9.13 HYBRID NOTATION

Lisp normally prints things in list notation, not dot notation. But as we have seen, some cons cell structures such as (A . B) cannot be written without dots. Lisp's policy is to print dots only when necessary. It never prints a dot unless the cons cell chain ends in a non-NIL atom. Its output is thus a hybrid of pure list and pure dot notations, called **hybrid notation**. Here are some examples of the differences between pure dot notation and hybrid notation:

Dot Notation	Hybrid Notation
(A . NIL)	(A)
(A . B)	(A . B)
(A . (B . NIL))	(A B)
(A . (B . C))	(A B . C)
(A . (B . (C . D)))	(A B C . D)
((A . NIL) . (B . (C . D)))	((A) B C . D)

EXERCISE

9.15. Write HYBRID-PRIN1. Here is how the function should decide whether to print a dot or not. If the cdr part of the cons cell is a list, HYBRID-PRIN1 continues to print in list notation. If the cdr part is NIL, HYBRID-PRIN1 should print a right parenthesis. If the cdr part is something else, such as a symbol, HYBRID-PRIN1 should print a dot, the symbol, and a right parenthesis. You will probably find it useful to define a subfunction to print cdrs of lists, as these always begin with a space, whereas the cars always begin with a left parenthesis. Test your function on the examples in the preceding table.

FUNCTIONS COVERED IN ADVANCED TOPICS

Lisp 1.5 output primitives: TERPRI, PRIN1, PRINC, PRINT.

10

Assignment

10.1 INTRODUCTION

We saw in Chapter 5 that the SETF macro changes the value of a variable; this is called *assignment*. We have avoided assignment as much as possible in this book, using it only at the top-level read-eval-print loop to set up global variables. We have not yet learned to use SETF inside of functions.

There are good reasons to avoid assignment when first learning to program. Assignment is easily misused, leading to functions that are hard to understand and debug. If your first programming language was BASIC, Pascal, Modula, or C, all of which are heavily dependent on assignment, you might be surprised to see how many interesting Lisp programs don't use assignment at all. Lisp provides a richer set of control structures than those languages (such as LET, and the applicative operators), which often makes assignment unnecessary.

There are, however, occasions where it is appropriate to use assignment in Lisp. This chapter introduces some standard techniques for programming with assignment, and some useful built-in assignment forms in addition to SETF. Assignment is frequently used in combination with iterative control structures, which are discussed in the following chapter.

10.2 UPDATING A GLOBAL VARIABLE

Suppose we are operating a lemonade stand, and we want to keep track of how many glasses have been sold so far. We keep the number of glasses sold in a global variable, *TOTAL-GLASSES*, which we will initialize to zero this way:

```
(setf *total-glasses* 0)
```

There is a convention in Common Lisp that global variables should have names that begin and end with an asterisk.[*] It's permissible to ignore the asterisk convention when performing some quick calculations with global variables at top level, but when you write a *program* to manipulate global variables, you should adhere to it. Therefore, we'll call our global variable *TOTAL-GLASSES*.

Now, every time we sell some lemonade, we have to update this variable. We also want to report back how many glasses have been sold so far. Here is a function to do that:

```
(defun sell (n)
  "Ye Olde Lemonade Stand:  Sales by the Glass."
  (setf *total-glasses* (+ *total-glasses* n))
  (format t
    "~&That makes ~S glasses so far today."
    *total-glasses*))

> (sell 3)
That makes 3 glasses so far today.
NIL

> (sell 2)
That makes 5 glasses so far today.
NIL
```

Notice that the SELL function contains two forms in its body. The first form updates the variable *TOTAL-GLASSES*. The second form prints a message about how many glasses have been sold so far. SELL returns NIL because that is the result returned by FORMAT.

---

[*]Note to instructors: In ANSI Common Lisp one should also declare globals with DEFVAR, which proclaims them special, but many implementations do not strictly enforce this. We omit discussion of DEFVAR here to avoid introducing the concept of dynamic scoping until Chapter 14.

EXERCISE

10.1. Suppose we had forgotten to set *TOTAL-GLASSES* to zero before calling SELL for the first time. What would happen? Suppose we had initialized *TOTAL-GLASSES* to the symbol FOO instead of to zero. When would the error become apparent?

10.3 STEREOTYPICAL UPDATING METHODS

SETF can assign any value to any variable. A very common use of assignment is to update a variable, in other words, the variable's old value is used to compute what its new value should be. The lemonade stand is a typical example of updating a variable. Many, perhaps most uses of assignment are of this form. Common Lisp provides built-in macros for expressing the most common update cases more concisely than with SETF. We will consider two of these: updating a counter by incrementing or decrementing it, and updating a list by adding or deleting an element at the front.

10.3.1 The INCF and DECF Macros

Instead of incrementing a numeric variable by writing, say, (SETF A (+ A 5)), you can write (INCF A 5). INCF and DECF are special assignment macros for incrementing and decrementing variables. If the increment/decrement value is omitted, it defaults to one.

```
> (setf a 2)
2

> (incf a 10)
12

> (decf a)
11
```

EXERCISE

10.2. Rewrite the lemonade stand SELL function to use INCF instead of SETF.

10.3.2 The PUSH and POP Macros

When adding an element to a list by consing it onto the front, such as (SETF X (CONS 'FOO X)), you can express your intent more elegantly by writing (PUSH 'FOO X). The name "push" comes from classical computer science terminology for **pushdown stacks**, or "stacks" for short. A stack is analogous to a spring-loaded stack of dishes in a cafeteria. When you push a dish onto the stack, it becomes the new topmost element. When you pop the topmost dish off of the stack, the dish below becomes the new topmost element. Let's try using PUSH to build a stack of dishes:

```
(setf mystack nil)

(push 'dish1 mystack)   ⇒   (dish1)

(push 'dish2 mystack)   ⇒   (dish2 dish1)

(push 'dish3 mystack)   ⇒   (dish3 dish2 dish1)
```

DISH3 is now at the top of the stack. (Reading a list from left to right is like reading a stack from top to bottom.) Since each call to PUSH results in an assignment, the variable MYSTACK will always be updated to point to the newest cons cell, at the head of the list. When we start popping dishes off the stack, the first dish to come off will be DISH3. Lisp provides a POP macro to update a variable by setting it to the REST of the list to which it was originally pointing.

```
mystack   ⇒   (dish3 dish2 dish1)

(pop mystack)   ⇒   dish3

mystack   ⇒   (dish2 dish1)
```

Notice that the result POP returns is the element that was formerly the top of the stack. That element is popped off the stack as a side effect. The following two forms are equivalent:

```
(pop mystack)

(let ((top-element (first mystack)))
  (setf mystack (rest mystack))
  top-element)
```

The LET expression first remembers the top of the stack in the local variable TOP-ELEMENT. Then in the body it pops the stack by setting

MYSTACK to (REST MYSTACK). Finally it returns the value of TOP-ELEMENT, which is the element it just popped.

PUSH and POP should really be called PUSHF and POPF for consistency with the other assignment forms. Their names don't end in ''F'' for historical reasons: They were invented before SETF, and hence, before there was an ''F'' convention. By the way, the name SETF stands for ''set field.''

Here is an example of programming with PUSH:

```
(setf *friends* nil)

(defun meet (person)
  (cond ((equal person (first *friends*))
         'we-just-met)
        ((member person *friends*)
         'we-know-each-other)
        (t (push person *friends*)
           'pleased-to-meet-you)))

> (meet 'fred)
PLEASED-TO-MEET-YOU

> (meet 'cindy)
PLEASED-TO-MEET-YOU

> (meet 'cindy)
WE-JUST-MET

> (meet 'joe)
PLEASED-TO-MEET-YOU

> (meet 'fred)
WE-KNOW-EACH-OTHER

> *friends*
(JOE CINDY FRED)
```

EXERCISES

10.3. Modify the MEET function to keep a count of how many people have been met more than once. Store this count in a global variable.

10.4. Write a function FORGET that removes a person from the *FRIENDS* list. If the person wasn't on the list in the first place, the function should complain.

10.3.3 Updating Local Variables

Assignment should not be used indiscriminately. For example, it is usually considered inelegant to change the value of a local variable; one should just bind a new local variable with LET instead. (There are exceptions, of course.) It is even less elegant to modify a variable that appears in a function's argument list; doing this makes the function hard to understand. Consider the following code written in very bad style:

```
(defun bad-style (n)
  (format t "~&N is ~S." n)
  (decf n 2)
  (format t "~&Now N is ~S." n)
  (decf n 2)
  (format t "~&Now N is ~S." n)
  (list 'result 'is (* n (- n 1))))

> (bad-style 9)
N is 9.
Now N is 7.
Now N is 5.
(RESULT IS 20)
```

This code can be cleaned up by introducing some extra variables and replacing the DECF expressions with a LET* form, which binds variables sequentially (page 144). When all assignments have been eliminated, we are assured that the value of a variable will never change once it is created. Programs written in this **assignment-free** style are easy to understand, and very elegant.

```
(defun good-style (n)
  (let* ((p (- n 2))
         (q (- p 2)))
    (format t "~&N is ~S." n)
    (format t "~&P is ~S." p)
    (format t "~&Q is ~S." q)
    (list 'result 'is (* q (- q 1)))))

> (good-style 9)
N is 9.
P is 7.
Q is 5.
(RESULT IS 20)
```

There are some occasions when it is more convenient to assign to a local variable instead of LET-binding it. The following is an example. Note that each variable is bound to NIL initially, and then is assigned a new value just once. This form of "disciplined" assignment is not bad style; it is quite different from the assignment occurring in the BAD-STYLE function.

```
(defun get-name ()
  (let ((last-name nil)
        (first-name nil)
        (middle-name nil)
        (title nil))
    (format t "~&Last name? ")
    (setf last-name (read))
    (format t "~&First name? ")
    (setf first-name (read))
    (format t "~&Middle name or initial? ")
    (setf middle-name (read))
    (format t "~&Preferred title? ")
    (setf title (read))
    (list title first-name middle-name last-name)))
```

```
> (get-name)
Last name? higginbotham
First name? waldo
Middle name or initial? j
Preferred title? admiral
(ADMIRAL WALDO J HIGGINBOTHAM)
```

10.4 WHEN AND UNLESS

WHEN and UNLESS are conditional forms that are useful when you need to evaluate more than one expression when a test is true. Their syntax is:

(WHEN *test*
  body)

(UNLESS *test*
  body)

WHEN first evaluates the test form. If the result is NIL, WHEN just returns NIL. If the result is non-NIL, WHEN evaluates the forms in its body and returns the value of the last one. UNLESS is similar, except it evaluates the forms in its body only if the test is false. For both of these conditionals, when the forms in the body are evaluated, the value of the last one is returned. Forms prior to the last one are only useful for side effects, such as i/o or assignment.

The only advantages of WHEN and UNLESS over COND are stylistic. They have a simpler and somewhat more pleasant syntax, and they need less indentation because their bodies are indented only two spaces. Here is an example of how WHEN and UNLESS can be useful. Suppose we want to write a function that takes two numbers as input and multiplies them. Suppose this function requires that its first input be odd and its second input even. If an input is of the wrong sort, the function can "fix" it by adding or subtracting one and printing a suitable warning message.

```
(defun picky-multiply (x y)
  "Computes X times Y.
   Input X must be odd; Y must be even."
  (unless (oddp x)
    (incf x)
    (format t
      "~&Changing X to ~S to make it odd." x))
  (when (oddp y)
    (decf y)
    (format t
      "~&Changing Y to ~S to make it even." y))
  (* x y))

> (picky-multiply 4 6)
Changing X to 5 to make it odd.
30

> (picky-multiply 2 9)
Changing X to 3 to make it odd.
Changing Y to 8 to make it even.
24
```

10.5 GENERALIZED VARIABLES

A **generalized variable** is any place a pointer may be stored. An ordinary variable like X or N contains a pointer to the object that is its value. But pointers can also be stored in other sorts of places, such as the car or cdr half of a cons cell. Assignment means replacing one pointer with another. So when we say the value of N is three, what we mean is that the variable named

N holds a pointer to the number three. An expression like (INCF N) replaces that pointer with a pointer to the number four.

The assignment macros we covered in this chapter can assign to generalized variables, meaning they can store pointers in many different places. The first argument to SETF, INCF, DECF, PUSH, or POP is treated as a place description. Consider this example:

```
(setf x '(jack benny was 39 for many years))

(setf (sixth x) 'several)

> x
(JACK BENNY WAS 39 FOR SEVERAL YEARS)

> (decf (fourth x) 2)
37

> x
(JACK BENNY WAS 37 FOR SEVERAL YEARS)
```

As you can see, SETF and related forms can accept place descriptions like (FOURTH X), and store new pointers in those places. For instance, the expression (FOURTH X) specifies a pointer that lives in the car of the fourth cons cell in the chain pointed to by X. This place can also be called the CAR of the CDDDR of X, as shown below:

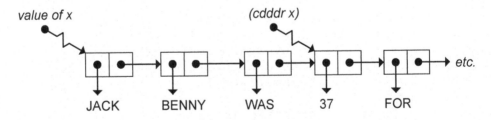

10.6 CASE STUDY: A TIC-TAC-TOE PLAYER

In this section we will write our first large program: a program that not only plays tic-tac-toe, but also explains the strategy behind each move. When writing a program this complex, it pays to take a few minutes at the outset to think about the overall design, particularly the data structures that will be used. Let's start by developing a representation for the board. We will number the squares on the tic-tac-toe board this way:

```
1 | 2 | 3
-----------
4 | 5 | 6
-----------
7 | 8 | 9
```

We will represent a board as a list consisting of the symbol BOARD followed by nine numbers that describe the contents of each position. A zero means the position is empty; a one means it is filled by an O; a ten means it is filled by an X. The function MAKE-BOARD creates a new tic-tac-toe board:

```
(defun make-board ()
  (list 'board 0 0 0 0 0 0 0 0 0))
```

Notice that if B is a variable holding a tic-tac-toe board, position one of the board can be accessed by writing (NTH 1 B), position two by (NTH 2 B), and so on. (NTH 0 B) returns the symbol BOARD.

Now let's write functions to print out the board. CONVERT-TO-LETTER converts a zero, one, or ten to a space, an O, or an X, respectively. It is called by PRINT-ROW, which prints out one row of the board. PRINT-ROW is in turn called by PRINT-BOARD.

```
(defun convert-to-letter (v)
  (cond ((equal v 1) "O")
        ((equal v 10) "X")
        (t " ")))

(defun print-row (x y z)
  (format t "~&   ~A | ~A | ~A"
          (convert-to-letter x)
          (convert-to-letter y)
          (convert-to-letter z)))

(defun print-board (board)
  (format t "~%")
  (print-row
    (nth 1 board) (nth 2 board) (nth 3 board))
  (format t "~&  -----------")
  (print-row
    (nth 4 board) (nth 5 board) (nth 6 board))
  (format t "~&  -----------")
  (print-row
    (nth 7 board) (nth 8 board) (nth 9 board))
  (format t "~%~%"))
```

```
> (setf b (make-board))
(BOARD 0 0 0 0 0 0 0 0 0)

> (print-board b)
```

```
       |   |
    - - - - - - - - - - -
       |   |
    - - - - - - - - - - -
       |   |
```

```
NIL
```

We can make a move by destructively changing one of the board positions from a zero to a one (for O) or a ten (for X). The variable PLAYER in MAKE-MOVE will be either one or ten, depending on who's moving.

```
(defun make-move (player pos board)
  (setf (nth pos board) player)
  board)
```

Let's make a sample board to test out these functions before proceeding further. We'll define variables *COMPUTER* and *OPPONENT* to hold the values ten and one (X and O), respectively, because this will make the example clearer.

```
(setf *computer* 10)

(setf *opponent* 1)

> (make-move *opponent* 3 b)       ;Put an O in square 3.
(BOARD 0 0 1 0 0 0 0 0 0)

> (make-move *computer* 5 b)       ;Put an X in square 5.
(BOARD 0 0 1 0 10 0 0 0 0)

> (print-board b)
```

```
       |   | O
    - - - - - - - - - - -
       | X |
    - - - - - - - - - - -
       |   |
```

```
NIL
```

For the program to select the best move, it must have some way of analyzing the board configuration. This is easy for tic-tac-toe. There are only eight ways to make three-in-a-row: three horizontally, three vertically, and two diagonally. We'll call each of these combinations a "triplet." We'll store a list of all eight triplets in a global variable *TRIPLETS*.

```
(setf *triplets*
  '((1 2 3) (4 5 6) (7 8 9)      ;Horizontal triplets.
    (1 4 7) (2 5 8) (3 6 9)      ;Vertical triplets.
    (1 5 9) (3 5 7)))            ;Diagonal triplets.
```

Now we can write a SUM-TRIPLET function to return the sum of the numbers in the board positions specified by that triplet. For example, the right diagonal triplet is (3 5 7). The sum of elements three, five, and seven of board B is eleven, indicating that there is one O, one X, and one blank (in some unspecified order) on that diagonal. If the sum had been twenty-one, there would be two Xs and one O; a sum of twelve would indicate one X and two Os, and so on.

```
(defun sum-triplet (board triplet)
  (+ (nth (first triplet) board)
     (nth (second triplet) board)
     (nth (third triplet) board)))

> (sum-triplet b '(3 5 7))      ;Left diagonal triplet.
11

> (sum-triplet b '(2 5 8))      ;Middle vertical triplet.
10

> (sum-triplet b '(7 8 9))      ;Bottom horizontal triplet.
1
```

To fully analyze a board we have to look at all the sums. The function COMPUTE-SUMS returns a list of all eight sums.

```
(defun compute-sums (board)
  (mapcar #'(lambda (triplet)
              (sum-triplet board triplet))
          *triplets*))

(compute-sums b)   ⇒   (1 10 0 0 10 1 10 11)
```

Notice that if player O ever gets three in a row, one of the eight sums will be three. Similarly, if player X manages to get three in a row, one of the eight sums will be 30. We can write a predicate to check for this condition:

```
(defun winner-p (board)
  (let ((sums (compute-sums board)))
    (or (member (* 3 *computer*) sums)
        (member (* 3 *opponent*) sums))))
```

We'll return to the subject of board analysis later. Let's look now at the basic framework for playing the game. The function PLAY-ONE-GAME offers the user the choice to go first, and then calls either COMPUTER-MOVE or OPPONENT-MOVE as appropriate, passing a new, empty board as input.

```
(defun play-one-game ()
  (if (y-or-n-p "Would you like to go first? ")
      (opponent-move (make-board))
      (computer-move (make-board))))
```

The OPPONENT-MOVE function asks the opponent to type in a move and checks that the move is legal. It then updates the board and calls COMPUTER-MOVE. But there are two special cases where we should not call COMPUTER-MOVE. First, if the opponent's move makes a three-in-a-row, the opponent has won and the game is over. Second, if there are no empty spaces left on the board, the game has ended in a tie. We assume that the opponent is O and the computer is X.

```
(defun opponent-move (board)
  (let* ((pos (read-a-legal-move board))
         (new-board (make-move
                      *opponent*
                      pos
                      board)))
    (print-board new-board)
    (cond ((winner-p new-board)
           (format t "~&You win!"))
          ((board-full-p new-board)
           (format t "~&Tie game."))
          (t (computer-move new-board)))))
```

A legal move is an integer between one and nine such that the corresponding board position is empty. READ-A-LEGAL-MOVE reads a Lisp object and checks whether it's a legal move using the <= (less than or equal) comparison predicate. If not, the function calls itself to read another move. Notice that the first two COND clauses each contain a test and two consequents. If the test is true, both consequents are evaluated, and the value of the last one (the recursive call) is returned.

```
(defun read-a-legal-move (board)
  (format t "~&Your move: ")
  (let ((pos (read)))
    (cond ((not (and (integerp pos)
                     (<= 1 pos 9)))
           (format t "~&Invalid  input.")
           (read-a-legal-move board))
          ((not (zerop (nth pos board)))
           (format t
             "~&That space is already occupied.")
           (read-a-legal-move board))
          (t pos))))
```

The BOARD-FULL-P predicate is called by OPPONENT-MOVE to test if there are no more empty spaces left on the board:

```
(defun board-full-p (board)
  (not (member 0 board)))
```

The COMPUTER-MOVE function is similar to OPPONENT-MOVE, except the player is X instead of O, and instead of reading a move from the keyboard, we will call CHOOSE-BEST-MOVE. This function returns a list of two elements. The first element is the position in which to place an X. The second element is a string explaining the strategy behind the move.

```
(defun computer-move (board)
  (let* ((best-move (choose-best-move board))
         (pos (first best-move))
         (strategy (second best-move))
         (new-board (make-move
                      *computer* pos board)))
    (format t "~&My move: ~S" pos)
    (format t "~&My strategy: ~A~%" strategy)
    (print-board new-board)
    (cond ((winner-p new-board)
           (format t "~&I win!"))
          ((board-full-p new-board)
           (format t "~&Tie game."))
          (t (opponent-move new-board)))))
```

Now we're almost ready to play our first game. Our first version of CHOOSE-BEST-MOVE will have only one strategy: Pick a legal move at random. The function RANDOM-MOVE-STRATEGY returns a list whose first element is the move, and whose second element is a string explaining the strategy behind the move. The function PICK-RANDOM-EMPTY-

POSITION picks a random number from one to nine. If that board position is empty, it returns that move. Otherwise, it calls itself recursively to try another random number.

```
(defun choose-best-move (board)            ;First version.
  (random-move-strategy board))

(defun random-move-strategy (board)
  (list (pick-random-empty-position board)
        "random move"))

(defun pick-random-empty-position (board)
  (let ((pos (+ 1 (random 9))))
    (if (zerop (nth pos board))
        pos
        (pick-random-empty-position board)))))
```

You can try playing a few games with the program to see how it feels. Pretty soon you'll notice that the random move strategy isn't very good near the end of the game; sometimes it causes the program to make moves that are downwright stupid. Consider this example:

```
(setf b '(board 10 10  0
                  0  0  0
                  1  1  0))
```

```
> (print-board b)

   X | X |
   -----------
     |   |
   -----------
   O | O |

NIL

> (computer-move b)
My move: 4
My strategy: random move

   X | X |
   -----------
   X |   |
   -----------
   O | O |
```

```
Your move: 9

   X | X |
  - - - - - - - - - -
   X |   |
  - - - - - - - - - -
   O | O | O
```

You win!
NIL

The computer already had two in a row; it could have won by putting an X in position three. But instead it picked a move at random and ended up putting an X in position four, which did no good at all because that vertical triplet was already blocked by the O at position seven.

To make our program smarter, we can program it to look for two-in-a-row situations. If there are two Xs in a row, it should fill in the third X to win the game. Otherwise, if there are two Os in a row, it should put an X there to block the opponent from winning.

```
(defun make-three-in-a-row (board)
  (let ((pos (win-or-block board
                           (* 2 *computer*))))
    (and pos (list pos "make three in a row"))))

(defun block-opponent-win (board)
  (let ((pos (win-or-block board
                           (* 2 *opponent*))))
    (and pos (list pos "block opponent"))))

(defun win-or-block (board target-sum)
  (let ((triplet (find-if
                   #'(lambda (trip)
                       (equal (sum-triplet board
                                            trip)
                              target-sum))
                   *triplets*)))
    (when triplet
      (find-empty-position board triplet))))

(defun find-empty-position (board squares)
  (find-if #'(lambda (pos)
               (zerop (nth pos board)))
           squares))
```

Both MAKE-THREE-IN-A-ROW and BLOCK-OPPONENT return NIL if they cannot find a move that fits their respective strategies. Now we need to revise CHOOSE-BEST-MOVE to prefer these two more clever strategies to the random move strategy. We introduce an OR into the body of CHOOSE-BEST-MOVE so that it will try its strategies one at a time until one of them returns a non-NIL move.

```
(defun choose-best-move (board)          ;Second version.
  (or (make-three-in-a-row board)
      (block-opponent-win board)
      (random-move-strategy board)))
```

This new strategy makes for a more interesting game. The computer will defend itself when it is obvious the opponent is about to win, and it will take advantage of the opportunity to win when it has two in a row.

```
> (play-one-game)
Would you like to go first? y
Your move: 1

    O |   |
   -----------
      |   |
   -----------
      |   |

My move: 5
My strategy: random move

    O |   |
   -----------
      | X |
   -----------
      |   |

Your move: 2

    O | O |
   -----------
      | X |
   -----------
      |   |
```

```
My move: 3
My strategy: block opponent

    O | O | X
   -----------
      | X |
   -----------
      |   |
```

```
Your move: 4

    O | O | X
   -----------
    O | X |
   -----------
      |   |
```

```
My move: 7
My strategy: make three in a row

    O | O | X
   -----------
    O | X |
   -----------
    X |   |
```

```
I win!
NIL
```

SUMMARY

The SETF macro can assign any value to any variable. "Updating" a variable means computing its new value based on its old value. Two stereotypical forms of updating are incrementing or decrementing a numeric variable (for which INCF and DECF may be used), and adding or removing an element from the front of a list (for which PUSH and POP may be used.) Most updates are performed on global variables. Changing the value of a local variable is usually considered bad programming style; it is better to bind a new variable with LET instead.

A generalized variable is any place a pointer may be stored. All of the assignment macros discussed in this chapter operate on generalized variables, not just ordinary variables.

Assignment is used only sparingly in Lisp programs. LET, applicative operators, and efficient tail-recursive functions, which most other languages lack, make assignment unnecessary in many cases. Assignment-free programs are considered very elegant.

REVIEW EXERCISES

10.5. Rewrite the following ugly function to use good Lisp style.

```
(defun ugly (x y)
  (when (> x y)
    (setf temp y)
    (setf y x)
    (setf x temp))
  (setf avg (/ (+ x y) 2.0))
  (setf pct (* 100 (/ avg y)))
  (list 'average avg 'is
        pct 'percent 'of 'max y))

(ugly 20 2)  ⇒
  (average 11.0 is 55.0 percent of max 20)
```

10.6. Suppose the variable X is NIL. What will its value be after evaluating (PUSH X X) three times?

10.7. What is wrong with the expression (SETF (LENGTH X) 3)?

FUNCTIONS COVERED IN THIS CHAPTER

Generalized assignment macros: SETF, INCF, DECF, PUSH, POP.

Conditionals: WHEN, UNLESS.

Lisp Toolkit: BREAK and ERROR

The BREAK and ERROR functions are useful for debugging, and for making programs more resistant to bugs. BREAK was introduced in the Chapter 8 toolkit section, but its full capabilities were not presented there. Both BREAK and ERROR take a format control string as their first argument. Additional arguments, if any, are used as arguments to the format directives such as ~S that appear in the control string.

BREAK prints out the message generated by the format control string, and then causes Lisp to enter the debugger. After you are done using the debugger, you can continue executing your program where it left off by issuing a debugger command called something like GO, PROCEED, or RESTART. (Debuggers are notoriously implementation dependent, so the precise command to use depends on which brand of Lisp you're running.) The BREAK function returns NIL, and evaluation proceeds with the next form.

Here's an example of using BREAK to debug a function. This function is supposed to take a selling price and a commission rate as input, figure the commission on the sale, print out a message, and then return either RICH or POOR depending on whether the commission was greater than 100 dollars. Sometimes, though, it returns NIL. This is a bug.

```
(defun analyze-profit (price commission-rate)
  (let* ((commission (* price commission-rate))
         (result
            (cond ((> commission 100) 'rich)
                  ((< commission 100) 'poor))))
    (format t "~&I predict you will be: ~S"
            result)
    result))


> (analyze-profit 1600 0.15)
I predict you will be: RICH
RICH


> (analyze-profit 3100 0.02)
I predict you will be: POOR
POOR


> (analyze-profit 2000 0.05)
I predict you will be: NIL
NIL
```

To debug the function, we begin by inserting a call to BREAK in the body. Then we can use the debugger to examine the control stack and check the values of local variables.

```
(defun analyze-profit (price commission-rate)
  (let* ((commission (* price commission-rate))
         (result
           (cond ((> commission 100) 'rich)
                 ((< commission 100) 'poor))))
    (break "Value of RESULT is ~S" result)
    (format t "~&I predict you will be: ~S"
            result)
    result))
```

```
> (analyze-profit 2000 0.05)
Value of RESULT is NIL
Entering the debugger:

Debug> price
2000

Debug> commission-rate
0.05

Debug> commission
100.0
```

Now the cause of the error is apparent: When the commission is exactly equal to 100, neither COND clause has a true test, so COND returns NIL. The solution is to replace the second test expression with T.

The ERROR function takes the same arguments as BREAK: a format control string followed by some optional arguments whose values will be printed by the format directives. One difference between ERROR and BREAK is that ERROR never returns: You can never continue from an ERROR. Second, ERROR merely reports the error and stops the program, it doesn't necessarily put you into the debugger, although in most implementations it will.

Programs can be made more resistant to bugs by inserting ''sanity checks'': expressions that check to make sure everything is normal, and call ERROR if something is wrong. For example, this version of the AVERAGE function checks to make sure its inputs are both numbers:

```
(defun average (x y)
  (unless (and (numberp x) (numberp y))
    (error "Arguments must be numbers:  ~S, ~S"
      x y))
  (/ (+ x y) 2.0))
```

Common Lisp provides several other functions for reporting errors. WARN prints a warning message but does not stop the program from running. CERROR signals a "continuable error." The user is told of the error and given the option to continue execution. These functions, and the new Common Lisp "condition system" that allows you to signal and trap arbitrary error conditions, will not be covered in this book. Check your reference manual for details.

Keyboard Exercise

This keyboard exercise requires you to add some additional strategies to the tic-tac-toe playing program discussed earlier. The first strategy we'll consider is called the squeeze play, which can be made using either of the two diagonal triplets. Suppose the opponent, O, goes first, and after three moves the board looks like this:

At this point, O has initiated a squeeze play. If X responds by choosing a corner, O can force a win by taking the remaining corner. Suppose, for example, that X chooses three (the upper-right corner), and O then takes seven (the lower-left corner.) The board looks like this:

```
  O |   | X
 -----------
    | X |
 -----------
  O |   | O
```

Now X is doomed because O can make three-in-a-row two different ways: either vertically or horizontally. No matter what move X chooses next, O is going to win.

The proper defense against a squeeze play is for X to choose a side square (two, four, six, or eight) instead of a corner. This forces O to take the opposite side square to block X, and the danger is past. Here's an example:

```
 O |   |
-----------
   | X |
-----------
   |   | O
```
Opponent sets up a squeeze play.

```
 O |   |
-----------
   | X | X
-----------
   |   | O
```
6: X defends by choosing a side square.

```
 O |   |
-----------
 O | X | X
-----------
   |   | O
```
4: O is forced to block X.

```
 O |   |
-----------
 O | X | X
-----------
 X |   | O
```
7: X is forced to block O.

```
 O |   | O
-----------
 O | X | X
-----------
 X |   | O
```
3: O is again forced to block X.

After two more moves, the game ends in a tie.

A second offensive strategy we want to guard against is called ''two on one.'' Like the squeeze play, it can be set up using either diagonal triplet. In a two-on-one strategy, the opponent takes the center square, the computer takes a corner, and the opponent takes the opposite corner, like this:

Now it's X's turn to move. If X takes a side square, O can force a win by taking a corner, like this:

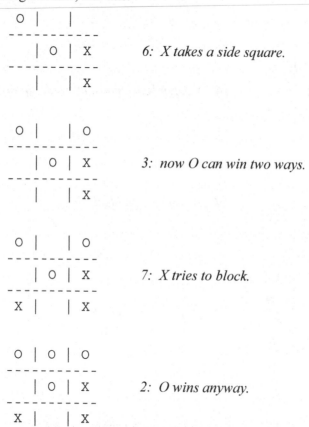

The only defense against a two-on-one attack is for X to take a corner instead of a side square.

EXERCISE

10.8. Type in the tic-tac-toe program as it appears in this book, with the second version of CHOOSE-BEST-MOVE. Try out the program by playing a few games before proceeding further.

a. Set up a global variable named *CORNERS* to hold a list of the four corner positions. Set up a global variable named *SIDES* to hold a list of the four side squares. Note that (FIND-EMPTY-POSITION BOARD *SIDES*) will return an empty side square, if there are any.

b. Write a function BLOCK-SQUEEZE-PLAY that checks the diagonals for an O-X-O pattern and defends by suggesting a side square as the best move. Your function should return NIL if there is no squeeze play in progress. Otherwise, it should return a list containing a move number and a string explaining the strategy behind the move. Test your function by calling it on a sample board.

c. Write a function BLOCK-TWO-ON-ONE that checks the diagonals for an O-O-X or X-O-O pattern and defends by suggesting a corner as the best move. Your function should return NIL if there is no two-on-one threat to which to respond. Otherwise, it should return a list containing a move and a strategy description.

d. Modify the CHOOSE-BEST-MOVE function so it that tries these two defensive strategies before choosing a move at random.

e. If the computer goes first, then after the opponent's first move there may be an opportunity for the computer to set up a squeeze play or two-on-one situation to trap the opponent. Write functions to check the diagonals and suggest an appropriate attack if the opportunity exists. Modify the CHOOSE-BEST-MOVE function to include these offensive strategies in its list of things to try.

10   Advanced Topics

10.7   DO-IT-YOURSELF LIST SURGERY

You can use SETF on generalized variables to manipulate pointers directly. For example, suppose we want to turn a chain of three cons cells into a chain of two cons cells by ''snipping out'' the middle cell. In other words, we want to change the cdr of the first cell so it points directly to the third cell. Here's how to do it:

```
(defun snip (x)
  (setf (cdr x) (cdr (cdr x))))

> (setf a '(no down payment))
(NO DOWN PAYMENT)

> (setf b (cdr a))
(DOWN PAYMENT)

> (snip a)
(PAYMENT)

> a
(NO PAYMENT)

> b
(DOWN PAYMENT)
```

Notice that the value of B was unchanged by SNIP. Only the cdr of the first cell in the chain has changed, as shown in Figure 10-1.

```
(setf a '(no down payment))

(setf b (cdr a))
```

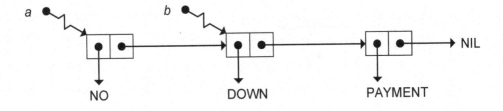

```
(snip a)  ⇒  (no payment)
```

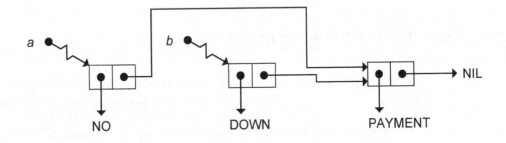

Figure 10-1 Effects of list surgery on the list (NO DOWN PAYMENT).

We can use SETF to create the following circular structure.

```
> (setf circ (list 'foo))
(FOO)

> (setf (cdr circ) circ)
(FOO FOO FOO FOO ...)
```

Here is what the circular list CIRC really looks like:

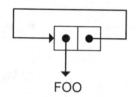

FOO

Modifying lists by directly changing the pointers in their cons cells is known as **list surgery**. List surgery is useful in large, complex programs because it can be much faster to change a few pointers than to build a whole new list. This also reduces the program's memory requirements (or causes it to garbage collect less frequently). Advanced Common Lisp programming includes lots of list surgery, but for beginners this isn't necessary. Many of the most common list surgery operations are already built in to Common Lisp, as we'll see in the next section.

10.8 DESTRUCTIVE OPERATIONS ON LISTS

Destructive list operations are those that change the contents of a cons cell. These operations are "dangerous" because they can create circular structures that may be hard to print, and because their effect on shared structures may be hard to predict. But destructive functions are also powerful and efficient tools. By convention, most of them have names that begin with N. (Like the CAR/CDR convention and the "F" convention, this arose essentially by accident but remains by virtue of brevity and usefulness.)

10.8.1 NCONC

NCONC (pronounced "en-konk," derived from "concatenate") is a destructive version of APPEND. While APPEND creates a new list for its result, NCONC physically changes the last cons cell of its first input to point to its second input. Example:

```
> (setf x '(a b c))
(A B C)

> (setf y '(d e f))
(D E F)
```

```
> (append x y)          Doesn't change X or Y, but
(A B C D E F)           result shares structure with Y.
```

```
> x                     X is unchanged.
(A B C)
```

```
> (nconc x y)           NCONC alters the list (A B C).
(A B C D E F)
```

```
> x                     X's value has changed.
(A B C D E F)
```

```
> y                     Y's has not.
(D E F)
```

If the first input to NCONC is NIL, it just returns its second input. For that reason, one shouldn't assume that (NCONC X Y) will alter the value of X. If X is NIL, its value will be unchanged. Therefore, one should always use SETF to store the result of NCONC into X, just in case.

```
> (setf x nil)
NIL

> (setf y '(no luck today))
(NO LUCK TODAY)

> (nconc x y)
(NO LUCK TODAY)

> x
NIL

> (setf x (nconc x y))
(NO LUCK TODAY)
```

```
> x
(NO LUCK TODAY)
```

The NCONC function actually accepts any number of inputs, and destructively concatenates all of them to form one long cons cell chain. We can write our own version of NCONC that takes exactly two lists as input and makes the cdr of the last cell of the first list point to the second list. It then returns a pointer to the beginning of the first list. One tricky point: If the first list is NIL, MY-NCONC should return the second list just as APPEND would do.

```
(defun my-nconc (x y)
  (cond ((null x) y)
        (t (setf (cdr (last x)) y)
           x)))
```

10.8.2 NSUBST

NSUBST is a destructive version of SUBST. It modifies a list by changing the pointers in the cars of some cells.

```
> (setf tree '(i say (e i (e i) o)))
(I SAY (E I (E I) O))

> (nsubst 'a 'e tree)
(I SAY (A I (A I) O))

> tree
(I SAY (A I (A I) O))

> (nsubst 'cheery '(a i) tree :test #'equal)
(I SAY (A I CHEERY O))
```

In the last example, since we were searching the tree for the list (A I), we had to tell NSUBST to use EQUAL as the equality test. The default test, EQL, would not have worked.

10.8.3 Other Destructive Functions

Many other Common Lisp functions have destructive counterparts. There are NREVERSE, NUNION, NINTERSECTION, and NSET-DIFFERENCE, for example. There are only two exceptions to the "N" naming convention.

APPEND was the very first Lisp function to have a destructive counterpart. Its destructive version was called NCONC, for "concatenate." (There was also a CONC function, but its use was obscure and it disappeared in later dialects.) It was many years later that NCONC gave rise to the "N" convention for indicating destructive functions. That's why there is no NAPPEND. The only other exception to the "N" convention is REMOVE. Its destructive counterpart is called DELETE, again for historical reasons. (DELETE was invented after NCONC, but before the "N" convention was established, so no one thought of the name NREMOVE.) The "N" is commonly held to stand for "noncopying" or "nonconsing."

10.9 PROGRAMMING WITH DESTRUCTIVE OPERATIONS

One place where destructive operations are especially useful is in making small changes to complex list structures, such as the MAKE-MOVE function in the tic-tac-toe program. Here's another example. Suppose we have the following table stored in the global variable *THINGS*:

```
((object1 large green shiny cube)
 (object2 small red dull metal cube)
 (object3 red small dull plastic cube))
```

How might we change the symbol OBJECT1 to FROB? The expression (ASSOC 'OBJECT1 *THINGS*) will return the list (OBJECT1 LARGE GREEN SHINY CUBE). We can use SETF on this list to physically change it by storing into the car half of the first cons cell. Since this is a destructive operation on the list, the value of *THINGS* will change as well. Let's go ahead and write a general function for renaming objects:

```
(defun rename (obj newname)
  (setf (car (assoc obj *things*)) newname))

> (rename 'object1 'frob)
FROB

> *things*
((FROB LARGE GREEN SHINY CUBE)
 (OBJECT2 SMALL RED DULL METAL CUBE)
 (OBJECT3 RED SMALL DULL PLASTIC CUBE))
```

We can use NCONC, another destructive operation, to add a new property to an object already in *THINGS*.

```
(defun add (obj prop)
  (nconc (assoc obj *things*) (list prop)))

> (assoc 'object2 *things*)
(OBJECT2 SMALL RED DULL METAL CUBE)

> (add 'object2 'sharp-edged)
(OBJECT2 SMALL RED DULL METAL CUBE SHARP-EDGED)

> *things*
((FROB LARGE GREEN SHINY CUBE)
 (OBJECT2 SMALL RED DULL METAL CUBE SHARP-EDGED)
 (OBJECT3 RED SMALL DULL PLASTIC CUBE))
```

EXERCISES

10.9. Write a destructive function CHOP that shortens any non-NIL list to a list of one element. (CHOP '(FEE FIE FOE FUM)) should return (FEE).

10.10. Write a function NTACK that *destructively* tacks a symbol onto a list. (NTACK '(FEE FIE FOE) 'FUM) should return (FEE FIE FOE FUM).

10.11. Draw the cons cell structure that results from the following sequence of operations:

```
(setf x '(a b c))

(setf (cdr (last x)) x)
```

10.12. Suppose the variable H is set to the list (HI HO). What is the critical difference between the results of (APPEND H H) and (NCONC H H)?

10.10 SETQ AND SET

In earlier Lisp dialects, where SETF and generalized variables were not available, the assignment function was called SETQ. The SETQ special function is still around today. Its syntax is the same as the SETF macro, and it can be used to assign values to ordinary (but not generalized) variables.

```
> (setq x '(slings and arrows))
(SLINGS AND ARROWS)
```

If you read older Lisp books you will notice that assignment is done with SETQ rather than SETF. Modern Common Lisp programmers use SETF for all forms of assignment, whether they are storing into an ordinary variable such as X or a generalized variable such as (SECOND X). SETQ is today considered archaic. Internally, though, most Lisp implementations turn a SETF into a SETQ if the assignment is to an ordinary variable, so you may see some references to SETQ in debugger output.

The SET function, like SETQ, comes from the earliest Lisp dialect, Lisp 1.5. SET evaluates both its arguments; the first argument must evaluate to a symbol. Because Common Lisp uses lexical scoping while Lisp 1.5 did not, the meaning of SET has changed somewhat. In Common Lisp, SET stores a value in the value cell of a symbol, meaning it assigns to the global variable named by the symbol, even if a local variable exists with the same name.[**] The SYMBOL-VALUE function returns the contents of a symbol's value cell. Here is an example of the use of SET and SYMBOL-VALUE:

```
(setf duck 'donald)       The global DUCK.

(defun test1 (duck)       A local DUCK.
  (list duck
        (symbol-value 'duck)))

(test1 'huey)  ⇒  (huey donald)

(defun test2 (duck)       Another local DUCK.
  (set 'duck 'daffy)      Change the global DUCK.
  (list duck
        (symbol-value 'duck)))

(test2 'huey)  ⇒  (huey daffy)

duck  ⇒  daffy
```

[**]For dynamically scoped variables, discussed in Chapter 14, SET assigns to the currently accessible dynamic variable with that name.

11

Iteration and Block Structure

11.1 INTRODUCTION

The word "iterate" means to repeat, or to do something over and over. Recursion and applicative operators are repetitive, but **iteration** (also known as "looping") is the simplest repetitive control structure. Virtually all programming languages include some way to write iterative expressions.

Iteration in Lisp is more sophisticated than in most other languages. Lisp provides powerful iteration constructs called DO and DO*, as well as simple ones called DOTIMES and DOLIST.

In this chapter we will also learn about "block structure," a concept borrowed from the Algol family of languages, which includes Pascal, Modula, and Ada. We will see how to group Lisp expressions into blocks, how to give names to the blocks, and why this is useful.

11.2 DOTIMES AND DOLIST

The simplest iterative forms are DOTIMES and DOLIST. Both are macro functions, meaning they don't evaluate all their arguments. They have the same syntax:

```
(DOTIMES (index-var n [result-form])
    body)
```

```
(DOLIST  (index-var  list  [result-form])
   body)
```

DOTIMES evaluates the forms in its body *n* times, while stepping an index variable from zero through *n* - 1. It then returns the value of *result-form*, which defaults to NIL if omitted. (The *result-form* is shown in brackets above because it's optional.) Here is an example of DOTIMES counting from zero up to three. The index variable is named I. Notice that the result returned by DOTIMES is NIL.

```
> (dotimes (i 4)
    (format t "~&I is ~S." i))
I is 0.
I is 1.
I is 2.
I is 3.
NIL
```

DOLIST has the same syntax as DOTIMES, but instead of counting, it steps the index variable through the elements of a list. In the following example the value returned by DOLIST is the symbol FLOWERS.

```
> (dolist (x '(red blue green) 'flowers)
    (format t "~&Roses are ~S." x))
Roses are RED.
Roses are BLUE.
Roses are GREEN.
FLOWERS
```

11.3 EXITING THE BODY OF A LOOP

The RETURN function can be used to exit the body of an iteration form immediately, without looping any further. RETURN takes one input: the value to return as the result of the iteration form. When RETURN is used to force an exit from an iteration form, the *result-form* expression, if any, is ignored.

Here is an iterative function called FIND-FIRST-ODD that returns the first odd number in a list. It uses DOLIST to loop through the elements of the list, and RETURN to exit the loop as soon as an odd number is found. If the list contains no odd numbers, then when the loop is finished, DOLIST will return NIL. An interesting point about FIND-FIRST-ODD is that the body of the loop contains two forms instead of one. Loop bodies may contain any number of forms.

```
(defun find-first-odd (list-of-numbers)
  (dolist (e list-of-numbers)
    (format t "~&Testing ~S..." e)
    (when (oddp e)
      (format t "found an odd number.")
      (return e))))
```

> (find-first-odd '(2 4 6 7 8))   ;*Will never reach 8.*
Testing 2...
Testing 4...
Testing 6...
Testing 7...found an odd number.
7

> (find-first-odd '(2 4 6 8 10))
Testing 2...
Testing 4...
Testing 6...
Testing 8...
Testing 10...
NIL

The following is an example where specifying an explicit *result-form* with DOLIST is useful. The function CHECK-ALL-ODD uses DOLIST to verify that all elements are odd. If so, DOLIST returns the symbol T at the completion of the loop. If any nonodd element is found, the function immediately returns from the loop with a value of NIL.

```
(defun check-all-odd (list-of-numbers)
  (dolist (e list-of-numbers t)
    (format t "~&Checking ~S..." e)
    (if (not (oddp e)) (return nil))))
```

> (check-all-odd '(1 3 5))
Checking 1...
Checking 3...
Checking 5...
T

> (check-all-odd '(1 3 4 5))
Checking 1...
Checking 3...
Checking 4...
NIL

EXERCISES

11.1. Write an iterative version of the MEMBER function, called IT-MEMBER. It should return T if its first input appears in its second input; it need not return a sublist of its second input.

11.2. Write an iterative version of ASSOC, called IT-ASSOC.

11.3. Write a recursive version of CHECK-ALL-ODD. It should produce the same messages and the same result as the preceding iterative version.

11.4 COMPARING RECURSIVE AND ITERATIVE SEARCH

For searching a flat list, iteration is simpler to use than recursion. It may also be more efficient, depending on the implementation. Compare these two versions of FIND-FIRST-ODD, which have been simplified by omitting the FORMAT expression:

```
(defun rec-ffo (x)
  "Recursively find first odd number in a list."
  (cond ((null x) nil)
        ((oddp (first x)) (first x))
        (t (rec-ffo (rest x)))))

(defun it-ffo (list-of-numbers)
  "Iteratively find first odd number in a list."
  (dolist (e list-of-numbers)
    (if (oddp e) (return e))))
```

There are a couple of small advantages to the iterative version. First, the termination test is implicit: DOLIST always stops when it gets to the end of the list. In the recursive version we have to write a COND clause to explicitly check for this. Second, in the iterative version the variable E names successive elements of the list, which is most convenient. In the recursive version, X names successive RESTs of the list, so we have to remember to write (FIRST X) to refer to the elements themselves, and we have to explicitly compute (REST X) with each recursive call.

In other situations recursion can be simpler and more natural than iteration. For example, you can easily search a tree with CAR/CDR recursion. There is no equally elegant way to do this iteratively. Iterative solutions exist, but they are ugly.

11.5 BUILDING UP RESULTS WITH ASSIGNMENT

In Chapter 8 we saw various ways to repetitively build up a result, such as a list, via recursive calls. In iterative programs results are built up via repetitive assignment. We'll first see how to do this in the body of a DOTIMES or DOLIST using explicit assignments such as SETF. Later in the chapter you'll see how assignments can be made implicitly, with DO.

Let's start by using DOTIMES to compute the factorial function. First we create an auxiliary variable PROD with initial value one. We will repetitively update this value in the body of the DOTIMES, and then return the final value of PROD as the result of the DOTIMES. Since the index variable I varies from zero to N−1 rather than from one to N, we must add one to I each time we reference its value in the body. Thus, (IT-FACT 5) counts from zero up to four, but it multiples PROD by the numbers one through five.

```
(defun it-fact (n)
  (let ((prod 1))
    (dotimes (i n prod)
      (setf prod (* prod (+ i 1))))))
```

```
(it-fact 5)  ⇒  120
```

Here is another use of explicit assignment: to write an iterative set intersection function. The variable ELEMENT is bound to successive elements of the set X. If ELEMENT is a member of the set Y, it gets pushed onto RESULT-SET; otherwise it doesn't. When all the elements of X have been processed, DOLIST returns the value of RESULT-SET.

```
(defun it-intersection (x y)
  (let ((result-set nil))
    (dolist (element x result-set)
      (when (member element y)
        (push element result-set)))))
```

```
> (it-intersection '(f a c e)
                   '(c l o v e))
(E C)
```

EXERCISES

11.4. Write an iterative version of LENGTH, called IT-LENGTH.

11.5. Write an iterative version of NTH, called IT-NTH.

11.6. Write an iterative version of UNION, called IT-UNION. Your function need not return its result in the same order as the built-in UNION function.

11.6 COMPARING DOLIST WITH MAPCAR AND RECURSION

MAPCAR is the simplest way to apply a function to every element of a list. Consider the problem of squaring a list of numbers. The applicative version is clearly simpler than the recursive version:

```
(defun app-square-list (list-of-numbers)
  (mapcar #'(lambda (n) (* n n))
          list-of-numbers))

(app-square-list '(1 2 3 4 5))  ⇒  (1 4 9 16 25)

(defun rec-square-list (x)
  (cond ((null x) nil)
        (t (cons (* (first x) (first x))
                 (rec-square-list (rest x))))))
```

The MAPCAR operator not only takes care of traveling down the input list and stopping when it gets to the end, but also takes care of consing the result list. All of this must be handled explicitly in the recursive version. If we use DOLIST to write an iterative solution, the termination test will be handled automatically, but we still have to build up the result with an explicit assignment. Here is a first attempt at a solution:

```
(defun faulty-it-square-list (list-of-numbers)
  (let ((result nil))
    (dolist (e list-of-numbers result)
      (push (* e e) result))))

> (faulty-it-square-list '(1 2 3 4 5))
(25 16 9 4 1)
```

The function's result is faulty: It's backwards. This is typical for an iterative solution. Since the function proceeds through the input list from left to right, and pushes each result onto the front of the result list, the result list ends up backwards. The square of the first number in the input list is the last number in the result list, and so on. We can fix this by writing (REVERSE RESULT) as the *result-form* of the DOLIST.

```
(defun it-square-list (list-of-numbers)
  (let ((result nil))
    (dolist (e list-of-numbers (reverse result))
      (push (* e e) result))))
```

```
> (it-square-list '(1 2 3 4 5))
(1 4 9 16 25)
```

If you've been reading the Advanced Topics sections, you'll understand why experienced Lisp programmers prefer to use the destructive function NREVERSE at the end of an iteration instead of using REVERSE. If you've been skipping these sections, don't worry about it.

EXERCISES

11.7. Why did the IT-INTERSECTION function return elements in reverse order from the order they appeared in its first input? How can you correct this?

11.8. Write an iterative version of REVERSE, called IT-REVERSE.

11.7 THE DO MACRO

DO is the most powerful iteration form in Lisp. It can bind any number of variables, like LET; it can step any number of index variables any way you like; and it allows you to specify your own test to decide when to exit the loop. Because it is so powerful, the syntax of DO is rather complicated:

```
(DO ( (var1  init1  [update1])
      (var2  init2  [update2])
        . . . )
    (test  action-1  . . .  action-n)
  body)
```

First, each variable in the DO's variable list is assigned its initial value. Then the test form is evaluated. If the result is true, DO evaluates the termination actions and returns the value of the last one. Otherwise DO evaluates the forms in its body in order. The body may contain RETURNs which force the DO to return immediately rather than iterate further. When DO reaches the end of the body, it begins the next iteration of the loop. First, each variable in the variable list is updated by setting it to the value of its update expression. (The update expression may be omitted, in which case the variable is left unchanged.) When all the variables have been updated, the termination test is tried again, and if it is true, DO evaluates the termination actions. Otherwise it goes on to evaluate the body again.

Here is a function called LAUNCH written with DO. Notice that it uses only one index variable, CNT, which it decrements from N down to zero. It is possible to write LAUNCH using DOTIMES instead, but it would be a little bit awkward because DOTIMES steps the index variable in the "wrong" direction.

```
(defun launch (n)
  (do ((cnt n (- cnt 1)))
      ((zerop cnt) (format t "Blast off!"))
    (format t "~S..." cnt)))
```

```
> (launch 10)
10...9...8...7...6...5...4...3...2...1...Blast off!
NIL
```

EXERCISES

11.9. Show how to write CHECK-ALL-ODD using DO.

11.10. Show how to write LAUNCH using DOTIMES.

Here is an implementation of COUNT-SLICES using DO. (COUNT-SLICES was introduced in Chapter 8.) This loop uses two index variables. CNT starts at zero and is used to build up the result. Z steps through successive RESTs of the loaf.

```
(defun count-slices (loaf)
  (do ((cnt 0 (+ cnt 1))
       (z loaf (rest z)))
      ((null z) cnt)))
```

This DO has an empty body: All the computation is done by expressions in the variable list. Suppose we evaluate (COUNT-SLICES '(X X)). When we enter the DO, CNT is initialized to zero and Z is initialized to (X X). Now comes the termination test: Since Z is not NIL, the loop does not terminate. The body is empty, so DO goes to update its variables. CNT is set to the value of (+ CNT 1), which is one. Z is set to (REST Z), which is the list (X). Now DO tries the termination test again. Z is still not NIL, so we iterate once more. This time CNT is set to two, and Z is set to NIL. Now the termination test is true. The expression to be evaluated and returned when the loop terminates is CNT, so DO returns two.

11.8 ADVANTAGES OF IMPLICIT ASSIGNMENT

DO has several advantages over DOTIMES and DOLIST. It can step the index variables any way you like, so it can count down instead of up, for example. DO can also bind multiple variables. This makes it easy to build up a result in the variable list of the DO; there is no need for a surrounding LET and an explicit SETF, as with the simpler iteration forms DOTIMES and DOLIST. Here is a version of the factorial function written with DO.

```
(defun fact (n)
  (do ((i n (- i 1))
       (result 1 (* result i)))
      ((zerop i) result)))
```

This version of FACT counts down rather than up, and it makes use of the parallel binding property of DO. When we compute (FACT 5), initially I is set to five and RESULT to one. When it's time to update the variables, the expression (- I 1) evaluates to four, and (* RESULT I) evaluates to five. Only after both update expressions have been evaluated are the variables themselves changed: I is set to four and RESULT is set to five. The next time through the loop, (- I 1) evaluates to three, and (* RESULT I) evaluates to 5×4 or 20. And so on. See the following table for the rest.

I	RESULT	(- I 1)	(* RESULT I)
5	1	4	5
4	5	3	20
3	20	2	60
2	60	1	120
1	120	0	120
0	120		

Both COUNT-SLICES and FACT have empty bodies. This is often the most compelling reason to use DO. You can make the assignments implicit by doing all the work in update expressions in the variable list, so you never have to write a PUSH or SETF. Functions written in this style are considered very elegant.

Sometimes, though, it is better not to try to do all the work in the update expressions. This is especially true when the update is conditional. Consider this version of IT-INTERSECTION, which has a null body:

```
(defun it-intersection (x y)
  (do ((x1 x (rest x1))
       (result nil (if (member (first x1) y)
                       (cons (first x1) result)
                       result)))
      ((null x1) result)))
```

This version is complicated because the DO wants to update RESULT every time through the loop, but we only want the value to change when (FIRST X) is a member of Y. A simpler version can be written by omitting the update expression for RESULT in the variable list. Instead we perform the update with a conditional PUSH in the body:

```
(defun it-intersection (x y)
  (do ((x1 x (rest x1))
       (result nil))
      ((null x1) result)
    (when (member (first x1) y)
      (push (first x1) result))))
```

If all you need to do is iterate over the elements of a list, DOLIST is more concise than DO. But DO is more general. For example, we can use DO to iterate over several lists at the same time, as in FIND-MATCHING-ELEMENTS. This function compares corresponding elements from two lists until it finds two that are equal, such as the third element of the lists (B I R D) and (C A R P E T).

```
(defun find-matching-elements (x y)
  "Search X and Y for elements that match."
  (do ((x1 x (rest x1))
       (y1 y (rest y1)))
      ((or (null x1) (null y1)) nil)
    (if (equal (first x1)
               (first y1))
        (return (first x1)))))
```

```
> (find-matching-elements
    '(b i r d)
    '(c a r p e t))
R
```

11.9 THE DO* MACRO

Here is FIND-FIRST-ODD written with DO. It follows the usual convention: A variable X is stepped through successive RESTs of the input. Within the body, we write (FIRST X) to refer to elements of the input.

```
(defun ffo-with-do (list-of-numbers)
   (do ((x list-of-numbers (rest x)))
       ((null x) nil)
     (if (oddp (first x)) (return (first x))))))
```

The DO* macro has the same syntax as DO, but it creates and updates the variables sequentially like LET*, rather than all at once like LET. One advantage of DO* in a function like FIND-FIRST-ODD is that it allows us to define a second index variable to hold the successive elements of a list, while the first index variable holds the successive cdrs:

```
(defun ffo-with-do* (list-of-numbers)
   (do* ((x list-of-numbers (rest x))
         (e (first x) (first x)))
        ((null x) nil)
     (if (oddp e) (return e))))
```

Notice that the index variable E uses the expression (FIRST X) for both its initial value and its update value. This is necessary because if the update value were omitted, the value of E would not change each time we went through the loop. It's also important that E appears *after* X in the variable list of the DO*, because E's value depends on X's value.

EXERCISES

11.11. Rewrite the following function to use DO* instead of DOLIST.

```
(defun find-largest (list-of-numbers)
   (let ((largest (first list-of-numbers)))
      (dolist (element (rest list-of-numbers)
                       largest)
         (when (> element largest)
            (setf largest element)))))
```

11.12. Rewrite the following function to use DO instead of DOTIMES.

```
(defun power-of-2 (n)    ;2 to the Nth power.
   (let ((result 1))
      (dotimes (i n result)
         (incf result result))))
```

11.13. Rewrite the following function using DOLIST instead of DO*.

```
(defun first-non-integer (x)
  "Return the first non-integer element of X."
  (do* ((z x (rest z))
        (z1 (first z) (first z)))
       ((null z) 'none)
    (unless (integerp z1)
      (return z1))))
```

11.14. Suppose we modified the function FFO-WITH-DO* above by just changing the DO* to a DO. What bug would this introduce?

11.15. The following version of the FFO-WITH-DO function has a much subtler bug in it. What is the bug? If you need a hint, try it on the list (2 4 6 7 8), and then on the list (2 4 6 7).

```
(defun ffo-with-do (x)
  (do ((z x (rest z))
       (e (first x) (first z)))
      ((null z) nil)
    (if (oddp e) (return e))))
```

11.10 INFINITE LOOPS WITH DO

You can make DO loop forever by specifying NIL as the termination test. One place where this is useful is a function that tries to read something specific from the keyboard, like a number. If the user types something other than a number, the function prints an error message and again waits for input. If the user does type a number, the function exits the loop using RETURN to return the number. Here's an example:

```
(defun read-a-number ()
  (do ((answer nil))
      (nil)
    (format t "~&Please type a number: ")
    (setf answer (read))
    (if (numberp answer)
        (return answer))
    (format t
      "~&Sorry, ~S is not a number.  Try again."
      answer)))
```

```
> (read-a-number)
Please type a number: foo
Sorry, FOO is not a number.  Try again.
Please type a number: (1 2 3)
Sorry, (1 2 3) is not a number.  Try again.
Please type a number: 37
37
```

11.11 IMPLICIT BLOCKS

In Common Lisp function bodies are contained in implicit **blocks**, and the function name also serves as the **block name**. A block is a sequence of expressions that can be exited at any point via the RETURN-FROM special function. In the following example the body of FIND-FIRST-ODD is a block named FIND-FIRST-ODD. The arguments to RETURN-FROM are a block name and a result expression; the block name is not evaluated, so it should not be quoted.

```
(defun find-first-odd (x)
  (format t "~&Searching for an odd number...")
  (dolist (element x)
    (when (oddp element)
      (format t "~&Found ~S." element)
      (return-from find-first-odd element)))
  (format t "~&None found.")
  'none)

> (find-first-odd '(2 4 6 7 8))
Searching for an odd number...
Found 7.
7

> (find-first-odd '(2 4 6 8 10))
Searching for an odd number...
None found.
NONE
```

In this example we used RETURN-FROM to exit the body of FIND-FIRST-ODD, not just the body of the DOLIST. RETURN-FROM returns from the closest enclosing block with the specified name. The bodies of looping forms such as DOTIMES, DOLIST, DO, and DO* are enclosed in implicit blocks named NIL. The expression (RETURN x) is actually just an abbreviation for (RETURN-FROM NIL x). So in the body of FIND-FIRST-

ODD, the RETURN-FROM is nested inside a block named NIL, which is in turn contained in a block named FIND-FIRST-ODD.

Here's an example where RETURN-FROM is needed, that does not involve iteration. The function SQUARE-LIST uses MAPCAR to square a list of numbers. However, if any of the elements turns out not to be a number, SQUARE-LIST returns the symbol NOPE instead of getting an error. The RETURN-FROM inside the lambda expression exits not only the lambda expression, but also the MAPCAR, and the body of SQUARE-LIST itself.

```
(defun square-list (x)
  (mapcar
    #'(lambda (e)
        (if (numberp e)
            (* e e)
            (return-from square-list 'nope)))
    x))
```

```
(square-list '(1 2 3 4 5))  ⇒  (1 4 9 16 25)
```

```
(square-list '(1 2 three four 5))  ⇒  NOPE
```

Besides the implicit blocks containing function bodies, blocks may also be defined explicitly via the BLOCK special function. This is only useful in advanced applications; we won't go into the details here.

SUMMARY

DOLIST and DOTIMES are the simplest iteration forms. DO and DO* are more powerful because they can step several variables at once using arbitrary update expressions and termination tests. But for simple problems like searching the elements of a list, DOLIST is more concise.

All the iteration forms make *implicit* assignments to their index variables. This is the cleanest type of assignment to use; you never actually have to write a SETF because the loop does the assignment for you. Sometimes, though, it is better to build up the result using explicit assignment in the loop body. This is especially true when we are using conditional assignment, as in the IT-INTERSECTION function.

Function names serve as implicit block names. We can therefore use RETURN-FROM to exit a function from anywhere in its body.

REVIEW EXERCISES

11.16. How do the variable lists of LET and DO differ?

11.17. What value is returned by the following expression? (This is a trick question.)

```
(dotimes (i 5 i)
   (format t "~&I = ~S" i))
```

11.18. Rewrite the DOTIMES expression in the preceding problem using DO. Does this help explain the value DOTIMES returns?

11.19. Does switching the order of entries in the variable list of a DO expression make a difference? Why?

11.20. If a loop uses only one index variable, can DO and DO* be used interchangeably?

11.21. One way to compute Fib(5) is to start with Fib(0) and Fib(1), which we know to be one, and add them together, giving Fib(2). Then add Fib(1) and Fib(2) to get Fib(3). Add Fib(2) and Fib(3) to get Fib(4). Add Fib(3) and Fib(4) to get Fib(5). This is an iterative method involving no recursion; we merely have to keep around the last two values of Fib to compute the next one. Write an iterative version of FIB using this technique.

FUNCTIONS COVERED IN THIS CHAPTER

Iteration macros: DOTIMES, DOLIST, DO, DO*.

Special functions for block structure: BLOCK, RETURN-FROM.

Ordinary function for exiting a block named NIL: RETURN.

Keyboard Exercise

In this keyboard exercise we will explore some properties of single- and double-stranded DNA, or *deoxyribonucleic acid*. DNA, and the related molecule RNA, make up the genetic material found in viruses and every type of cell, from bacteria to people. A strand of DNA is very much like a chain of cons cells; the elements of the chain are of four types, corresponding to the four **bases** adenine, thymine, guanine, and cytosine. We will represent a strand of DNA by a list of bases. The list (A G G T C A T T G) corresponds to a strand that is nine bases long; the first base being adenine and the next two guanine. Here is a schematic diagram of the strand:

Each of the four bases has a complement with which it can form a pair. Adenine pairs with thymine, while guanine pairs with cytosine. Two single strands of DNA can combine to form double-stranded DNA (whose shape is the famous "double helix") when each of their corresponding bases are complementary. The strand (A G G T C A T T G) and the strand (T C C A G T A A C) are complementary, for example. Double-stranded DNA looks like this:

```
  !   !   !   !   !   !   !   !   !
  A   G   G   T   C   A   T   T   G
  .   .   .   .   .   .   .   .   .

  .   .   .   .   .   .   .   .   .
  T   C   C   A   G   T   A   A   C
  !   !   !   !   !   !   !   !   !
```

EXERCISE

11.22. Write iterative solutions to all parts of this exercise that require repetitive actions.

a. Write a function COMPLEMENT-BASE that takes a base as input and returns the matching complementary base. (COMPLEMENT-BASE 'A) should return T; (COMPLEMENT-BASE 'T) should return A; and so on.

b. Write a function COMPLEMENT-STRAND that returns the complementary strand of a sequence of single-stranded DNA. (COMPLEMENT-STRAND '(A G G T)) should return (T C C A).

c. Write a function MAKE-DOUBLE that takes a single strand of DNA as input and returns a double-stranded version. We will represent double-stranded DNA by making a list of each base and its complement. (MAKE-DOUBLE '(G G A C T)) should return ((G C) (G C) (A T) (C G) (T A)).

d. One of the important clues to DNA's double-stranded nature was the observation that in naturally occurring DNA, whether from people, animals, or plants, the observed percentage of adenine is always very close to that of thymine, while the observed percentage of guanine is very close to that of cytosine. Write a function COUNT-BASES that counts the number of bases of each type in a DNA strand, and returns the result as a table. Your function should work for both single- and double-stranded DNA. Example: (COUNT-BASES '((G C) (A T) (T A) (T A) (C G))) should return ((A 3) (T 3) (G 2) (C 2)), whereas (COUNT-BASES '(A G T A C T C T)) should return ((A 2) (T 3) (G 1) (C 2)). In the latter case the percentages are not equal because we are working with only a single strand. What answer do you get if you apply COUNT-BASES to the corresponding double-stranded sequence?

e. Write a predicate PREFIXP that returns T if one strand of DNA is a prefix of another. To be a prefix, the elements of the first strand must exactly match the corresponding elements of the second, which may be longer. Example: (G T C) is a prefix of (G T C A T), but not of (A G G T C).

f. Write a predicate APPEARSP that returns T if one DNA strand appears anywhere within another. For example, (C A T) appears in (T C A T G) but not in (T C C G T A). *Hint:* If *x* appears in *y*, then *x* is a either a prefix of *y*, or of (REST *y*), or of (REST (REST *y*)), and so on.

g. Write a predicate COVERP that returns T if its first input, repeated some number of times, matches all of its second input. Example: (A G C) covers (A G C A G C A G C) but not (A G C T T G). You may assume that neither strand will be NIL.

h. Write a function PREFIX that returns the leftmost N bases of a DNA strand. (PREFIX 4 '(C G A T T A G)) should return (C G A T). Do not confuse the function PREFIX with the predicate PREFIXP.

i. Biologists have found that portions of some naturally occurring DNA strands consist of many repetitions of a short "kernel" sequence. Write a function KERNEL that returns the shortest prefix of a DNA strand that can be repeated to cover the strand. (KERNEL

'(A G C A G C A G C)) should return (A G C). (KERNEL '(A A A
A A)) should return (A). (KERNEL '(A G G T C)) should return (A
G G T C), because in this case only a single repetition of the entire
strand will cover the strand. *Hint:* To find the kernel, look at
prefixes of increasing length until you find one that can be repeated
to cover the strand.

j. Write a function DRAW-DNA that takes a single-stranded DNA
sequence as input and draws it along with its complementary strand,
as in the diagram at the beginning of this exercise.

Lisp Toolkit: TIME

The TIME macro function tells you how long it took to evaluate an
expression. It may also tell you how much memory was used during the
evaluation, and other useful things. The exact details of what TIME measures
and how the information is displayed are implementation dependent. TIME is
useful for gauging the efficiency of programs, for example, to compare two
solutions to a problem to see which is faster, or to see how much slower a
function runs when given a larger input. Here is an example:

```
(defun addup (n)
  "Adds up the first N integers"
  (do ((i 0 (+ i 1))
       (sum 0 (+ sum i)))
      ((> i n) sum)))

> (time (addup 1000))        Input is one thousand.
Evaluation took:
  0.83 seconds of real time,
  0.65625 seconds of user run time,
  81 page faults, and
  48208 bytes consed.
500500
```

```
> (time (addup 10000))    Input is ten thousand.
Evaluation took:
  6.909996 seconds of real time,
  6.484375 seconds of user run time,
  217 page faults, and
  480208 bytes consed.
50005000
```

As you can see, when the input to ADDUP was increased from 1000 to 10,000, the user run time and total bytes consed also increased by a factor of ten. But the number of page faults increased by a factor of just 2.6.

11   Advanced Topics

11.12 PROG1, PROG2, AND PROGN

PROG1, PROG2, and PROGN are three very simple functions. They all take an arbitrary number of expressions as input and evaluate the expressions one at a time. PROG1 returns the value of the first expression; PROG2 returns the value of the second; PROGN returns the value of the last expression.

```
> (prog1 (setf x 'foo)
         (setf x 'bar)
         (setf x 'baz)
         (format t "~&X is ~S" x))
X is BAZ
FOO

> (prog2 (setf x 'foo)
         (setf x 'bar)
         (setf x 'baz)
         (format t "~&X is ~S" x))
X is BAZ
BAR
```

```
> (progn (setf x 'foo)
         (setf x 'bar)
         (setf x 'baz)
         (format t "~&X is ~S" x))
X is BAZ
NIL
```

These forms are used infrequently today. They were important in earlier versions of Lisp, in which the body of a function could contain at most one expression and a COND clause could contain at most one consequent.

One place where PROGN is still useful is in the true-part and false-part of an IF. If you want to evaluate several expressions in the true-part or false-part, you must group them together using something like PROGN, BLOCK, or a LET.

The effects of PROG1 and PROG2 can easily be achieved with LET. For example, (POP X) is equivalent to both of the following expressions:

```
(prog1
  (first x)
  (setf x (rest x)))

(let ((old-top (first x)))
  (setf x (rest x))
  old-top)
```

Today, the second is generally considered easier to read and understand.

11.13 OPTIONAL ARGUMENTS

Common Lisp functions can be written to accept optional arguments, keyword arguments or any number of arguments, by putting special symbols called **lambda-list keywords** in the argument list. For example, variables following an &OPTIONAL lambda-list keyword name optional arguments. The following function accepts one required argument X and one optional argument Y. If an optional argument is unsupplied, it defaults to NIL.

```
(defun foo (x &optional y)
  (format t "~&X is ~S" x)
  (format t "~&Y is ~S" y)
  (list x y))
```

```
> (foo 3 5)
X is 3
Y is 5
(3 5)

> (foo 4)
X is 4
Y is NIL
(4 NIL)
```

We don't have to use NIL as the default value for unsupplied arguments. It is possible to specify what default value to use by replacing the optional argument name in the lambda list with a list of form *(name default)*. In the following function DIVIDE-CHECK, the default value for the divisor is two. (REM, called by DIVIDE-CHECK, is a built-in function that returns the remainder of dividing one number by another.)

```
(defun divide-check (dividend &optional (divisor 2))
  (format t "~&~S ~A divide evenly by ~S"
    dividend
    (if (zerop (rem dividend divisor)) "does"
        "does not")
    divisor))

> (divide-check 27 3)
27 does divide evenly by 3
NIL

> (divide-check 27)
27 does not divide evenly by 2
```

11.14   REST ARGUMENTS

The variable following an &REST lambda-list keyword will be bound to a list of the remaining arguments to a function. This allows the function to accept an unlimited number of arguments, as + and FORMAT do. Here's a function that takes an unlimited number of arguments and returns their average. It includes a division by 1.0 to ensure that the result is in floating point form.

```
(defun average (&rest args)
  (/ (reduce #'+ args)
     (length args)
     1.0))
```

```
(average 1 2 3 4 5)    ⇒    3.0

(average 3 5 11 19)    ⇒    9.5

(average)    ⇒    Error! division by zero.
```

One place where you must be careful about using an &REST argument is in a recursive function. With the first call, the function's arguments are collected into a list. If the function then calls itself recursively on the cdr of that list, it will be processing a list of a list, rather than the original list. Here is an example: a function FAULTY-SQUARE-ALL that is supposed to return a list of the squares of all its arguments:

```
(defun faulty-square-all (&rest args)
  (if (null args) nil
      (cons (* (first args) (first args))
            (faulty-square-all (cdr args)))))

(dtrace faulty-square-all)

> (faulty-square-all 1 2 3 4 5)
----Enter FAULTY-SQUARE-ALL
|      ARGS = (1 2 3 4 5)
|    ----Enter FAULTY-SQUARE-ALL
|    |      ARGS = ((2 3 4 5))

Error in function *.
Argument (2 3 4 5) is not a NUMBER.
```

We can correct the problem by using APPLY to make the recursive call. With APPLY, the value of (CDR ARGS) is treated as a list of arguments to the recursive call, not as a single argument.

```
(defun square-all (&rest args)
  (if (null args) nil
      (cons (* (first args) (first args))
            (apply #'square-all (cdr args)))))

(square-all 1 2 3 4 5)    ⇒    (1 4 9 16 25)
```

The PROG1, PROG2, and PROGN functions can be defined using the &REST lambda-list keyword as follows:

```
(defun my-prog1 (x &rest ignore) x)
```

```
(defun my-prog2 (x y &rest ignore) y)

(defun my-progn (&rest x)
  (car (last x)))
```

The built-in versions of PROG1, PROG2, and PROGN don't bother to create a list of their arguments, because they only need to return one value.

11.15 KEYWORD ARGUMENTS

In previous Advanced Topics sections we've seen several functions that accept keyword arguments, such as MEMBER and FIND-IF. For example, when you want MEMBER to use EQUAL as the equality test, you write:

```
(member x y :test #'equal)
```

Keyword arguments are useful when a function accepts a large number of optional arguments. By using keywords, we avoid having to memorize an order for these optional arguments; all we have to remember are their names. You can create your own functions that accept keyword arguments by using the &KEY lambda-list keyword. As with &OPTIONAL, default values can be supplied if desired. Here is a function MAKE-SUNDAE that accepts up to five keyword arguments:

```
(defun make-sundae (name &key (ice-cream 'vanilla)
                              (syrup 'hot-fudge)
                              nuts
                              cherries
                              whipped-cream)
  (list 'sundae
        (list 'for name)
        (list ice-cream 'with syrup 'syrup)
        (list 'toppings '=
              (remove nil
                      (list (and nuts 'nuts)
                            (and cherries 'cherries)
                            (and whipped-cream
                                 'whipped-cream)))))))

> (make-sundae 'john)
(SUNDAE (FOR JOHN)
        (VANILLA WITH HOT-FUDGE SYRUP)
        (TOPPINGS = NIL))
```

```
> (make-sundae 'cindy
               :syrup 'strawberry
               :nuts t
               :cherries t)
(SUNDAE (FOR CINDY)
        (VANILLA WITH STRAWBERRY SYRUP)
        (TOPPINGS = (NUTS CHERRIES)))
```

Keywords such as :CHERRIES always evaluate to themselves; that's why they don't need to be quoted. Notice that we use the keyword :CHERRIES when calling MAKE-SUNDAE, but in the argument list and body of MAKE-SUNDAE we use the ordinary symbol CHERRIES. This is an important distinction. Inside MAKE-SUNDAE, CHERRIES is just another variable. The only thing special about it is the way it gets its value. Just as an &REST variable is treated specially, variables defined with &KEY get their values in a special way: When calling MAKE-SUNDAE, we specify a value for CHERRIES by preceding the value with the :CHERRIES keyword.

11.16 AUXILIARY VARIABLES

The &AUX lambda-list keyword is used to define auxiliary local variables. You can specify just the variable name, in which case the variable is created with an initial value of NIL, or you can use a list of form (*var expression*). In the latter case *expression* is evaluated, and the result serves as the initial value for the variable. Here is an example of the use of an auxiliary variable LEN to hold the length of a list:

```
(defun average (&rest args
                &aux (len (length args)))
  (/ (reduce #'+ args) len 1.0))
```

The &AUX keyword accomplishes the same thing as the LET* special function: Both create new local variables using sequential binding. The choice of which to use is purely a matter of taste.

FUNCTIONS COVERED IN ADVANCED TOPICS

PROG1, PROG2, PROGN.

Lambda-list keywords: &OPTIONAL, &REST, &KEY, &AUX.

12

Structures and The Type System

12.1 INTRODUCTION

Common Lisp includes many built-in datatypes, which together form a **type system**. The types we've covered so far are numbers (of several varieties), symbols, conses, strings, function objects, and stream objects. These are the basic datatypes, but there are quite a few more.

The Common Lisp type system has two important properties. First, types are *visible:* They are described by Lisp data structures (symbols or lists), and there are built-in functions for testing the type of an object and for returning a type description of an object. Second, the type system is *extensible:* Programmers can create new types at any time.

Structures are an example of a programmer-defined datatype. After covering the basics of the type system, this chapter explains how new structure types are defined and how structures may be created and modified.

The Common Lisp Object System (CLOS) provides an advanced programmer-defined datatype facility that supports "object-oriented programming." We will not cover CLOS in this book. For our purposes, structures will suffice.

12.2 TYPEP AND TYPE-OF

The TYPEP predicate returns true if an object is of the specified type. Type specifiers may be complex expressions, but we will only deal with simple cases here.

```
(typep 3 'number)  ⇒  t

(typep 3 'integer)  ⇒  t

(typep 3 'float)  ⇒  nil

(typep 'foo 'symbol)  ⇒  t
```

Figure 12-1 shows a portion of the Common Lisp type hierarchy. This diagram has many interesting features. T appears at the top of the hierarchy, because all objects are instances of type T, and all types are subtypes of T. Type COMMON includes all the types that are built in to Common Lisp. Type NULL includes only the symbol NIL. Type LIST subsumes the types CONS and NULL. NULL is therefore a subtype of both SYMBOL and LIST. STRING is a subtype of VECTOR, which is a subtype of ARRAY. Arrays are discussed in Chapter 13.

The TYPE-OF function returns a type specifier for an object. Since objects can be of more than one type (for example, 3 is a number, an integer, and a fixnum; NIL is both a symbol and a list), the exact result returned by TYPE-OF is implementation dependent. Here are some typical examples:

```
(type-of 'aardvark)  ⇒  symbol

(type-of 3.5)  ⇒  short-float

(type-of '(bat breath))  ⇒  cons

(type-of "Phooey")  ⇒  (simple-string 6)
```

The type specifier (SIMPLE-STRING 6) describes a fixed-length character string with six elements. Some Lisp implementations might return just SIMPLE-STRING, or STRING, or (VECTOR STRING-CHAR). The relationship between strings and vectors will be explained in Chapter 13.

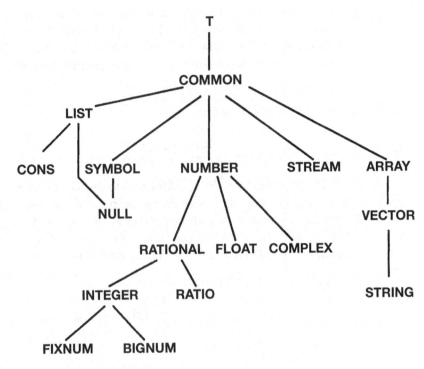

Figure 12-1 A portion of the Common Lisp type hierarchy.

12.3 DEFINING STRUCTURES

Structures are programmer-defined Lisp objects with an arbitrary number of named components. Structure types automatically become part of the Lisp type hierarchy. The DEFSTRUCT macro defines new structures and specifies the names and default values of their components. For example, we can define a structure called STARSHIP like this:

```
(defstruct starship
  (name nil)
  (speed 0)
  (condition 'green)
  (shields 'down))
```

This DEFSTRUCT form defines a new type of object called a STARSHIP whose components are called NAME, SPEED, CONDITION, and SHIELDS. STARSHIP becomes part of the system type hierarchy and can be referenced by such functions as TYPEP and TYPE-OF.

The DEFSTRUCT macro function also does several other things. It defines a constructor function MAKE-STARSHIP for creating new structures of this type. When a new starship is created, the name component will default to NIL, the speed to zero, the condition to GREEN, and the shields to DOWN.

```
> (setf s1 (make-starship))
#S(STARSHIP NAME NIL
            SPEED 0
            CONDITION GREEN
            SHIELDS DOWN)
```

The #S notation is the standard way to display structures in Common Lisp. The list following the #S contains the type of the structure followed by an alternating sequence of component names and values. Do not be misled by the use of parentheses in #S notation: Structures are not lists. Ordinary list operations like CAR and CDR will not work on structures.

```
s1  ⇒  #s(starship name nil
                   speed 0
                   condition green
                   shields down)
```

```
> (car s1)
Error: CAR of non-list object:  #S(STARSHIP ...)
```

Although new instances are usually created by calling the constructor function MAKE-STARSHIP, it is also possible to type in STARSHIP objects directly to the read-eval-print loop, using #S notation. Notice that the structure must be quoted to prevent its evaluation.

```
> (setf s2 '#s(starship speed (warp 3)
                        condition red
                        shields up))
#S(STARSHIP NAME NIL
            SPEED (WARP 3)
            CONDITION RED
            SHIELDS UP)
```

12.4 TYPE PREDICATES FOR STRUCTURES

Another side effect of DEFSTRUCT is that it creates a type predicate for the structure based on the structure name. In this case the predicate is called STARSHIP-P.

```
(starship-p s2)  ⇒  t

(starship-p 'foo)  ⇒  nil
```

Since the type name STARSHIP is fully integrated into the type system, it can be used with TYPEP and will be returned by TYPE-OF.

```
(typep s1 'starship)  ⇒  t

(type-of s2)  ⇒  starship
```

12.5 ACCESSING AND MODIFYING STRUCTURES

When a new structure is defined, DEFSTRUCT creates **accessor functions** for each of its components. For example, it creates a STARSHIP-SPEED accessor for retrieving the SPEED component of a starship.

```
(starship-speed s2)  ⇒  (warp 3)

(starship-shields s2)  ⇒  up
```

These accessor functions can also serve as place descriptions to SETF and the other generalized assignment operators.

```
> s1
#S(STARSHIP NAME NIL
            SPEED 0
            CONDITION GREEN
            SHIELDS DOWN)

 (setf (starship-name s1) "Enterprise")

(incf (starship-speed s1))  ⇒  1

> s1
#S(STARSHIP NAME "Enterprise"
            SPEED 1
            CONDITION GREEN
            SHIELDS DOWN)
```

Using these accessor functions, we can easily write our own functions to manipulate structures in interesting ways. For example, the ALERT function below causes a starship to raise its shields, and in addition raises the condition level to be at least YELLOW.

```
(defun alert (x)
  (setf (starship-shields x) 'up)
  (if (equal (starship-condition x) 'green)
      (setf (starship-condition x) 'yellow))
  'shields-raised)

(alert s1)  ⇒  shields-raised
```

```
s1  ⇒  #s(starship name "Enterprise"
                    speed 1
                    condition yellow
                    shields up)
```

An experienced Lisp programmer would prefer to use a more descriptive name than X for the argument to ALERT. Since ALERT expects its argument to be a starship, why not use that name in the argument list? The result would look like this:

```
(defun alert (starship)
  (setf (starship-shields starship) 'up)
  (if (equal (starship-condition starship) 'green)
      (setf (starship-condition starship) 'yellow))
  'shields-raised)
```

On the other hand, a few programmers find this writing style confusing, because it uses the symbol STARSHIP as both a local variable name and as a type name. If you fall into this category, you might prefer to use an abbreviated form for the variable name, such as STRSHIP.

12.6 KEYWORD ARGUMENTS TO CONSTRUCTOR FUNCTIONS

When a new structure instance is created, we aren't required to use the default values for the components. We can specify different values by supplying them as keyword arguments in the call to the constructor. (See Advanced Topics section 6.14 for an explanation of keywords and keyword arguments.) Here's an example using the MAKE-STARSHIP constructor:

```
> (setf s3 (make-starship :name "Reliant"
                          :shields 'damaged))
#S(STARSHIP NAME "Reliant"
            SPEED 0
            CONDITION GREEN
            SHIELDS DAMAGED)
```

12.7 CHANGING STRUCTURE DEFINITIONS

If you redefine a structure type using DEFSTRUCT to change the names or orderings of components, you should throw away all the old structures of that type; the accessor functions may no longer work properly on them, and there may be other problems as well. For example, having stored a starship named Reliant in S3, if we redefine STARSHIP, the value of S3 will become a strange object and the fields will be all mixed up.

```
> (defstruct starship
    (captain nil)
    (name nil)
    (shields 'down)
    (condition 'green)
    (speed 0))
STARSHIP


> s3
#S(STARSHIP CAPTAIN "Reliant"
            NAME 0
            SHIELDS GREEN
            CONDITION DAMAGED)


> (starship-speed s3)
Error:  vector index out of bounds
in #S(STARSHIP CAPTAIN "Reliant" ...)
```

To correct the problem, we simply need to rebuild the structure using the redefined constructor function MAKE-STARSHIP.

```
> (setf s3 (make-starship :captain "Benson"
                          :name "Reliant"
                          :shields 'damaged))
#S(STARSHIP CAPTAIN "Benson"
            NAME "Reliant"
            SHIELDS DAMAGED
            CONDITION GREEN
            SPEED 0)


> (starship-speed s3)      ;Now it works correctly.
0
```

SUMMARY

Common Lisp contains many built-in datatypes; only the basic ones are discussed in this book. The Common Lisp type system is both visible and extensible. Users can extend the type system by defining new structure types.

DEFSTRUCT defines structure types. The structure definition includes the names of all the components, and optionally specifies default values for them. If no default is given for a component, NIL is used. DEFSTRUCT also automatically defines a constructor function for the type (such as MAKE-STARSHIP) and a type predicate (such as STARSHIP-P).

REVIEW EXERCISES

12.1. Describe the roles of the symbols CAPTAIN, :CAPTAIN, and STARSHIP-CAPTAIN in the starship example.

12.2. Is (STARSHIP-P 'STARSHIP) true?

12.3. What are the values of (TYPE-OF 'MAKE-STARSHIP), (TYPE-OF #'MAKE-STARSHIP), and (TYPE-OF (MAKE-STARSHIP))?

FUNCTIONS COVERED IN THIS CHAPTER

Structure-defining macro: DEFSTRUCT.

Type system functions: TYPEP and TYPE-OF.

Lisp Toolkit: DESCRIBE and INSPECT

DESCRIBE is a function that takes any kind of Lisp object as input and prints an informative description of it. Many Lisp systems come with online documentation that can be conveniently accessed this way. DESCRIBE is also a good way to see how Lisp systems work internally, since you can describe symbols like CONS, NIL, and DEFUN and learn interesting things.

The exact output produced by DESCRIBE depends on which implementation of Common Lisp you are using. Here are some typical examples. As a beginning Lisper you probably won't understand all the details of what DESCRIBE is telling you, but puzzling them out with the help of a manual (and DESCRIBE too) can be fun.

```
> (describe 7)
7 is a FIXNUM.
It is a prime number.

> (describe 'fred)
FRED is an internal symbol in package USER.

> (describe t)
T is an external symbol in package LISP.
It is a constant.  Its value is T.

> (describe 'cons)
CONS is an external SYMBOL in package LISP.
   It can be called with these arguments: (x y)
   Function documentation:
      Returns a list with x as the CAR and y as
      the CDR.
```

DESCRIBE is particularly useful for displaying structures. In most implementations of Common Lisp, DESCRIBE shows the fields of the structure in a more readable format than the #S notation Lisp uses by default.

```
> (describe s1)
#S(STARSHIP ...) is a structure of type STARSHIP.
   NAME            "Enterprise"
   SPEED           1
   CONDITION       YELLOW
   SHIELDS         UP
```

Another tool that can be fun to experiment with is called INSPECT. If your computer has a mouse and a window system, INSPECT may let you inspect the components of an object by pointing to them with the mouse. Try defining a simple function like HALF, then do (INSPECT 'HALF) to see how function definitions are stored internally.

Different Lisp implementations provide different sorts of inspectors. You will need to look in the manual for the particular Lisp you are using to learn how to use its inspector effectively.

Keyboard Exercise

In this keyboard exercise we will implement a **discrimination net**. Discrimination nets are networks of yes and no questions used for problem-solving tasks, such as diagnosing automotive engine trouble. Here are two examples of dialogs with a car diagnosis net:

```
> (run)
Does the engine turn over? no
Do you hear any sound when you turn the key? no
Is the battery voltage low? no
Are the battery cables dirty or loose? yes
Clean the cables and tighten the connections.
NIL

> (run)
Does the engine turn over? yes
Does the engine run for any length of time? no
Is there gas in the tank? no
Fill the tank and try starting the engine again.
NIL
```

Figure 12-2 shows a portion of the discrimination net that generated this dialog. The net consists of a series of nodes. Each node has a name (a symbol), an associated question (a string), a "yes" action, and a "no" action. The yes and no actions may either be the names of other nodes to go to, or they may be strings that give the program's diagnosis. Since in the latter case there is no new node to which to go, the program stops after displaying the string.

Figure 12-3 shows how the net will be created. Note that the tree of questions is incomplete. If we follow certain paths, we may end up trying to go to a node that hasn't been defined yet, as shown in the following. In that case the program just prints a message and stops.

```
> (run)
Does the engine turn over? yes
Will the engine run for any period of time? yes
Node ENGINE-RUNS-AT-LEAST-BRIEFLY not yet defined.
NIL
```

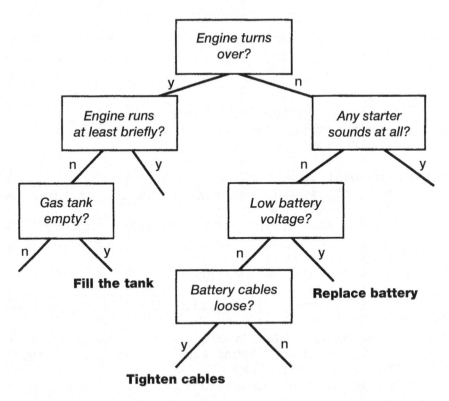

Figure 12-2 A portion of a discrimination net for solving automotive diagnosis problems.

EXERCISE

12.4. In this exercise you will create a discrimination net for automotive diagnosis that mimics the behavior of the system shown in the preceding pages.

   a. Write a DEFSTRUCT for a structure called NODE, with four components called NAME, QUESTION, YES-CASE, and NO-CASE.

   b. Define a global variable *NODE-LIST* that will hold all the nodes in the discrimination net. Write a function INIT that initializes the network by setting *NODE-LIST* to NIL.

   c. Write ADD-NODE. It should return the name of the node it added.

   d. Write FIND-NODE, which takes a node name as input and returns the node if it appears in *NODE-LIST*, or NIL if it doesn't.

```
(add-node 'start
          "Does the engine turn over?"
          'engine-turns-over
          'engine-wont-turn-over)

(add-node 'engine-turns-over
          "Will the engine run for any period of time?"
          'engine-will-run-briefly
          'engine-wont-run)

(add-node 'engine-wont-run
          "Is there gas in the tank?"
          'gas-in-tank
          "Fill the tank and try starting the engine again.")

(add-node 'engine-wont-turn-over
          "Do you hear any sound when you turn the key?"
          'sound-when-turn-key
          'no-sound-when-turn-key)

(add-node 'no-sound-when-turn-key
          "Is the battery voltage low?"
          "Replace the battery"
          'battery-voltage-ok)

(add-node 'battery-voltage-ok
          "Are the battery cables dirty or loose?"
          "Clean the cables and tighten the connections."
          'battery-cables-good)
```

Figure 12-3 Lisp code to create the automotive diagnosis network.

e. Write PROCESS-NODE. It takes a node name as input. If it can't find the node, it prints a message that the node hasn't been defined yet, and returns NIL. Otherwise it asks the user the question associated with that node, and then returns the node's yes action or no action depending on how the user responds.

f. Write the function RUN. It maintains a local variable named CURRENT-NODE, whose initial value is START. It loops, calling PROCESS-NODE to process the current node, and storing the value returned by PROCESS-NODE back into CURRENT-NODE. If the value returned is a string, the function prints the string and stops. If the value returned is NIL, it also stops.

g. Write an interactive function to add a new node. It should prompt the user for the node name, the question, and the yes and no actions. Remember that the question must be a string, enclosed in double quotes. Your function should add the new node to the net.

h. If the engine will run briefly but then stalls when it's cold, it is possible that the idle rpm is set too low. Write a new node called ENGINE-WILL-RUN-BRIEFLY to inquire whether the engine stalls when cold but not when warm. If so, have the net go to another node where the user is asked whether the cold idle speed is at least 700 rpm. If it's not, tell the user to adjust the idle speed.

12   Advanced Topics

12.8   PRINT FUNCTIONS FOR STRUCTURES

It is often convenient to invent specialized notations for printing structures. For example, we may not want to see all the fields of a starship object whenever it is printed; we may be satisfied to just see the name. The convention for printing abbreviated structure descriptions in Common Lisp is to make up a notation beginning with "#<" and ending with ">" that includes the type of the structure plus whatever identifying information is desired. For example, we might choose to print starships this way:

```
#<STARSHIP Enterprise>
```

The first step in customizing the way starship objects print is to write our own print function. It must take three inputs: the object being printed, the stream on which to print it, and a number (called depth) that Common Lisp uses to limit the depth of nesting when printing complex structures. We will ignore the depth argument in this book, but our function must still accept three arguments to work correctly. Here it is:

```
(defun print-starship (x stream depth)
  (format stream "#<STARSHIP ~A>"
    (starship-name x)))
```

We can test this function by calling it with a starship as first input. We'll use T for the second input (T refers to the default output stream, which is the console), and a depth of zero.

```
> (print-starship s1 t 0)
#<STARSHIP Enterprise>
```

Now to make Lisp call this function whenever it tries to print a starship, we must include the print function as an option to the DEFSTRUCT:

```
> (defstruct (starship
                (:print-function print-starship))
    (captain nil)
    (name nil)
    (shields 'down)
    (condition 'green)
    (speed 0))
STARSHIP

> (setf s4 (make-starship :name "Reliant"))
#<STARSHIP Reliant>

> (starship-shields s4)
GREEN

> (format t "~&This is ~S leaving orbit." s4)
This is #<STARSHIP Reliant> leaving orbit.
NIL
```

Print functions are especially useful when a structure contains other structures as components and we want to suppress most of the detail. They are almost essential when there are circular pointers between structures. For instance, every captain has a ship, and every ship a captain. If the structures for Kirk and the Enterprise point to each other, then when either one is printed, Lisp could enter an infinite loop, or else be forced to use the rather unaesthetic #1# notation to correctly express the circularity. If the print functions for the STARSHIP and CAPTAIN structures display only the NAME fields, we will have a concise notation for these objects in which the circularities are not evident.

EXERCISE

12.5. Create a defstruct for CAPTAIN with fields NAME, AGE, and SHIP. Make a structure describing James T. Kirk, captain of the Enterprise, age 35. Make the Enterprise point back to Kirk through its CAPTAIN component. Notice that when you print Kirk, you see his ship as well. Now define a print function for CAPTAIN that displays only the name, such as #<CAPTAIN "James T. Kirk">.

12.9 EQUALITY OF STRUCTURES

The EQUAL function does not treat two distinct structures as equal even if they have the same components. For example:

```
> (setf s5 (make-starship))
#S(STARSHIP NAME NIL
            SPEED 0
            CONDITION GREEN
            SHIELDS DOWN)

> (setf s6 (make-starship))
#S(STARSHIP NAME NIL
            SPEED 0
            CONDITION GREEN
            SHIELDS DOWN)

> (equal s5 s6)
NIL

> (equal s6 s6)
T
```

However, the EQUALP function will treat two structures as equal if they are of the same type and all their components are equal.

```
> (equalp s5 s6)
T

> (equalp s5 '#s(starship name nil
                          speed 0
                          condition green
                          shields down))
T
```

EQUALP also differs from EQUAL in ignoring case distinctions when comparing characters.

```
(equal "enterprise" "Enterprise")  ⇒  nil

(equalp "enterprise" "Enterprise")  ⇒  t
```

12.10 INHERITANCE FROM OTHER STRUCTURES

Structure types can be organized into a hierarchy using the :INCLUDE option to DEFSTRUCT. For example, we could define a structure type SHIP whose components are NAME, CAPTAIN, and CREW-SIZE. Then we could define STARSHIP as a type of SHIP with additional components WEAPONS and SHIELDS, and SUPPLY-SHIP as a type of SHIP with an additional component called CARGO.

```
(defstruct ship
  (name nil)
  (captain nil)
  (crew-size nil))

(defstruct (starship (:include ship))
  (weapons nil)
  (shields nil))

(defstruct (supply-ship (:include ship))
  (cargo nil))
```

The fields of a STARSHIP structure include all the components of SHIP. Thus, when we make a starship, its first three components will be NAME, CAPTAIN, and CREW-SIZE. The same holds for supply ships.

```
> (setf z1 (make-starship
              :captain "James T. Kirk"))
#S(STARSHIP NAME NIL
            CAPTAIN "James T. Kirk"
            CREW-SIZE NIL
            WEAPONS NIL
            SHIELDS NIL)

> (setf z2 (make-supply-ship
              :captain "Harry Mudd"))
#S(SUPPLY-SHIP NAME NIL
               CAPTAIN "Harry Mudd"
               CREW-SIZE NIL
               CARGO NIL)
```

The Enterprise is both a ship and a starship, so both type predicates will return true.

```
> (ship-p z1)
T

> (starship-p z1)
T

> (supply-ship-p z1)
NIL
```

Finally, note that the accessor functions for ships also apply to all subtypes of ship, which include starships and supply ships. Thus we can access the captain of the Enterprise using either SHIP-CAPTAIN or STARSHIP-CAPTAIN, but not SUPPLY-SHIP-CAPTAIN.

```
> (ship-captain z1)
"James T. Kirk"

> (starship-captain z1)
"James T. Kirk"

> (supply-ship-captain z1)
Error: #S(STARSHIP NAME NIL ...) is not
of type SUPPLY-SHIP.
```

13

Arrays, Hash Tables, and Property Lists

13.1 INTRODUCTION

This chapter briefly covers three distinct datatypes: arrays, hash tables, and property lists. Arrays are used very frequently in other programming languages, but not so often in Lisp. The reason is that most languages have such an impoverished set of datatypes that arrays must be used for many applications where lists, structures, or hash tables would be preferable.

Property lists are the oldest of the three datatypes discussed in this chapter; they were part of the original Lisp dialect, Lisp 1.5. In modern Lisp programming they have largely been replaced by hash tables, but they're still worth understanding.

13.2 CREATING AN ARRAY

An array is a contiguous block of storage whose elements are named by numeric subscripts. In this book we will consider only one-dimensional arrays, which are called **vectors**. (It's only a minor step from vectors to matrices and higher dimensional arrays; see your reference manual for details.) The components of a vector of length n are numbered zero through $n-1$. Let's create our first vector and store it in the variable MY-VEC:

```
(setf my-vec '#(tuning violin 440 a))
```

Do not let the #() notation confuse you into thinking that arrays are lists. A list is a chain of cons cells. An array is not a chain; it is a contiguous block of storage. The vector #(TUNING VIOLIN 440 A) is represented this way in memory:

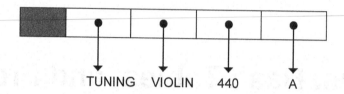

The shaded portion of the array is called an **array header**. It contains useful information about the array, such as its length and number of dimensions, which Lisp uses whenever you access the array's elements. As you might expect, basic list operations such as CAR and CDR do not work on arrays, since arrays are not cons cells.

```
> my-vec
#(TUNING VIOLIN 440 A)

> (car my-vec)
Error: #(TUNING VIOLIN 440 A) is not a list.
```

Because storage in arrays is contiguous, we can access each element of an array as fast as any other element. With lists, we have to follow a chain of pointers to get from one cons cell to the next, so depending on the length of the list, it can take much, much longer to access the last element than the first. Efficient access is the prime advantage arrays have over lists. Another advantage is that in most implementations, an array uses only half as much memory as a list of equal length. But lists also have some advantages over arrays. Lists of arbitrary length are easily built up element by element, either recursively or iteratively. It is not as easy to grow an array one element at a time.* Another advantage of lists is that they can share structure in ways that are impossible for arrays, but we won't get into the details of that in this book.

---

*Note to instructors: You can of course use arrays with fill pointers, but you can only add elements at one end, and the maximum length must be fixed in advance. Or you can use adjustable arrays, but repeated calls to ADJUST-ARRAY are very expensive.

13.3 PRINTING ARRAYS

To be able to see the elements of an array, we must set the global variable
PRINT-ARRAY to T. This assures that vectors will be printed in the same
#(*thing1 thing2...*) notation we use to type them in. If *PRINT-ARRAY* is
NIL, vectors and arrays will print in a more concise implementation-dependent
form using #< > notation, in which their individual elements are suppressed.

```
> (setf *print-array* nil)
NIL

> my-vec
#<Vector {204844}>

> (setf *print-array* t)
T

> my-vec
#(TUNING VIOLIN 440 A)
```

13.4 ACCESSING AND MODIFYING ARRAY ELEMENTS

The vector we stored in MY-VEC has four elements, numbered zero, one, two,
and three. The AREF function is used to access the elements of an array by
number, just as NTH is used to access the elements of lists.

```
> (aref my-vec 1)
VIOLIN
```

AREF is also understood as a place name by SETF; this is how one stores
new values in an array. Let's make a fresh array and store some items in it.

```
> (setf a '#(nil nil nil nil nil))
#(NIL NIL NIL NIL NIL)

> (setf (aref a 0) 'foo)
FOO

> (setf (aref a 1) 37)
37

> (setf (aref a 2) 'bar)
BAR
```

```
> a
#(FOO 37 BAR NIL NIL)

> (aref a 1)
37
```

Many functions we originally learned to use on lists are actually designed to work on **sequences**, which include both lists and vectors. Some examples of sequence functions are LENGTH, REVERSE, and FIND-IF.

```
> (length a)
5

> (reverse a)
#(NIL NIL BAR 37 FOO)

> (find-if #'numberp a)
37
```

On the other hand, some functions work only on lists. Besides the obvious CAR and CDR, there are MEMBER and the other set functions, plus SUBST and SUBLIS, and destructive list functions like NCONC (described in Advanced Topics section 10.8.) But destructive sequence functions, like NREVERSE, work on either lists or vectors.

13.5 CREATING ARRAYS WITH MAKE-ARRAY

The Lisp function MAKE-ARRAY creates and returns a new array. The length of the array is specified by the first argument. The initial contents of the array are undefined. Some Common Lisp implementations initialize array elements to zero; others use NIL. To be safe, you should not rely on array elements having any particular initial value unless you have specified one explicitly.

MAKE-ARRAY accepts several keyword arguments. The :INITIAL-ELEMENT keyword specifies one initial value to use for all the elements of the array.

```
> (make-array 5 :initial-element 1)
#(1 1 1 1 1)
```

The :INITIAL-CONTENTS keyword specifies a list of values for initializing the respective elements of an array. The list must be exactly as long as the array.

```
> (make-array 5 :initial-contents '(a e i o u))
#(A E I O U)
```

If you do not use one of these keywords when calling MAKE-ARRAY, the initial contents of the array will be unpredictable.

13.6 STRINGS AS VECTORS

Strings are actually a special type of vector. Thus, such functions as LENGTH, REVERSE, and AREF, which work on vectors, also work on strings. Remember that vectors are indexed starting from 0, not 1.

```
(length "Cockatoo")  ⇒  8

(reverse "Cockatoo")  ⇒  "ootakcoC"

(aref "Cockatoo" 3)  ⇒  #\k
```

The elements of a string are called **character objects**. For example, #\k denotes the character object known as lowercase ''k.'' Characters are yet another datatype, distinct from symbols and numbers. Character objects do not need to be quoted because they evaluate to themselves, just as numbers do.

```
#\k  ⇒  #\k

(type-of #\k)  ⇒  character
```

Since SETF understands AREF as a place name, you can destructively modify strings with SETF. You must only store character objects in the string, though, or an error will result.

```
> (setf pet "Cockatoo")
"Cockatoo"

> (setf (aref pet 5) #\p)
#\p

> pet
"Cockapoo"

> (setf (aref pet 6) 'cute)
Error: CUTE is not of type CHARACTER.
```

13.7 HASH TABLES

A **hash table** offers essentially the same functionality as an association list. You supply a key, which may be any sort of object, and Lisp gives you back the item associated with that key. The advantage of hash tables is that they are implemented using special **hashing algorithms** that allow Lisp to look things up much faster than it can look them up in an association list. Hashing is fast in part because hash tables are implemented using vectors rather than cons cell chains.

Association lists still have some advantages over hash tables. They are easier to create and manipulate because they are ordinary list structures. Hash tables use implementation-dependent representations that are not directly visible to the user. So if you want utter simplicity, use an association list. If you're willing to trade some simplicity for efficiency, use a hash table.

Hash tables cannot be typed in from the keyboard the way vectors can. They can only be created by the MAKE-HASH-TABLE function. In the default kind of hash table, EQL is used to compare the keys of items that are stored. It is also possible to create hash tables that use EQ or EQUAL. Hash table objects are printed in an implementation-dependent manner that usually does not show you the elements. The following example is typical:

```
> (setf h (make-hash-table))
#<EQL Hash table 5173142>

> (type-of h)
HASH-TABLE
```

The GETHASH function looks up a key in a hash table. The key can be any sort of object. GETHASH is understood as a place specification by SETF, so it can also be used to store into the hash table.

```
> (setf (gethash 'john h)
        '(attorney (16 maple drive)))
(ATTORNEY (16 MAPLE DRIVE))

> (setf (gethash 'mary h)
        '(physician (23 cedar court)))
(PHYSICIAN (23 CEDAR COURT))

> (gethash 'john h)
(ATTORNEY (16 MAPLE DRIVE))
T
```

```
> (gethash 'bill h)
NIL
NIL

> h
#<EQL Hash table 5173142>
```

GETHASH returns two values instead of one. The first value is the item associated with the key, or NIL if the key was not found in the hash table. The second value is T if the key was found in the hash table, or NIL if it was not found. The reason for this second value is to distinguish a key that appears in the table with an associated item of NIL from a key that does not appear at all. You can safely ignore the second return value; we will not make use of multiple return values in this book.

DESCRIBE will tell you useful things about a hash table, such as the number of **buckets** it has. A bucket is a group of entries. The more buckets there are, the fewer entries will be assigned to the same bucket, so retrievals will be faster. But the price of this speed is an increase in the amount of memory the hash table uses. INSPECT can be used to look at the entries of a hash table.

```
> (describe h)
#<EQL Hash Table 5173142> is a HASH-TABLE.
It currently has 2 entries and 65 buckets.
```

13.8 PROPERTY LISTS

In Lisp, every symbol has a **property list**. Property lists provide basically the same facilities as association lists and hash tables: You can store a value in a property list under a given key (called an **indicator**), and later look things up in the property list by supplying the indicator. Property lists are organized as lists of alternating indicators and values, like this:

```
(ind-1 value-1 ind-2 value-2 ...)
```

Property lists are very old; they were part of the original Lisp 1.5. They are included here for the sake of completeness; for most applications it is better to use an association list or hash table. Many Lisp implementations use the property lists of symbols for their own purposes. For example, if you look on the property list of CONS or COND you may see some system-specific information. Users are free to put their own properties on the property list, but it is a very bad idea to tamper with the properties your Lisp puts there.

The GET function retrieves a property of a symbol given the indicator. SETF understands GET as a place description; that is how new properties are stored on the property list. Let's give the symbol FRED a property called SEX with value MALE, a property called AGE with value 23, and a property called SIBLINGS with value (GEORGE WANDA).

```
(setf (get 'fred 'sex) 'male)

(setf (get 'fred 'age) 23)

(setf (get 'fred 'siblings) '(george wanda))

> (describe 'fred)
FRED is a SYMBOL.
Its SIBLINGS property is (GEORGE WANDA).
Its AGE property is 23.
Its SEX property is MALE.
```

The actual property list of FRED looks like this:

```
(siblings (george wanda) age 23 sex male)
```

Retrieving one of FRED's properties is easy: We just use GET to search the property list. *Note:* GET uses the EQ function to check for equality, so property indicators must not be numbers. Normally they are symbols.

```
(get 'fred 'age)   ⇒   23

(get 'fred 'favorite-ice-cream-flavor)   ⇒   nil
```

As you can see, when a symbol does not have the specified property, GET normally returns NIL. However, GET also accepts a third argument that it will return instead of NIL if it can't find the property it was asked to look up. This is one way to distinguish a symbol having a property FOO with value NIL from a symbol that does not have a FOO property at all. For example, we may know that Mabel is an only child (her SIBLINGS property is NIL), but Clara's siblings may not be recorded.

```
(setf (get 'mabel 'siblings) nil)

(get 'mabel 'siblings 'unknown)   ⇒   nil

(get 'clara 'siblings 'unknown)   ⇒   unknown
```

The value of a property can be changed at any time. Suppose FRED has a birthday:

```
(incf (get 'fred 'age))  ⇒  24
```

```
(get 'fred 'age)  ⇒  24
```

The SYMBOL-PLIST function returns a symbol's property list. It is discussed in more detail in Advanced Topics section 13.10.

```
> (symbol-plist 'fred)
(siblings (george wanda) age 24 sex male)
```

We can remove a property entirely using a function called REMPROP. The value returned by REMPROP is implementation dependent. It will be non-NIL if the property was found on the property list, or NIL if the property was not found. As a side effect, both the property name and the associated value are removed from the property list.

```
> (remprop 'fred 'age)
(AGE 24 SEX MALE)          ;Implementation-dependent value.
```

```
(get 'fred 'age)  ⇒  nil
```

13.9 PROGRAMMING WITH PROPERTY LISTS

Suppose we are building a database about the characters in a story, and one of the facts we want to record is meetings between the characters. We can store a list of names under the HAS-MET property of each individual. A name should not appear on the list more than once, in other words, the list should be a set. The easiest way to do this is to write a function called ADDPROP to add an element to a set stored under a property name. Here is the definition of ADDPROP:

```
(defun addprop (sym elem prop)
  (pushnew elem (get sym prop)))
```

PUSHNEW is a generalized assignment operator like PUSH, but it first checks to make sure the element is not a member of the list, so it is useful for adding an element to a set.

Using our ADDPROP function we can easily write a function to record meetings:

```
(defun record-meeting (x y)
  (addprop x y 'has-met)
  (addprop y x 'has-met)
  t)
```

This function makes use of the fact that ''has-met'' is a symmetric relation, in other words, if *x* has met *y*, then *y* has also met *x*.

```
> (symbol-plist 'little-red)
NIL

> (record-meeting 'little-red 'wolfie)
T

> (symbol-plist 'little-red)
(HAS-MET (WOLFIE))

> (symbol-plist 'wolfie)
(HAS-MET (LITTLE-RED))

> (record-meeting 'wolfie 'grandma)
T

> (symbol-plist 'wolfie)
(HAS-MET (GRANDMA LITTLE-RED))
```

EXERCISES

13.1. Write a function called SUBPROP that deletes an element from a set stored under a property name. For example, if the symbol ALPHA has the list (A B C D E) as the value of its FOOPROP property, doing (SUBPROP 'ALPHA 'D 'FOOPROP) should leave (A B C E) as the value of ALPHA's FOOPROP property.

13.2. Write a function called FORGET-MEETING that forgets that two particular persons have ever met each other. Use SUBPROP in your solution.

13.3. Using SYMBOL-PLIST, write your own version of the GET function.

13.4. Write a predicate HASPROP that returns T or NIL to indicate whether a symbol has a particular property, independent of the value of that property. *Note:* If symbol A has a property FOO with value NIL, (HASPROP 'A 'FOO) should still return T.

SUMMARY

Arrays are a kind of sequence, as are lists. One-dimensional arrays are called vectors. Strings are vectors of characters. Arrays can be created with MAKE-ARRAY, and their elements accessed with the AREF function. Many functions that work on lists also work on arrays, such as LENGTH, REVERSE, and FIND-IF.

Hash tables offer essentially the same functionality as association lists. Hash tables provide for very efficient lookup of items, because they don't search the table sequentially the way ASSOC does. Instead they use a hashing algorithm to compute a subscript, which is used to access a vector.

Property lists are attached to symbols, and are used by some Lisp systems to store implementation dependent information. They are used infrequently in modern Lisp programming. Hash tables are preferred over both property lists and association lists when efficient access is important.

REVIEW EXERCISES

13.5. Give one advantage of arrays over lists.

13.6. Give one advantage of lists over arrays.

13.7. Which requires more cons cells: a property list, or an association list of dotted pairs?

FUNCTIONS COVERED IN THIS CHAPTER

Array functions: MAKE-ARRAY, AREF.

Printer switch: *PRINT-ARRAY*.

Hash table functions: MAKE-HASH-TABLE, GETHASH.

Property list functions: GET, SYMBOL-PLIST, REMPROP.

Array Keyboard Exercise

Let's find out how random your Lisp's random number generator is. In this exercise we will produce a histogram plot of 200 random values between zero and ten. We will use an array to keep track of how many times we encounter each value. Here is an example of how the program will work:

```
> (new-histogram 11)      ;Eleven bins: 0 to 10.
T

> (dotimes (i 200)
    (record-value (random 11)))
NIL

> (print-histogram)
 0 [ 14] **************
 1 [ 18] ******************
 2 [ 19] *******************
 3 [  8] ********
 4 [ 21] *********************
 5 [ 13] *************
 6 [ 17] *****************
 7 [ 23] ***********************
 8 [ 18] ******************
 9 [ 25] *************************
10 [ 24] ************************
    200 total
NIL
```

The RANDOM function returns a random integer from zero up to, but not including, its argument. Thus (RANDOM 11) returns a number from zero to ten. In the histogram display, the first number on each line is the value we're counting. The next number, in brackets, is how many instances of that value have been seen. The remainder of the line contains one asterisk for each instance. The last line gives the total number of points recorded so far.

EXERCISE

13.8. Follow the steps below to create a histogram-drawing program. Your functions should not assume that the histogram will have exactly eleven bins. In other words, don't use eleven as a constant in your program; use (LENGTH *HIST-ARRAY*) instead. That way your program will be able to generate histograms of any size.

a. Write expressions to set up a global variable *HIST-ARRAY* that holds the array of counts, and a global variable *TOTAL-POINTS* that holds the number of points recorded so far.

b. Write a function NEW-HISTOGRAM to initialize these variables appropriately. It should take one input: the number of bins the histogram is to have.

c. Write the function RECORD-VALUE that takes a number as input.
    If the number is between zero and ten, it should increment the
    appropriate element of the array, and also update *TOTAL-
    POINTS*. If the input is out of range, RECORD-VALUE should
    issue an appropriate error message.

d. Write a function PRINT-HIST-LINE that takes a value from zero to
    ten as input, looks up that value in the array, and prints the
    corresponding line of the histogram. To get the numbers to line up
    in columns properly, you will need to use the format directives ~2S
    to display the value and ~3S to display the count. You can use a
    DOTIMES to print the asterisks.

e. Write the function PRINT-HISTOGRAM.

Hash Table Keyboard Exercise

A **cryptogram** is a type of puzzle that requires the solver to decode a message.
The code is known as a **substitution cipher** because it consists of substituting
one letter for another throughout the message. For example, if we substitute J
for F, T for A, and W for L, the word ''fall'' would be encoded as JTWW.
Here is an actual cryptogram for you to solve:

```
zj ze kljjls jf slapzi ezvlij pib kl jufwxuj p hffv jupi jf
enlpo pib slafml pvv bfwkj
```

The purpose of this keyboard exercise is not to solve cryptograms by hand,
but to write a program to help you solve them. Here is how our cryptogram-
solving program will start out. The cryptogram is represented as a list of
strings. All letters should be lowercase.

```
(setf crypto-text
  '("zj ze kljjls jf slapzi ezvlij pib kl jufwxuj p hffv jupi jf"
    "enlpo pib slafml pvv bfwkj"))
```

```
> (solve crypto-text)
--------------------
zj ze kljjls jf slapzi ezvlij pib kl jufwxuj p hffv jupi jf


enlpo pib slafml pvv bfwkj

--------------------
Substitute which letter?
```

When tackling a new cryptogram, it helps to look at the shortest words first. In English there are only two one-letter words, ''I'' and ''a,'' so the tenth word of the cryptogram, P, must be one of those. Suppose we guess that P deciphers to A. Beneath each P in the text we write an A.

```
Substitute which letter? p
What does 'p' decipher to? a
--------------------
zj ze kljjls jf slapzi ezvlij pib kl jufwxuj p hffv jupi jf
                      a            a           a    a

enlpo pib slafml pvv bfwkj
  a   a          a
--------------------
Substitute which letter?
```

Next we might look at all the two-letter words and guess that Z deciphers to I. Beneath each Z in the message we write an I.

```
Substitute which letter? z
What does 'z' decipher to? i
--------------------
zj ze kljjls jf slapzi ezvlij pib kl jufwxuj p hffv jupi jf
i  i                ai i      a             a        a

enlpo pib slafml pvv bfwkj
  a   a          a
--------------------
Substitute which letter?
```

An important constraint on cryptograms that helps to make them solvable is that no letter can decipher to more than one thing, and no two letters can decipher to the same thing. Our program must check to ensure that this constraint is obeyed by any solution we generate.

```
Substitute which letter? z
'z' has already been deciphered as 'i'!
--------------------
zj ze kljjls jf slapzi ezvlij pib kl jufwxuj p hffv jupi jf
i  i            ai  i    a           a          a

enlpo pib slafml pvv bfwkj
  a  a         a
--------------------
Substitute which letter? k
What does 'k' decipher to? a
But 'p' already deciphers to 'a'!
--------------------
zj ze kljjls jf slapzi ezvlij pib kl jufwxuj p hffv jupi jf
i  i            ai  i    a           a          a

enlpo pib slafml pvv bfwkj
  a  a         a
--------------------
Substitute which letter?
```

At some point we may want to take back a substitution. Suppose that after deciphering P and Z we decide that P shouldn't really decipher to A after all. The program must allow for this:

```
Substitute which letter? undo
Undo which letter? p
--------------------
zj ze kljjls jf slapzi ezvlij pib kl jufwxuj p hffv jupi jf
i  i                  i    i

enlpo pib slafml pvv bfwkj


--------------------
Substitute which letter?
```

The process continues until we have solved the cryptogram.

EXERCISE

13.9. Set up the global variable CRYPTO-TEXT as shown. Then build the cryptogram-solving tool by following these instructions:

  a. Each letter in the alphabet has a corresponding letter to which it deciphers, for example, P deciphers to A. As we solve the cryptogram we will store this information in two hash tables called

ENCIPHER-TABLE and *DECIPHER-TABLE*. We will use *DECIPHER-TABLE* to print out the deciphered cryptogram. We need *ENCIPHER-TABLE* to check for two letters being deciphered to the same thing, for example, if P is deciphered to A and then we tried to decipher K to A, a look at *ENCIPHER-TABLE* would reveal that A had already been assigned to P. Similarly, if P is deciphered to A and then we tried deciphering P to E, a look at *DECIPHER-TABLE* would tell us that P had already been deciphered to A. Write expressions to initialize these global variables.

b. Write a function MAKE-SUBSTITUTION that takes two character objects as input and stores the appropriate entries in *DECIPHER-TABLE* and *ENCIPHER-TABLE* so that the first letter deciphers to the second and the second letter enciphers to the first. This function does *not* need to check if either letter already has an entry in these hash tables.

c. Write a function UNDO-SUBSTITUTION that takes one letter as input. It should set the *DECIPHER-TABLE* entry of that letter, and the *ENCIPHER-TABLE* entry of the letter it deciphered to, to NIL.

d. Look up the documentation for the CLRHASH function, and write a function CLEAR that clears the two hash tables used in this problem.

e. Write a function DECIPHER-STRING that takes a single encoded string as input and returns a new, partially decoded string. It should begin by making a new string the same length as the input, containing all spaces. Here is how to do that, assuming the variable LEN holds the length:

```
(make-string len :initial-element #\Space)
```

Next the function should iterate through the elements of the input string, which are character objects. For each character that deciphers to something non-NIL, that value should be inserted into the corresponding position in the new string. Finally, the function should return the new string. When testing this function, make sure its inputs are all lowercase.

f. Write a function SHOW-LINE that displays one line of cryptogram text, with the deciphered text displayed beneath it.

g. Write a function SHOW-TEXT that takes a cryptogram (list of strings) as input and displays the lines as in the examples at the beginning of this exercise.

h. Type in the definition of GET-FIRST-CHAR, which returns the first character in the lowercase printed representation of an object.

```
(defun get-first-char (x)
  (char-downcase
    (char (format nil "~A" x) 0)))
```

i. Write a function READ-LETTER that reads an object from the keyboard. If the object is the symbol END or UNDO, it should be returned as the value of READ-LETTER. Otherwise READ-LETTER should use GET-FIRST-CHAR on the object to extract the first character of its printed representation; it should return that character as its result.

j. Write a function SUB-LETTER that takes a character object as input. If that character has been deciphered already, SUB-LETTER should print an error message that tells to what the letter has been deciphered. Otherwise SUB-LETTER should ask ''What does *(letter)* decipher to?'' and read a letter. If the result is a character and it has not yet been enciphered, SUB-LETTER should call MAKE-SUBSTITUTION to record the substitution. Otherwise an appropriate error message should be printed.

k. Write a function UNDO-LETTER that asks ''Undo which letter?'' and reads in a character. If that character has been deciphered, UNDO-LETTER should call UNDO-SUBSTITUTION on the letter. Otherwise an appropriate error message should be printed.

l. Write the main function SOLVE that takes a cryptogram as input. SOLVE should perform the following loop. First it should display the cryptogram. Then it should ask ''Substitute which letter?'' and call READ-LETTER. If the result is a character, SOLVE should call SUB-LETTER; if the result is the symbol UNDO, it should call UNDO-LETTER; if the result is the symbol END, it should return T; otherwise it should issue an error message. Then it should go back to the beginning of the loop, unless the value returned by READ-LETTER was END.

m. P deciphers to A, and Z deciphers to I. Solve the cryptogram.

Lisp Toolkit: ROOM

Lisp systems tend to use a lot of memory. When they run out, they try to get more. There are several ways Lisp might get more memory. First, it can try to reclaim any previously allocated storage that is no longer in use, such as cons cells to which nothing points anymore. This process is called **garbage collection**. Some Lisps garbage collect continuously, but most have to stop what they're doing, garbage collect, and then resume. The pause for a garbage collection is usually only a few seconds, but if your Lisp is garbage collecting frequently, these pauses can be annoying.

Although all Lisp implementations include a garbage collector, it is not part of the Common Lisp standard, so there is no standard way to modify a garbage collector's parameters or otherwise interact with it. In many implementations, though, there is a built-in function called GC that causes Lisp to garbage collect immediately. It usually prints some sort of informative message afterwards.

```
> (gc)
Garbage collection complete.
Approximately 303,008 bytes have been reclaimed.
NIL
```

Another way Lisp tries to obtain memory is by asking the operating system for more when it runs out. If you install more memory chips in your computer, your Lisp may not have to garbage collect as frequently, and may therefore run faster. The ROOM function prints a summary of Lisp's current memory usage, so you can tell how much memory has been allocated. Since each Lisp implementation manages its memory differently, the details of the display ROOM produces will differ. A typical example follows. This Lisp is using a total of 6.7 megabytes of memory.

If you're using a workstation with virtual memory, when Lisp needs more memory, it will start using up more of your disk for swap space. But if the disk is full, Lisp will run out of swap space. If there is a danger of the disk filling up, it is better to garbage collect more frequently than to increase virtual memory size. You can set limits on the maximum amount of memory your Lisp is allowed to use, but each implementation handles this a different way. See your user's manual for details.

```
> (room)
        Type      | Dynamic |   Static   | Read-Only |    Total
------------------|---------|------------|-----------|----------
Bignum            |     528 |         16 |       596 |     1,140
Ratio             |       0 |          0 |         8 |         8
Single-Float      |       0 |          0 |         0 |         0
Long-Float        |      36 |          0 |     2,592 |     2,628
Complex           |       0 |          0 |         0 |         0
String            |  19,008 |  1,130,416 |    22,772 | 1,172,196
Bit-Vector        |       0 |        456 |         0 |       456
Integer-Vector    |  31,880 |  3,706,124 |    13,092 | 3,751,096
General-Vector    |   8,740 |    421,540 |    72,772 |   503,052
Array             |       0 |        244 |         0 |       244
Function          |   3,924 |     21,948 |   415,040 |   440,912
Symbol            |   7,520 |    322,700 |       360 |   330,580
List              |  22,992 |    398,708 |   152,816 |   574,516
------------------|---------|------------|-----------|----------
  Totals:         |  94,628 |  6,002,152 |   680,048 = 6,776,828
NIL
```

13 Advanced Topics

13.10 PROPERTY LIST CELLS

Recall that a symbol is composed of five pointers. So far we've seen three of them: the symbol name, the value cell, and the function cell. The property list cell is another of these components. Every symbol has a property list, although it may be NIL. In contrast, not every symbol has a function definition in its function cell, or a value in its value cell.

Suppose we establish a property list for the symbol CAT-IN-HAT. The SYMBOL-PLIST function can be used to access the property list we have created.

```
(setf (get 'cat-in-hat 'bowtie) 'red)

(setf (get 'cat-in-hat 'tail) 'long)

> (symbol-plist 'cat-in-hat)
(TAIL LONG BOWTIE RED)
```

The structure of the symbol CAT-IN-HAT now looks like this:

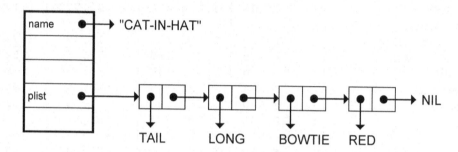

SETF understands SYMBOL-PLIST as a place name, so it is possible to give a symbol a new property list using SETF. Replacing the contents of a symbol's property list cell is dangerous, though, because it could wipe out important properties that Lisp itself had stored on the property list.

One reason property lists are today considered archaic is that they are global data structures: A symbol has only one property list, and it is accessible everywhere. If we use hash tables to store our information, we can keep several of them around at the same time, representing different sets of facts. Each hash table is independent, so changes made to one will not affect the others.

13.11 MORE ON SEQUENCES

The COERCE function can be used to convert a sequence from one type to another. If we coerce a string to a list, we can see the individual character objects. Conversely, we can use COERCE to turn a list of characters into a string.

```
> (coerce "Cockatoo" 'list)
(#\C #\o #\c #\k #\a #\t #\o #\o)

> (coerce '(#\b #\i #\r #\d) 'string)
"bird"
```

```
> (coerce '(foo bar baz) 'vector)
#(FOO BAR BAZ)
```

Yet another way to make a string is to make a vector with MAKE-ARRAY, using the :ELEMENT-TYPE keyword to specify that this vector holds only objects of type STRING-CHAR. (STRING-CHAR is a subtype of CHARACTER.) Vectors of STRING-CHARs are strings.

```
> (make-array 3 :element-type 'string-char
               :initial-contents '(#\M #\o #\m))
"Mom"
```

Most of the applicative operators, such as FIND-IF and REDUCE, work on any type of sequence, not just lists. MAPCAR is specific to lists, but there is also a general mapping function, MAP, that works on sequences of any type. The first input to MAP specifies the type of the result, the second input is the mapping function, and the remaining inputs are sequences to be mapped over. MAP stops when it reaches the end of any of the input sequences.

```
> (map 'list #'+
    '(1 2 3 4)
    '#(10 20 30 40))
(11 22 33 44)

> (map 'list #'list
    '(a b c)
    '#(1 2 3)
    "xyz")
((A 1 #\x) (B 2 #\y) (C 3 #\z))
```

If MAP is given NIL as a first argument, it returns NIL instead of constructing a sequence from the results of the mapping. This is useful if you want to apply a function to every element of a sequence only for its side effect.

```
> (map nil #'print "a b")
#\a
#\Space
#\b
NIL
```

FUNCTIONS COVERED IN ADVANCED TOPICS

Sequence functions: MAP, COERCE.

14

Macros and Compilation

14.1 INTRODUCTION

Macro functions, or **macros** for short, are a way to extend the syntax of Lisp. In this chapter we will use evaltrace diagrams and a little tool called PPMX (defined in the Lisp Toolkit section) to see how macros work. There will be a few references to material in previous Advanced Topics sections, but you'll be told where to look if you haven't read those sections before.

In the second half of the chapter we'll take a look at compilation. If you decide one of your programs runs too slowly, compiling it is an easy way to make it faster. The compiler translates Lisp programs into machine language programs, which can result in a 10 to 100 times speedup.

14.2 MACROS AS SHORTHAND

Think of macros as the computer equivalent of shorthand. Anything you write in shorthand can also be written in plain English; it just takes longer. Similarly, Common Lisp macros don't let you say anything that can't be expressed with ordinary functions, but they do help you to say things more concisely. A good example is INCF. It is quicker to write (INCF A) than (SETF A (+ A 1)).

Some macros are very clever, especially the generalized assignment macros like SETF and INCF. They are able to interpret arbitrarily complex place descriptions as generalized variable references. When you write an expression like

```
(incf (aref (nth array-num *list-of-arrays*)
            (first subscripts)))
```

you're relying on the cleverness of INCF to figure out what this place description means.

Macros can generate complicated programs from simple instructions. The DEFSTRUCT macro, for example, turns a structure definition for STARSHIPs into a long stream of instructions for supporting the STARSHIP datatype. These include function definitions for MAKE-STARSHIP and STARSHIP-P, and accessor functions for all the STARSHIP's components, such as STARSHIP-NAME. Not only would it be a lot of work to type in all these definitions by hand, but some of what DEFSTRUCT produces is implementation dependent. For example, the instructions for entering STARSHIP as a part of the Common Lisp type hierarchy differ from one Common Lisp implementation to the next. They involve functions and variables that aren't part of the Common Lisp standard, and probably aren't even documented by the Lisp vendor. The DEFSTRUCT macro allows Lisp vendors to hide these messy details from their customers by providing an agreed-upon, standard way to define structures that works in every Common Lisp implementation.

14.3 MACRO EXPANSION

If you write something in shorthand, eventually it will have to be "expanded" into plain English to understand and act on it. Lisp automatically expands macro calls for the same reason. A macro is actually a special shorthand-expanding function that does not evaluate its arguments. Its job is to look at its arguments and produce an expression that Lisp *can* evaluate. In the case of (INCF A), the INCF macro is called on the (unevaluated) argument A. It constructs an expression such as (SETQ A (+ A 1)), which it returns. The exact expression INCF constructs is implementation dependent, but it will look something like this SETQ. Lisp then evaluates the expression, and increments the value of A.

Recall from Section 10.10 that the SETQ special function performs assignment on ordinary variables. When you use SETF to assign to an ordinary variable, the SETF macro actually expands into a call to SETQ.

In evaltrace notation, macro expansion is shown by a dotted line. The expression the macro returns is evaluated normally, shown by a thin solid line in the following diagram:

```
····> (incf a)
         ┏→ Enter INCF macro with input A
         ┗→ Macro expansion:  (SETQ A (+ A 1))
     ┝→ (setq a (+ a 1))
         ┏→ (+ a 1)
         ┗→ 5
         set A to 5
     ┕→ 5
```

If you want to look at macro expansions on the computer, you can use a little tool called PPMX, defined in the Lisp Toolkit section of this chapter. The name PPMX stands for ''Pretty Print Macro eXpansion.'' Some Lisp editors also provide commands for displaying macro expansions; see your user's manual.

```
> (ppmx (incf a))
Macro expansion:
(SETQ A (+ A 1))
```

In some Lisp implementations INCF expands differently. For example, it might expand into a LET expression that creates a local variable to hold the value of (+ A 1), and then stores that value back into A. This may seem a rather indirect approach to incrementing A, but remember that INCF is designed to handle much more complex cases involving generalized assignment. In those cases a LET may really be necessary.

```
> (ppmx (incf a))
Macro expansion:
(LET ((#:G0144 (+ A 1)))
  (SETQ A #:G0144))
```

In the example above, #:G0144 is an internal symbol, called a **gensym**. It was automatically generated by INCF to serve as a local variable name. Gensyms are guaranteed not to conflict with the names of any of your variables. For reasons we won't go into here, #:G0144 is a different symbol than G0144. You cannot type this symbol from the keyboard, so it will never conflict with any variable in your program, even if you happen to choose the name G0144.

EXERCISES

14.1. Use PPMX to find the expression to which (POP X) expands.

14.2. Use PPMX to see to what expression the following DEFSTRUCT expands. (The results will be highly implementation dependent.)

```
(defstruct starship
  (name nil)
  (condition 'green))
```

14.4 DEFINING A MACRO

Macros are defined with DEFMACRO. Its syntax is similar to DEFUN. Let's define a simplified version of INCF to increment ordinary variables. Our macro will take a variable name as input and construct an expression to increment that variable by one.

```
(defmacro simple-incf (var)
  (list 'setq var (list '+ var 1)))

(setf a 4)

> (simple-incf a)
5

> (ppmx (simple-incf a))
Macro expansion:
(SETQ A (+ A 1))
```

Another way to see how SIMPLE-INCF works is to trace it with DTRACE. (If you're using the standard TRACE supplied with your Lisp implementation instead of DTRACE, you may be unable to trace macros.)

```
(dtrace simple-incf)

> (simple-incf a)
----Enter SIMPLE-INCF macro
|     Form = (SIMPLE-INCF A)
\--SIMPLE-INCF expanded to (SETQ A (+ A 1))
6
```

It's fine to use DTRACE on macros you write yourself, but in some Lisp implementations it may be inadvisable to trace important built-in macros, like SETF. If tracing these macros causes problems, use PPMX instead. One advantage of PPMX over tracing is that the result of the macroexpansion is only printed, not evaluated. PPMX allows you to experiment with macro-expanding arbitrary expressions without worrying about their causing an evaluation error.

Now let's modify SIMPLE-INCF to accept an optional second argument specifying the amount by which to increment the variable. We do this with the &OPTIONAL lambda-list keyword. (Optional arguments were explained in Advanced Topics section 11.13.) The default amount to increment the variable will be one.

```
(defmacro simple-incf (var &optional (amount 1))
   (list 'setq var (list '+ var amount)))

(setf a 5)

(setf b 2)

> (ppmx (simple-incf b (* 3 a)))
Macro expansion:
(SETQ B (+ B (* 3 A)))

> (simple-incf b (* 3 a))
17
```

Macros do not evaluate their arguments, so the inputs to SIMPLE-INCF are the symbol B and the list (* 3 A), not the numbers 2 and 15. An evaltrace diagram shows how SIMPLE-INCF computes the macro expansion, which Lisp then evaluates.

```
···> (simple-incf b (* 3 a))
       ▶ Enter SIMPLE-INCF macro with inputs B and (* 3 A)
          create variable VAR with value B
          create variable AMOUNT with value (* 3 A)
          ┌→ (list 'setq var ...)
          └→ Result of LIST is (SETQ B (+ B (* 3 A)))
       ▶ Macro expansion:  (SETQ B (+ B (* 3 A)))
   ┌→ (setq b (+ b (* 3 a)))
      ┌→ (+ b (* 3 a))
      └→ 17
      set B to 17
   └→ 17
```

Let's now consider why INCF has to be a macro rather than a function. Suppose we try to make an INCF function, using DEFUN. We'll call it FAULTY-INCF.

```
(defun faulty-incf (var)
  (setq var (+ var 1)))
```

Since FAULTY-INCF is a function, it evaluates its arguments, and it is not expected to return an expression for Lisp to evaluate. It can just go ahead and do the incrementing itself. But since its arguments are evaluated, there is a problem. Let's see what happens:

```
(setf a 7)

> (faulty-incf a)
8

> (faulty-incf a)
8

> a
7
```

The input to FAULTY-INCF is the number seven. FAULTY-INCF creates a local variable named VAR to hold its input, and then it increments VAR by one. It doesn't know anything about the variable A, because its argument was evaluated before the function was entered. An evaltrace diagram makes this clear.

```
┌→ (faulty-incf a)
│   A evaluates to 7
├→ Enter FAULTY-INCF with input 7
│     create variable VAR with value 7
│     ┌→ (setq var (+ var 1))
│     │     ┌→ (+ var 1)
│     │     └→ 8
│     │   set VAR to 8
│     └→ 8
└→ Result of FAULTY-INCF is 8
```

We might try quoting the variable A when passing it to the FAULTY-INCF function. Of course we'll have to modify the definition of FAULTY-

INCF, because its input will no longer be a number. But for reasons that will be explained in the Advanced Topics section, this won't work either. SIMPLE-INCF *must* be written as a macro. This doesn't invalidate what was said earlier about macros being no more than shorthand; we are still free to write a SETQ expression instead of using SIMPLE-INCF. SETQ is not a macro: It is a **special function**. The difference is explained in the next section.

EXERCISE

14.3. Write a SET-NIL macro that sets a variable to NIL.

14.5 MACROS AS SYNTACTIC EXTENSIONS

Since the purpose of macros is to extend the syntax of the language, Lisp does not treat a macro call like an ordinary function call. There are three important differences between ordinary functions and macro functions:

1. The arguments to ordinary functions are always evaluated; the arguments to macro functions are not evaluated.

2. The result of an ordinary function can be anything at all; the result returned by a macro function must be a valid Lisp expression.

3. After a macro function returns an expression, that expression is immediately evaluated. The results returned by ordinary functions do not get evaluated.

In addition to macros, Common Lisp also includes a small number of special functions. Some examples are SETQ, IF, LET, and BLOCK. Special functions are the lowest level building blocks of Common Lisp; they are responsible for things like assignment, scoping, and basic control structure such as blocks and loops. Like macros, special functions do not evaluate their arguments, but they also don't return expressions to be evaluated. They are primitives that do very special things. You cannot write new special functions; only a Lisp implementor can do that.

Returning to our discussion of macros as shorthand, we should say that anything that can be done with a macro can also be done without macros, by using a combination of ordinary Common Lisp functions, special functions, and in some cases, implementation dependent functions.

14.6 THE BACKQUOTE CHARACTER

SIMPLE-INCF constructed a Lisp expression by combining two calls to LIST, some quoted symbols, and the values of the variables VAR and AMOUNT. This approach works well enough when the expression is small, but when macros must produce large, complicated expressions, it is awkward to construct them bit by bit. What we need instead is a way to write a template for the expression the macro is to return. Then all the macro has to do is fill in the blanks. The backquote character provides such a facility.

The backquote character (`` ` ``) is analogous to quote, in that both are used to quote lists. However, inside a backquoted list, any expression that is preceded by a comma is considered to be "unquoted," meaning the value of the expression rather than the expression itself is used.

```
(setf name 'fred)

> '(this is ,name from pittsburgh)
(THIS IS FRED FROM PITTSBURGH)

> '(i gave ,name about ,(* 25 8) dollars)
(I GAVE FRED ABOUT 200 DOLLARS)
```

We can use backquote to write a more concise version of the SIMPLE-INCF macro:

```
(defmacro simple-incf (var &optional (amount 1))
  '(setq ,var (+ ,var ,amount)))

> (ppmx (simple-incf fred-loan (* 25 8)))
(SETQ FRED-LOAN (+ FRED-LOAN (* 25 8)))
```

EXERCISE

14.4. Write a macro called SIMPLE-ROTATEF that switches the value of two variables. For example, if A is two and B is seven, then (SIMPLE-ROTATEF A B) should make A seven and B two. Obviously, setting A to B first, and then setting B to A won't work. Your macro should expand into a LET expression that holds on to the original values of the two variables and then assigns them their new values in its body.

A very common use of macros is to avoid having to quote arguments. The macro expands into an ordinary function call with quoted versions of the arguments filled in where needed. You can use backquote to generate expressions with quotes in them by including the quotes as part of the template, like this:

```
'(setf foo 'bar)  ⇒  (setf foo 'bar)
```

In the example below, TWO-FROM-ONE is a macro that takes a function name and another object as arguments; it expands into a call to the function with two arguments, both of which are the quoted object.

```
(defmacro two-from-one (func object)
  '(,func ',object  ',object))

> (two-from-one cons aardvark)
(AARDVARK . AARDVARK)

> (ppmx (two-from-one cons aardvark))
Macro expansion:
(CONS 'AARDVARK 'AARDVARK)
```

We place a comma before OBJECT because we want the value of that variable to be inserted into the list that backquote constructs; the quote before the comma also becomes part of the list. If we leave out the quote, the macro will expand to (CONS AARDVARK AARDVARK), which will cause an unassigned variable error unless AARDVARK has a value. If we leave out the comma instead of the quote, the macro will expand to (CONS 'OBJECT 'OBJECT).

EXERCISE

14.5. Write a macro SET-MUTUAL that takes two variable names as input and expands into an expression that sets each variable to the name of the other. (SET-MUTUAL A B) should set A to 'B, and B to 'A.

Let's try a more complex example of backquote. We'll write a macro SHOWVAR that displays the value of a variable, like this:

```
(defun f (x y)
  (showvar x)
  (showvar y)
  (* x y))

> (f 3 7)
The value of X is 3
The value of Y is 7
21
```

SHOWVAR must be a macro because it needs to know the name of the variable it's displaying, not just the value. Let's break the problem down a little. The message about X's value could be printed by the following expression. Notice that only the first instance of X is quoted.

```
(format t "~&The value of ~S is ~S" 'x x)
```

We can now easily abstract the template needed for the SHOWVAR macro. The combination of a quote followed by a comma may look strange, but you can see from the preceding example where the quote comes from.

```
(defmacro showvar (var)
  `(format t "~&The value of ~S is ~S"
     ',var
     ,var))
```

14.7 SPLICING WITH BACKQUOTE

Another feature of backquote is that if a template element is preceded by a comma and an at sign (, @), the value of that element is *spliced* into the result that backquote constructs rather than being inserted. (The value of the element must be a list.) If only a comma is used, the element would be inserted as a single object, resulting in an extra level of parentheses.

```
(setf name 'fred)

(setf address '(16 maple drive))

> `(,name lives at ,address now)          Inserting.
(FRED LIVES AT (16 MAPLE DRIVE) NOW)

> `(,name lives at ,@address now)         Splicing.
(FRED LIVES AT 16 MAPLE DRIVE NOW)
```

Here is an example of where splicing is useful. The SET-ZERO macro, to be defined later, takes any number of variables as input. It expands into an expression to set each of them to zero and also to return a message to that effect. Because the macro must generate several actions but can return only one value, it combines the actions into a single expression with PROGN.

```
> (set-zero a b c)
(ZEROED A B C)
```

```
> (ppmx (set-zero a b c))
Macro expansion:
(PROGN
   (SETF A 0)
   (SETF B 0)
   (SETF C 0)
   '(ZEROED A B C))
```

Here is the definition of SET-ZERO. It uses MAPCAR to construct a SETF expression for each variable in the argument list. The SETF expressions are then spliced into the body of the PROGN. Also, the final expression in the PROGN's body is a quoted list constructed by splicing. If there were a plain comma there instead of a comma and at sign combination, the result would be (ZEROED (A B C)).

```
(defmacro set-zero (&rest variables)
  `(progn ,@(mapcar #'(lambda (var)
                        (list 'setf var 0))
                    variables)
          '(zeroed ,@variables)))
```

EXERCISE

14.6. Write a macro called VARIABLE-CHAIN that accepts any number of inputs. The expression (VARIABLE-CHAIN A B C D) should expand into an expression that sets A to 'B, B to 'C, and C to 'D.

14.8   THE COMPILER

The compiler translates Lisp programs into machine language. This makes programs run faster: the typical speedup is a factor of 10 to 100. As a beginning Lisp programmer you are probably not writing very large programs, so speed may not be a concern. However, as you tackle more ambitious problems, you will eventually find yourself concerned with performance issues such as how fast a program runs and how much memory it uses. Compilation can reduce both figures.

There are two ways to use the compiler. You can compile a single function using COMPILE, or an entire file using COMPILE-FILE. Many Lisp-oriented editors provide ways for you to invoke the compiler with just a keystroke or two, so you may never need to call these functions explicitly.

Let's take a look at the effect of COMPILE on the running time of a simple function. This function returns the smallest integer larger than the square root of its input. It computes the result in a very tedious way, but that will help us measure the speedup achieved by compilation.

```
(defun tedious-sqrt (n)
  (dotimes (i n)
    (if (> (* i i) n) (return i))))
```

```
> (time (tedious-sqrt 5000000))
Evaluation took:
  1.169998 seconds of real time,
  0.953125 seconds of user run time,
  0.09375 seconds of system run time,
  69 page faults, and
  53816 bytes consed.
2237
```

We see that the square root of five million is between 2236 and 2237. We also see that the interpreted version of TEDIOUS-SQRT takes roughly 0.95 seconds of user run time on this example. You might want to choose a smaller argument if your machine is a lot slower than this. Now, let's try compiling TEDIOUS-SQRT and see how fast the compiled version runs.

```
> (compile 'tedious-sqrt)
TEDIOUS-SQRT
```

```
> (time (tedious-sqrt 5000000))
Evaluation took:
  0.04999995 seconds of real time,
  0.03125 seconds of user run time,
  0.0 seconds of system run time,
  1 page fault, and
  32 bytes consed.
2237
```

The compiled version took only .03125 seconds of user run time, making it 30 times faster than the interpreted version. It also consed only 32 bytes, while the interpreted version consed over 50K bytes. In this particular implementation, the consing is due to the * function's use of &REST to collect its arguments. This conses a list each time the function is called. The compiler turns calls to * into machine language *multiply* instructions, which eliminates both the cost of a function call and the accompanying consing.

14.9   COMPILATION AND MACRO EXPANSION

The Common Lisp standard permits macro calls to be replaced by their expansions at any time. In some Lisp implementations DEFUN does the macro expansion right away. In others the macro call gets replaced the first time the function is evaluated. In very simple implementations a macro call may never be replaced with the resulting expansion; instead the macro is expanded anew each time the expression is evaluated.

Since macro expansion can happen at any time, you should not write macros that produce side effects, such as assignments or i/o. But it's fine for the macro to expand into an *expression* that produces side effects.

```
(defmacro bad-announce-macro ()
   (format t "~%Hi mom!"))

(defun say-hi ()
   (bad-announce-macro))

> (compile 'say-hi)
Hi, mom!
SAY-HI

> (say-hi)
NIL
```

In the above example the macro was expanded as part of the process of compiling SAY-HI. So the compiler said ''Hi, mom!'' The result of the macro was NIL, so that's what got compiled into the body of SAY-HI. When we call the compiled SAY-HI function, it says nothing because the macro has been replaced with its expansion. The problem can be resolved by making the macro *return* the FORMAT expression instead of executing it.

```
(defmacro good-announce-macro ()
   `(format t "~%Hi mom!"))
```

14.10   COMPILING ENTIRE PROGRAMS

When you compile an entire program, it will generally be stored in a file. You can use the COMPILE-FILE function on the file. Some Lisp editors allow you to do this with an editor command. They may also allow you to compile the contents of an editor buffer without writing it to a file. See your user's manual for details.

Because of the way compilers work, you will need to follow a few simple rules for organizing your program. If you don't follow these rules, the compiler may produce error messages and not compile your program correctly.

First, if your program uses any global variables, the compiler may issue a warning message saying that the variable was "assumed to be SPECIAL." Special variables are explained in the Advanced Topics section. You can get rid of these warnings by declaring the variables with DEFVAR, DEFPARAMETER, or DEFCONSTANT. The declaration should occur early in the file, prior to any function that references those variables. You can also ignore the warnings if you choose.

Second, if your program contains macros, the macro definitions must be placed earlier in the file than any functions that reference them. Otherwise, if function FOO calls a macro BAR, Lisp may not realize when compiling FOO that it needs to treat the call to BAR as a macro call to be expanded. If FOO has been compiled incorrectly, most compilers will issue a warning when they find out that BAR is a macro.

Third, if your program redefines any built-in functions, the compiler may not handle it correctly. Be sure to use names that don't conflict with built-in functions. Online documentation can help you check for this.

14.11 CASE STUDY: FINITE STATE MACHINES

Finite state machines (FSMs) are a technique from theoretical computer science for describing how simple devices like vending machines or traffic lights work. In this section we will write a general purpose simulator for finite state machines to demonstrate how real Lisp programs are developed. To make the discussion more concrete we will focus on a particular machine to simulate, but our simulator will work for any finite state machine.

Consider a vending machine with two products: gum and mints. Gum costs 15 cents, and mints cost 20 cents. Any combination of nickels and dimes may be used to operate the machine; it will issue appropriate change automatically. If enough money has been put in, pressing the gum button or the mint button will deliver the desired product. Pressing the coin return lever at any time will return an amount equal to what has been put in so far.

The behavior of our vending machine can be formally described by the finite state machine shown in Figure 14-1. The machine is initially in a state called START. If it gets the symbol NICKEL as input, it goes "Clunk!" and

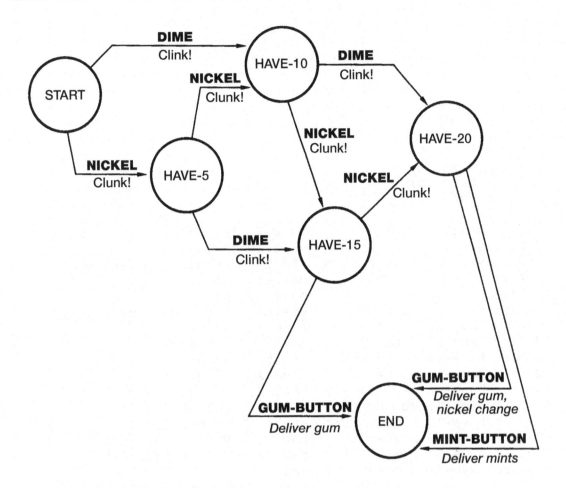

Figure 14-1 Finite state diagram for a vending machine.

moves to a state named HAVE-5. If it's in state HAVE-5 and it gets the symbol DIME as input, it goes "Clink!" and moves to state HAVE-15. In state HAVE-15, if it gets the input GUM-BUTTON, it delivers a packet of gum and goes to state END.

The machine has a total of six states: START, HAVE-5, HAVE-10, HAVE-15, HAVE-20, and END. (It's called a finite state machine precisely because the number of states is finite.) Each state is represented by a node in Figure 14-1, and each possible transition from one state to the next is represented by an arc (an arrow). The arc is labeled with the input needed to make the transition and the action the machine should take when it follows that transition. For example, the arc from HAVE-10 to HAVE-15 is labeled NICKEL / "Clunk!".

```
(defnode start)
(defnode have-5)
(defnode have-10)
(defnode have-15)
(defnode have-20)
(defnode end)

(defarc start    nickel      have-5  "Clunk!")
(defarc start    dime        have-10 "Clink!")
(defarc start    coin-return start   "Nothing to return.")
(defarc have-5   nickel      have-10 "Clunk!")
(defarc have-5   dime        have-15 "Clink!")
(defarc have-5   coin-return start   "Returned five cents.")
(defarc have-10  nickel      have-15 "Clunk!")
(defarc have-10  dime        have-20 "Clink!")
(defarc have-10  coint-return start  "Returned ten cents.")
(defarc have-15  nickel      have-20 "Clunk!")
(defarc have-15  dime        have-20 "Nickel change.")
(defarc have-15  gum-button  end     "Deliver gum.")
(defarc have-15  coin-return start "Returned fifteen cents.")
(defarc have-20  nickel      have-20 "Nickel returned.")
(defarc have-20  dime        have-20 "Dime returned.")
(defarc have-20  gum-button  end
                    "Deliver gum, nickel change.")
(defarc have-20  mint-button end     "Deliver mints.")
(defarc have-20  coin-return start   "Returned twenty cents.")
```

Figure 14-2 Node and arc definitions for the vending machine.

The complete definition of the vending machine is shown in Figure 14-2. The macros DEFNODE and DEFARC provide a convenient syntax for defining the finite state machine one part at a time. Here is a sample run of the FSM simulator so you can see the goal toward which we'll be working.

```
> (fsm)
State START.   Input: nickel
Clunk!
State HAVE-5.  Input: dime
Clink!
State HAVE-15.   Input: gum-button
Deliver gum.
NIL
```

We begin constructing our simulator by creating structures for nodes and arcs, using DEFSTRUCT. Each node has a name, a list of input arcs, and a list of output arcs. Each arc has a "from" node, a "to" node, a label, and an action. We also define print functions for these structures.

```
(defstruct (node (:print-function print-node))
  (name nil)
  (inputs nil)
  (outputs nil))

(defun print-node (node stream depth)
  (format stream "#<Node ~A>"
          (node-name node)))

(defstruct (arc (:print-function print-arc))
  (from nil)
  (to nil)
  (label nil)
  (action nil))

(defun print-arc (arc stream depth)
  (format stream "#<ARC ~A / ~A / ~A>"
          (node-name (arc-from arc))
          (arc-label arc)
          (node-name (arc-to arc))))
```

Now we need a global variable *NODES* to hold the list of nodes comprising the machine, and a global variable *ARCS* to hold the list of arcs. Another variable, *CURRENT-NODE*, keeps track of the machine's state. We declare these global variables with DEFVAR, explained in the Advanced Topics section. The INITIALIZE function sets these variables to NIL.

```
(defvar *nodes*)
(defvar *arcs*)
(defvar *current-node*)

(defun initialize ()
  (setf *nodes* nil)
  (setf *arcs* nil)
  (setf *current-node* nil))
```

The DEFNODE macro is a bit of "syntactic sugar" for defining new nodes. It simply puts a quote in front of its argument and calls the ADD-NODE function.

```
(defmacro defnode (name)
  '(add-node ',name))
```

ADD-NODE constructs a new node with the given name and adds it to the list kept in the global variable *NODES*. It uses NCONC (the destructive version of APPEND) so it can add the node to the *end* of the list. This assures that the nodes in *NODES* will appear in the order in which they were defined with DEFNODE, rather than in reverse order. ADD-NODE also returns the newly created node.

```
(defun add-node (name)
  (let ((new-node (make-node :name name)))
    (setf *nodes* (nconc *nodes* (list new-node)))
    new-node))
```

```
> (initialize)
NIL

> (defnode start)
#<Node START>

> (defnode have-5)
#<Node HAVE-5>
```

```
*nodes*  ⇒  (#<Node START> #<Node HAVE-5>)
```

FIND-NODE takes a node name as input and returns the corresponding node. If no node exists with that name, FIND-NODE signals an error.

```
(defun find-node (name)
  (or (find name *nodes* :key #'node-name)
      (error "No node named ~A exists." name)))
```

```
> (find-node 'have-5)
#<Node HAVE-5>

> (find-node 'have-6)
Error: No node named HAVE-6 exists.
```

The DEFARC macro provides a convenient syntax for defining arcs, and the ADD-ARC function does the real work. When an arc is created, it is added to the NODE-OUTPUTS list of the *from* node and the NODE-INPUTS list of the *to* node. It is also added to the list kept in the global variable *ARCS*.

```
(defmacro defarc (from label to &optional action)
  '(add-arc ',from ',label ',to  ',action))

(defun add-arc (from-name label to-name action)
  (let* ((from (find-node from-name))
         (to (find-node to-name))
         (new-arc (make-arc :from from
                            :label label
                            :to to
                            :action action)))
    (setf *arcs* (nconc *arcs* (list new-arc)))
    (setf (node-outputs from)
          (nconc (node-outputs from)
                 (list new-arc)))
    (setf (node-inputs to)
          (nconc (node-inputs to)
                 (list new-arc)))
    new-arc))
```

```
> (defarc start nickel have-5 "Clunk!")
#<ARC START / NICKEL / HAVE-5>
```

Now we can write the top-level function FSM. It takes an optional input specifying the initial state of the machine. The default initial state is START. FSM repeatedly calls the function ONE-TRANSITION to move to the next state. When the machine reaches a state with no output arcs (such as END), it stops. Notice that the DO has an empty variable list.

```
(defun fsm (&optional (starting-point 'start))
  (setf *current-node* (find-node starting-point))
  (do ()
      ((null (node-outputs *current-node*)))
    (one-transition)))
```

Finally, we write ONE-TRANSITION. It prompts for an input and makes the appropriate state transition by changing the value of *CURRENT-NODE*. If there is no legal transition from the current state given that input, it prints an error message and prompts for input again.

```lisp
(defun one-transition ()
  (format t "~&State ~A.  Input: "
          (node-name *current-node*))
  (let* ((ans (read))
         (arc (find ans
                    (node-outputs *current-node*)
                    :key #'arc-label)))
    (unless arc
      (format t "~&No arc from ~A has label ~A.~%"
              (node-name *current-node*) ans)
      (return-from one-transition nil))
    (let ((new (arc-to arc)))
      (format t "~&~A" (arc-action arc))
      (setf *current-node* new))))
```

```
> (fsm)
State START.  Input: dime
Clink!
State HAVE-10.  Input: quarter
No arc from HAVE-10 has label QUARTER.
State HAVE-10.  Input: dime
Clink!
State HAVE-20.  Input: dime
Dime  returned.
State HAVE-20.  Input: mint-button
Deliver mints.
NIL
```

Our simulator is not limited to simulating vending machines. Any device that can be described in a finite number of states and state transitions can be simulated by this program.

EXERCISE

14.7. Extend the vending machine example to sell chocolate bars for 25 cents. Make it accept quarters as well as nickels and dimes. When you put in a quarter it should go ''Ker-chunk!''

SUMMARY

Macros are Lisp's version of shorthand, with several uses. They allow programmers to define syntactic extensions to Lisp and to say things more concisely. They also help Lisp implementors hide messy implementation-specific details from their customers. Macros do not evaluate their arguments; they return Lisp expressions that are evaluated. New macros can be defined with DEFMACRO.

Like macros, special functions do not evaluate their inputs. But unlike macros, they do not return Lisp expressions that are to be evaluated. Special functions provide the primitives on which Lisp is built, such as assignment, conditionals, and block structure.

The backquote character constructs a list from a template. If a template element is preceded by a comma it will be evaluated; the value is then inserted into the list being constructed. Elements preceded by a comma and at sign combination are spliced into the list rather than inserted. Backquote is particularly useful in macros that construct complex expressions by filling in the blanks of a template.

REVIEW EXERCISES

14.8. Why is it unwise to write macros that have side effects?

14.9. Common Lisp contains exactly 24 built-in special functions. What are they? (*Hint:* Look in Chapter 5 of *Common Lisp: The Language*.)

14.10. How much faster do typical programs run after being compiled?

FUNCTIONS COVERED IN THIS CHAPTER

Macro definition: DEFMACRO.

Compiler: COMPILE, COMPILE-FILE.

Lisp Toolkit: PPMX

PPMX stands for "Pretty Print Macro eXpansion." It macroexpands its first argument (unevaluated) and prints the result. PPMX is not only useful for learning about built-in macros like SETF, it is also quite handy for debugging macros you write yourself if there is a problem with their expansion.

```
(defmacro ppmx (form)
  "Pretty prints the macro expansion of FORM."
  `(let* ((exp1 (macroexpand-1 ',form))
          (exp (macroexpand exp1))
          (*print-circle* nil))
     (cond ((equal exp exp1)
            (format t "~&Macro expansion:")
            (pprint exp))
           (t (format t "~&First step of expansion:")
              (pprint exp1)
              (format t "~%~%Final expansion:")
              (pprint exp)))
     (format t "~%~%")
     (values)))
```

If a macro expands into another macro call, PPMX shows both the result of the first expansion and the final expression derived when all macros have been expanded. For example, the LENGTHY-INCF macro below expands into a call to the SETF macro. SETF in turn expands into a call to the SETQ special function.

```
(defmacro lengthy-incf (var)
  `(setf ,var (+ ,var 1)))

> (ppmx (lengthy-incf a))
First step of expansion:
(SETF A (+ A 1))

Final expansion:
(SETQ A (+ A 1))
```

In some implementations, the DOTIMES macro expands into a call to the DO macro. In the example below, DO in turn expands into a more complex expression involving BLOCK, LET, TAGBODY, and GO. We will not cover tagbodies and GO in this book.

```
> (ppmx (dotimes (i n)
           (if (> (* i i) n) (return i))))
First step of expansion:
(DO ((I 0 (1+ I))
     (#:G6517 N))
    ((>= I #:G6517) NIL)
  (IF (> (* I I) N) (RETURN I)))

Final expansion:
(BLOCK NIL
  (LET ((I 0)
        (#:G6517 N))
    (TAGBODY (GO #:G6519)
     #:G6518 (IF (> (* I I) N) (RETURN I))
             (PSETQ I (1+ I))
     #:G6519 (UNLESS (>= I #:G6517) (GO #:G6518))
             (RETURN (PROGN NIL)))))
```

Keyboard Exercise

Our finite state machine simulator is called an "interpreter": It operates by interpreting the node and arc data structures as a machine description. A faster way to simulate a finite state machine is to write a specialized function for each node. The function takes as its argument a list of input symbols for the machine. It looks at the first symbol, decides on the appropriate state transition to make, and then calls the function corresponding to that state, passing it the REST of the input list.

This approach is faster because we don't have to call ASSOC or FIND-NODE. In fact, we don't reference the node and arc data structures as all. The speedup may be important if we are simulating a complex machine with many states, such as a piece of computer circuitry.

Since all the inputs must be supplied at once as a list, instead of prompting for them interactively, it is possible for the machine to run out of inputs before reaching an end state. In that case we simply return the name of the last state reached by the machine. Following is a function for simulating the machine when it is in state START.

```
(defun start (input-syms &aux
                  (this-input (first input-syms)))
    (cond ((null input-syms) 'start)
          ((equal this-input 'nickel)
           (format t "~&~A" "Clunk!")
           (have-5 (rest input-syms)))
          ((equal this-input 'dime)
           (format t "~&~A" "Clink!")
           (have-10 (rest input-syms)))
          ((equal this-input 'coin-return)
           (format t "~&~A" "Nothing to return.")
           (start (rest input-syms)))
          (t (error "No arc from ~A with label ~A."
                    'start this-input))))
```

Assuming all the other states had similar functions defined, we could write (START '(NICKEL DIME GUM-BUTTON)) to get some gum. The result would look like this:

```
> (start '(nickel dime gum-button))
Clunk!
Clink!
Deliver gum.
END
```

Writing a function for each state is tedious. It's much more convenient to define a machine with DEFNODE and DEFARC expressions. To get speed, though, we need to convert the nodes to functions. It would be good if we could get the computer to do this work for us.

EXERCISE

14.11. In this keyboard exercise we will write a compiler for finite state machines that turns each node into a function. The definition of the vending machine's nodes and arcs should already be loaded into your Lisp before beginning the exercise.

   a. Write a function COMPILE-ARC that takes an arc as input and returns a COND clause, following the example shown previously. Test your function on some of the elements in the list *ARCS*. (COMPILE-ARC (FIRST *ARCS*)) should return this list:

```
((equal this-input 'nickel)
 (format t "~&~A" "Clunk!")
 (have-5 (rest input-syms)))
```

b. Write a function COMPILE-NODE that takes a node as input and returns a DEFUN expression for that node. (COMPILE-NODE (FIND-NODE 'START)) should return the DEFUN shown previously.

c. Write a macro COMPILE-MACHINE that expands into a PROGN containing a DEFUN for each node in *NODES*.

d. Compile the vending machine. What does the expression (START '(DIME DIME DIME GUM-BUTTON)) produce?

14   Advanced Topics

14.12   THE &BODY LAMBDA-LIST KEYWORD

One reason people write macros is so they can add new bits of syntax to Lisp. For example, we can write a WHILE macro to provide the same control structure as WHILE loops in other languages.

```
(defmacro while (test &body body)
  `(do ()
       ((not ,test))
     ,@body))
```

The WHILE macro takes a test expression as its first argument, followed by zero or more body expressions to be evaluated if the test is true. The body expressions could be collected with &REST, but Common Lisp includes a special keyword, &BODY, to use when the remaining arguments to a macro form the body of some control structure. Some Lisp editors pay special attention to the &BODY keyword when indenting calls to macros. The use of

&BODY also signifies to human readers of the macro definition that the remaining arguments are a body of Lisp code.

The NEXT-POWER-OF-TWO function below uses a WHILE loop to repeatedly double the value of the variable I, starting from one, up to the first power of two that is greater than the input N.

```
(defun next-power-of-two (n &aux (i 1))
  (while (< i n)
    (format t "~&Not ~S" i)
    (setf i (* i 2)))
  i)

> (next-power-of-two 11)
Not 1
Not 2
Not 4
Not 8
16
```

For best style, this particular problem should be solved with DO instead of WHILE, to avoid explicit SETFs.

14.13 DESTRUCTURING LAMBDA LISTS

The MIX-AND-MATCH macro takes two pairs as input and returns an expression that produces four pairs:

```
(defmacro mix-and-match (p q)
  (let ((x1 (first p))
        (y1 (second p))
        (x2 (first q))
        (y2 (second q)))
    `(list '(,x1 ,y1)
           '(,x1 ,y2)
           '(,x2 ,y1)
           '(,x2 ,y2))))

> (mix-and-match (fred wilma) (barney betty))
((FRED WILMA) (FRED BETTY) (BARNEY WILMA)
 (BARNEY BETTY))
```

In this example we took apart the two inputs (FRED WILMA) and (BARNEY BETTY) manually, using a LET expression. But since macros don't evaluate their inputs, they are able to treat input expressions as list structures to be taken apart automatically. This is known as **destructuring**. You can specify how to destructure an input expression by replacing a variable in the macro's argument list with another whole argument list. For example, we can replace the variable P in MIX-AND-MATCH with the argument list (X1 Y1), and the variable Q with (X2 Y2). Here then is a version of MIX-AND-MATCH using destructuring:

```
(defmacro mix-and-match ((x1 y1) (x2 y2))
  '(list '(,x1 ,y1)
         '(,x1 ,y2)
         '(,x2 ,y1)
         '(,x2 ,y2)))
```

Destructuring is only available for macros, not ordinary functions. It is particularly useful for macros that define new bits of control structure with a complex syntax. The DOVECTOR macro that follows is modeled after DOTIMES and DOLIST. It steps an index variable through successive elements of a vector. The macro uses destructuring to pick apart the index variable name, the vector expression, and the result form.

```
(defmacro dovector ((var vector-exp
                         &optional result-form)
                    &body body)
  '(do* ((vec-dov ,vector-exp)
         (len-dov (length vec-dov))
         (i-dov 0 (+ i-dov 1))
         (,var nil))
        ((equal i-dov len-dov) ,result-form)
     (setf ,var (aref vec-dov i-dov))
     ,@body))

> (dovector (x '#(foo bar baz))
    (format t "~&X is ~S" x))
X is FOO
X is BAR
X is BAZ
NIL
```

You can see from the expansion of DOVECTOR why this macro is useful as a form of shorthand:

```
> (ppmx (dovector (x '#(foo bar baz))
        (format t "~&X is ~S" x)))
First step of expansion:
(DO* ((VEC-DOV '#(FOO BAR BAZ))
      (LEN-DOV (LENGTH VEC-DOV))
      (I-DOV 0 (+ I-DOV 1))
      (X NIL))
     ((EQUAL I-DOV LEN-DOV) NIL)
  (SETF X (AREF VEC-DOV I-DOV))
  (FORMAT T "~&X is ~S" X))


Final expansion:
(BLOCK NIL
  (LET* ((VEC-DOV '#(FOO BAR BAZ))
         (LEN-DOV (LENGTH VEC-DOV))
         (I-DOV 0)
         (X NIL))
    (TAGBODY
            (GO #:G955)
     #:G954 (SETF X (AREF VEC-DOV I-DOV))
            (FORMAT T "~&X is ~S" X)
            (SETQ I-DOV (+ I-DOV 1))
     #:G955
            (UNLESS (EQUAL I-DOV LEN-DOV)
              (GO #:G954))
            (RETURN NIL))))
```

The DOVECTOR expands into a DO* expression with local variables
VEC-DOV (to hold the vector) and LEN-DOV (to hold its length), and an
index variable called I-DOV. These names were chosen because they are
unlikely to conflict with any user variable names. If we had used VEC, LEN,
and I instead, they might prevent users from accessing some local variables of
their own with those names.[*]   The expansion also contains an explicit
assignment to the variable X in the body of the DO*. After the DOVECTOR
macro returns the DO* expression, it is further macro expanded by Lisp into a
combination of BLOCK, LET, TAGBODY, and GO. The DOVECTOR
expression is much nicer for humans to read than the macro expansion.

---

[*]Note to instructors: Of course there are better ways to prevent such name conflicts. We could use the
package system, or gensyms. But those are outside the scope of an introductory book.

14.14 MACROS AND LEXICAL SCOPING

Let's return to our consideration of FAULTY-INCF, an attempt to implement INCF as a function rather than a macro. Suppose we quote the variable before passing it to the function, by writing (FAULTY-INCF 'A). FAULTY-INCF needs to do two things: It must find out the current value of the variable, and it must replace that value with a new one.

In the case of global variables this is possible. Recall that a global lives in the value cell of the symbol that names it. We can use the built-in function SYMBOL-VALUE to access the value cell. We can store into this cell by using SETF or by using the built-in SET function discussed in Section 10.10. Here is our new version of FAULTY-INCF:

```
(defun faulty-incf (var)
  (set var (+ (symbol-value var) 1)))

(setf a 7)

> (faulty-incf 'a)
8

> (faulty-incf 'a)
9

> a
9
```

The function appears to work correctly, but it will only work for global variables. If we try to use it on a local variable, it will fail. SIMPLE-INCF works correctly for either local or global variables.

```
(defun test-simple (turnip)
  (simple-incf turnip))

(defun test-faulty (turnip)
  (faulty-incf 'turnip))

> (test-simple 37)
38

> (test-faulty 37)
Error: TURNIP unassigned variable.
```

In TEST-SIMPLE the SIMPLE-INCF macro expands into an expression that is then evaluated in the lexical context of TEST-SIMPLE. So the local variable TURNIP is lexically apparent, and there is no problem.

```
→ (test-simple 37)
→ Enter TEST-SIMPLE with input 37
    create variable TURNIP with value 37
    ┈> (simple-incf turnip)
        ┏→ Enter SIMPLE-INCF macro with input TURNIP
        ┗→ Macro expansion:  (SETQ TURNIP (+ TURNIP 1))
    → (setq turnip (+ turnip 1))
        ┌→ (+ turnip 1)
        │   TURNIP evaluates to 37
        └→ 38
       set TURNIP to 38
    └→ 38
→ Result of TEST-SIMPLE is 38
```

We can see the bug in FAULTY-INCF with an evaltrace diagram. Inside the body of the FAULTY-INCF function the only local variable visible is VAR. The heavy solid line surrounding the body indicates that the parent lexical context of FAULTY-INCF is the global context, so TEST-FAULTY's local variable TURNIP is not lexically accessible. There is no value assigned to the global variable TURNIP, so when SYMBOL-VALUE looks in the value cell it gets an unassigned variable error.

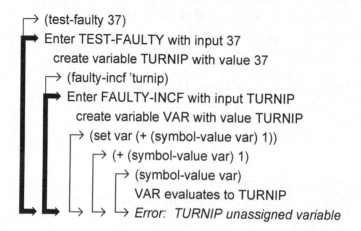

```
→ (test-faulty 37)
→ Enter TEST-FAULTY with input 37
    create variable TURNIP with value 37
    → (faulty-incf 'turnip)
    → Enter FAULTY-INCF with input TURNIP
       create variable VAR with value TURNIP
       → (set var (+ (symbol-value var) 1))
          → (+ (symbol-value var) 1)
             → (symbol-value var)
                VAR evaluates to TURNIP
             → Error:  TURNIP unassigned variable
```

14.15 HISTORICAL SIGNIFICANCE OF MACROS

One of the nice features of macros is that their syntax is identical to that of ordinary and special functions. This makes it easy for programmers to make syntactic extensions to Lisp in an invisible way: people who use the extensions can't tell that they are programmer defined rather than built in. In contrast, in languages like Pascal it is not possible to add new statement types, only new procedures. The only ways to extend the syntax of Pascal are to write a preprocessor or modify the compiler. Both approaches are impractical if you want to be able to combine extensions contributed by several programmers.

Many features of Common Lisp originated in earlier dialects as some programmer's private macro package. Examples include the SETF, DEFSTRUCT, and WITH-OPEN-FILE macros. Even DEFMACRO was originally an extension. (Although Lisp has had macros from the very beginning, before DEFMACRO came along they had to be defined in a more cumbersome way.)

Lisp has evolved continuously over its 30-year history, with many people contributing good ideas for extensions. This evolution would not have been possible without macros. Besides extending Lisp, macros can also be used to define entirely new languages. Specialized high-level languages for artificial intelligence programming are often built on top of Lisp this way. The figures in this book were created using a specialized graphics language implemented as Common Lisp macros.

14.16 DYNAMIC SCOPING

Throughout this book we have used lexical scoping for all variables. Lexical scoping means that in order for a function FOO to access a variable X, the definition of FOO must appear within the context where X is defined. If FOO is defined at top level with DEFUN, then it can only access global variables (plus whatever locals it defines itself.) But if a function is defined by a lambda expression appearing inside the body of another function BAR, then it can access BAR's local variables as well as its own. Functions defined outside of BAR cannot access any of BAR's variables.

The alternative to lexical scoping is called **dynamic** scoping. Prior to Common Lisp, dynamic scoping was the norm in Lisp. Lexical scoping was found only in two offshoot dialects called Scheme and T.

Dynamically scoped variables are also called **special** variables. When a variable name is declared to be special, that variable will not be local to any function; its value will be accessible anywhere. In contrast, lexically scoped variables are accessible only within the body of the form that defines them. One way to declare a variable name special is with the DEFVAR macro.

```
(defvar birds)
```

Let's compare the effects of lexical versus dynamic scoping of variables. We've declared BIRDS to be dynamically scoped. We'll use FISH as a lexically scoped variable, so it should not be DEFVARed. Each variable will be assigned an appropriate initial value below; then we'll write a function to reference the value of each variable.

```
(setf fish '(salmon tuna))

(setf birds '(eagle vulture))

(defun ref-fish ()
  fish)

(defun ref-birds ()
  birds)

(ref-fish) ⇒ (salmon tuna)

(ref-birds) ⇒ (eagle vulture)
```

Now to see the difference between the two scoping disciplines, we'll write functions that name their inputs FISH and BIRDS. First, we'll consider the familiar, lexically scoped case using FISH.

```
(defun test-lexical (fish)
  (list fish (ref-fish)))

> (test-lexical '(guppy minnow))
((GUPPY MINNOW) (SALMON TUNA))
```

In TEST-LEXICAL the expression FISH refers to the local variable FISH. This local variable is not visible to REF-FISH. The symbol FISH in the body of REF-FISH continues to refer to the global variable FISH. In the evaltrace diagram you can see that the body of REF-FISH is enclosed in a solid line, indicating that its parent lexical context is the global context. Since REF-FISH doesn't create a local variable of its own named FISH, any occurrence of FISH in its body is taken as a reference to the global variable.

the global variable FISH has value (SALMON TUNA)

→ (test-lexical '(guppy minnow))
→ Enter TEST-LEXICAL with input (GUPPY MINNOW)
   create local variable FISH with value (GUPPY MINNOW)
   → (list fish (ref-fish))
   FISH evaluates to (GUPPY MINNOW)
    → (ref-fish)
    → Enter REF-FISH
    FISH evaluates to (SALMON TUNA)
    → Result of REF-FISH is (SALMON TUNA)
   → ((GUPPY MINNOW) (SALMON TUNA))
→ Result of TEST-LEXICAL is ((GUPPY MINNOW) (SALMON TUNA))

In the dynamically scoped case, using BIRDS, the testing function looks identical to the previous one, but it behaves differently. This difference is due to the effect of the DEFVAR's declaring BIRDS to be special.

```
(defun test-dynamic (birds)
  (list birds (ref-birds)))

> (test-dynamic '(robin sparrow))
((ROBIN SPARROW) (ROBIN SPARROW))

> (ref-birds)
(EAGLE VULTURE)
```

When we enter the body of TEST-DYNAMIC, a new dynamic variable named BIRDS is created. From now until we leave the body, every use of BIRDS anywhere in the program will refer to this variable, even if it occurs in some other function outside of TEST-DYNAMIC. The global variable named BIRDS is inaccessible as long as this new dynamic variable is in existence. When TEST-DYNAMIC returns, the dynamic variable BIRDS that it created will cease to exist, and the name BIRDS will again be associated with the global variable BIRDS.

There is no special evaltrace notation for dynamic variables; you simply have to note whether a given name has been DEFVARed or not. Once it has, all variables with that name will be dynamically scoped.

the global variable BIRDS has value (EAGLE VULTURE)

```
→ (test-dynamic '(robin sparrow))
  Enter TEST-DYNAMIC with input (ROBIN SPARROW)
    create dynamic variable BIRDS with value (ROBIN SPARROW)
    → (list birds (ref-birds))
      BIRDS dynamically evaluates to (ROBIN SPARROW)
      → (ref-birds)
        Enter REF-BIRDS
        BIRDS dynamically evaluates to (ROBIN SPARROW)
        Result of REF-BIRDS is (ROBIN SPARROW)
    ↳ ((ROBIN SPARROW) (ROBIN SPARROW))
  Result of TEST-DYNAMIC is ((ROBIN SPARROW) (ROBIN SPARROW))
```

The rule for evaluating dynamically scoped variables is that when we hit a thick solid line, instead of jumping to the global lexical context, we just pass right on through, continuing to look for the creation of a variable with that name. We only use the global value if we make it all the way out to the global context, meaning no function presently has a variable with the same name as the global variable.

The term "dynamic binding" refers to the property that the name BIRDS in REF-BIRDS is not permanently associated with any one variable, the way FISH is associated with a global variable in REF-FISH. Instead, the connection between the name and the actual variable is made dynamically. When REF-BIRDS is called inside TEST-DYNAMIC, the symbol BIRDS refers to the dynamic variable BIRDS established by TEST-DYNAMIC. When REF-BIRDS is called at top level, the same symbol BIRDS is interpreted as a reference to the global variable BIRDS.

Dynamic scoping should be used sparingly. In earlier Lisp dialects where it was the default, its use caused quite a few program bugs where one function would accidentally modify a dynamic variable created by another. Lexical scoping protects a function's local variables from modification by other, unrelated functions. But there are some contexts where dynamic scoping is exactly the right thing to use. An example is given in Section 14.18.

14.17   DEFVAR, DEFPARAMETER, DEFCONSTANT

DEFVAR, DEFPARAMETER, and DEFCONSTANT all declare names to be special. DEFVAR is used for declaring variables whose values will change during the normal operation of the program. It accepts an optional initial variable value and a documentation string.

```
> (defvar *total-glasses* 0
    "Total glasses sold so far")
*TOTAL-GLASSES*
```

A curious fact about DEFVAR is that if the variable already has a value, DEFVAR will not change it. It only assigns the initial value if the variable has none.

```
> (defvar *total-glasses* 3)
*TOTAL-GLASSES*

> *total-glasses*
0
```

DEFPARAMETER has the same syntax as DEFVAR, but it is used to declare variables whose values will not change while the program runs. They hold ''parameter settings'' that tell the program how to behave. Another difference between DEFPARAMETER and DEFVAR is that DEFPARAMETER will assign a value to a variable even if it already has one.

```
> (defparameter *max-glasses* 500
    "Maximum number of glasses we can make")
*MAX-GLASSES*

> (defparameter *max-glasses* 300)
*MAX-GLASSES*

> *max-glasses*
300
```

DEFCONSTANT is used to define constants, which are guaranteed never to change. The convention in Lisp is to surround the names of special variables with an asterisk, but this does not apply to constants. It is an error to try to change the value of a constant, or to create a new variable with the same name as a constant. PI is a built-in constant in Common Lisp.

```
> (defconstant speed-of-light 299792500.0
```

```
     "Speed of light in meters per second")
SPEED-OF-LIGHT

> (setf speed-of-light 'very-fast)
Error: can't assign to SPEED-OF-LIGHT.
It's a constant.

> (let ((pi 'greek))
    (list pi 'salad))
Error: can't create a variable named PI.
It's a constant.
```

Declaring a quantity to be constant sometimes allows the compiler to generate more efficient machine language than if it were a variable. It also prevents someone from changing the value accidentally. Most implementations still permit you to change the value deliberately, though, by going through the debugger.

14.18 REBINDING SPECIAL VARIABLES

Much of Lisp's terminology for variables is a holdover from the days when dynamic scoping was the norm. For historical reasons some writers talk about "binding a variable" when they mean "creating a new variable." But people also say "unbound variable" when they mean "unassigned variable." Binding does not refer strictly to assignment; that is one of the major sources of terminological confusion in Lisp. Nonglobal lexical variables always have values, but it is possible for global or special variables to exist without a value. We won't get into the arcane details of that in this book.

We have avoided confusion so far by declining to use the term "binding" at all. In this final section we introduce the term "rebinding' to refer to the creation of a new special variable with the same name as the old one. While the new variable is in existence, all uses of that name anywhere in the program will refer to it (unless the name is rebound yet again), and the previous variable with that name will be inaccessible. Strictly speaking, we aren't rebinding any variable: We're dynamically rebinding the *name*, making it refer temporarily to a different variable.

Common Lisp contains quite a few built-in special variables. Some of these control the way input/output is handled. For example, the variable *PRINT-BASE* is used by FORMAT and other functions to determine the base in which numbers are to be printed. Normally they are printed in base ten. We can dynamically rebind *PRINT-BASE* to print numbers in other bases. Since it is already declared special, we don't have to DEFVAR it. To rebind it, we merely include it in the argument list of our function.

```
(defun print-in-base (*print-base* x)
  (format t "~&~D is written ~S in base ~D."
    x x *print-base*))

> (print-in-base 10 205)
205 is written 205 in base 10.
NIL

> (print-in-base 8 205)
205 is written 315 in base 8.
NIL

> (print-in-base 2 205)
205 is written 11001101 in base 2.
NIL
```

We can also rebind special variables using LET, as PPMX rebound the variable *PRINT-CIRCLE*. When a special variable is rebound, any assignments, no matter where they occur in the program, will affect the new variable, not the old one. In the following example, when BUMP-FOO is called in the body of the LET inside REBIND-FOO, it increments the dynamic variable named *FOO* that was established by the LET. When it is called outside of the LET, it increments the global variable *FOO*. If *FOO* had not been declared special, BUMP-VAR would always access the global *FOO*.

```
(defvar *foo* 2)

(defun bump-foo ()
  (incf *foo*))

(defun rebind-foo ()
  (bump-foo)
  (showvar *foo*)
  (let ((*foo* 100))
    (format t "~&Enter the LET...~%")
    (showvar *foo*)
    (incf *foo*)
    (showvar *foo*)
    (bump-foo)
    (showvar *foo*)
```

```
        (format t "~&Leave the LET.~%"))
    (bump-foo)
    (showvar *foo*))

> (rebind-foo)
The value of *FOO* is 3
Enter the LET...
The value of *FOO* is 100
The value of *FOO* is 101
The value of *FOO* is 102
Leave the LET.
The value of *FOO* is 4
NIL
```

Rebinding of special variables is most useful when different parts of a large program need to communicate with each other, and passing information via extra arguments to functions is impractical. Writing really large programs requires a different set of skills than what this book emphasizes; it is a good topic for an advanced Lisp course.

FUNCTIONS COVERED IN ADVANCED TOPICS

DEFMACRO: the &BODY lambda list keyword.

Declarations: DEFVAR, DEFPARAMETER, DEFCONSTANT.

Appendix A
The SDRAW Tool

The SDRAW tool provides three user-level functions: SDRAW, SDRAW-LOOP, and SCRAWL. SDRAW takes a list as input and draws the corresponding cons cell diagram on the display. SDRAW-LOOP implements a read-eval-draw loop similar to the normal read-eval-print loop. SCRAWL is used to interactively "crawl around" in list structure by taking successive CARs and CDRs. It uses SDRAW to display the current position in the list. See page 186 for examples.

The generic version of SDRAW shown here will work in any legal Common Lisp implementation. It "draws" cons cells by outputting an appropriate character sequence. More sophisticated versions of SDRAW for a variety of Common Lisp implementations are available on diskette from the publisher. Some of these versions draw cons cells using the IBM PC graphic character set. Others, designed for the X Windows system, use CLX functions to produce bitmapped graphics.

Two notes about the implementation: First, the software lives in package SDRAW and uses SHADOWING-IMPORT to inject the symbols SDRAW, SDRAW-LOOP, and SCRAWL into the USER package. This can be disabled by deleting the first four forms in the file. Second, the function SDL1 (part of SDRAW-LOOP) uses HANDLER-CASE to trap evaluation errors. HANDLER-CASE is part of the new condition system recently added to the Common Lisp standard. Not all implementations support HANDLER-CASE yet. If necessary you can replace it with IGNORE-ERRORS, or whatever the equivalent function is called in your implementation.

```
;;; -*- Mode: Lisp; Package: SDRAW -*-
;;;
;;; SDRAW - draws cons cell structures.
;;; From the book "Common Lisp:  A Gentle Introduction to
;;;      Symbolic Computation" by David S. Touretzky.
;;; The Benjamin/Cummings Publishing Co., 1989.
;;;
;;; User-level routines:
;;;    (SDRAW obj)   - draws obj on the terminal
;;;    (SDRAW-LOOP) - puts the user in a read-eval-draw loop
;;;    (SCRAWL obj) - interactively crawl around obj

(in-package "SDRAW")

(export '(sdraw::sdraw sdraw::sdraw-loop sdraw::scrawl))

(shadowing-import  '(sdraw::sdraw sdraw::sdraw-loop sdraw::scrawl)
                 (find-package "USER"))


;;;;;;;;;;;;;;;;;;;;;;;;;;;;;;;;;;;;;;;;;;;;;;;;;;;;;;;;;;;;;;;;;;;
;;;
;;; The parameters below are in units of characters (horizontal)
;;; and lines (vertical).  They apply to all versions of SDRAW,
;;; but their values may change if cons cells are being drawn as
;;; bit maps rather than as character sequences.

(defparameter *sdraw-display-width* 79.)
(defparameter *sdraw-horizontal-atom-cutoff* 79.)
(defparameter *sdraw-horizontal-cons-cutoff* 65.)

(defparameter *etc-string* "etc.")
(defparameter *circ-string* "circ.")
(defparameter *etc-spacing* 4.)
(defparameter *circ-spacing* 5.)

(defparameter *inter-atom-h-spacing* 3.)
(defparameter *cons-atom-h-arrow-length* 9.)
(defparameter *inter-cons-v-arrow-length* 3.)
(defparameter *cons-v-arrow-offset-threshold* 2.)
(defparameter *cons-v-arrow-offset-value* 1.)

(defparameter *sdraw-vertical-cutoff* 22.)
(defparameter *sdraw-num-lines* 25)
(defvar *line-endings* (make-array *sdraw-num-lines*))
```

```
;;;;;;;;;;;;;;;;;;;;;;;;;;;;;;;;;;;;;;;;;;;;;;;;;;;;;;;;;;;;;;;;;
;;;
;;; SDRAW and subordinate definitions.

(defun sdraw (obj)
  (fill *line-endings* most-negative-fixnum)
  (draw-structure (struct1 obj 0 0 nil))
  (values))

(defun struct1 (obj row root-col obj-memory)
  (cond ((atom obj)
             (struct-process-atom (format nil "~S" obj) row root-col))
            ((member obj obj-memory :test #'eq)
             (struct-process-circ row root-col))
            ((>= row *sdraw-vertical-cutoff*)
             (struct-process-etc row root-col))
            (t (struct-process-cons obj row root-col
                                    (cons obj obj-memory)))))

(defun struct-process-atom (atom-string row root-col)
  (let* ((start-col (struct-find-start row root-col))
          (end-col (+ start-col (length atom-string))))
    (cond ((< end-col *sdraw-horizontal-atom-cutoff*)
              (struct-record-position row end-col)
              (list 'atom row start-col atom-string))
            (t (struct-process-etc row root-col)))))

(defun struct-process-etc (row root-col)
  (let ((start-col (struct-find-start row root-col)))
    (struct-record-position
      row
      (+ start-col (length *etc-string*) *etc-spacing*))
    (list 'msg row start-col *etc-string*)))

(defun struct-process-circ (row root-col)
  (let ((start-col (struct-find-start row root-col)))
    (struct-record-position
      row
      (+ start-col (length *circ-string*) *circ-spacing*))
    (list 'msg row start-col *circ-string*)))
```

```lisp
(defun struct-process-cons (obj row root-col obj-memory)
  (let* ((cons-start (struct-find-start row root-col))
         (car-structure
          (struct1 (car obj)
                   (+ row *inter-cons-v-arrow-length*)
                   cons-start obj-memory))
         (start-col (third car-structure)))
    (if (>= start-col *sdraw-horizontal-cons-cutoff*)
        (struct-process-etc row root-col)
        (list 'cons row start-col car-structure
              (struct1 (cdr obj) row
                       (+ start-col *cons-atom-h-arrow-length*)
                       obj-memory)))))

(defun struct-find-start (row root-col)
  (max root-col (+ *inter-atom-h-spacing*
                   (aref *line-endings* row))))

(defun struct-record-position (row end-col)
  (setf (aref *line-endings* row) end-col))
```

```lisp
;;;;;;;;;;;;;;;;;;;;;;;;;;;;;;;;;;;;;;;;;;;;;;;;;;;;;;;;;;;;;;;;;;;;;;
;;;
;;; SDRAW-LOOP and subordinate definitions.

(defparameter *sdraw-loop-prompt-string* "S> ")

(defun sdraw-loop ()
  "Read-eval-print loop using sdraw to display results."
  (format t "~&Type any Lisp expression, or (ABORT) to exit.~%~%")
  (sdl1))
```

```
(defun sdl1 ()
  (loop
    (format t "~&~A" *sdraw-loop-prompt-string*)
    (let ((form (read)))
      (setf +++ ++
            ++  +
            +   -
            -   form)
      (let ((result (multiple-value-list
                      (handler-case (eval form)
                        (error (condx) condx)))))
        (typecase (first result)
          (error (display-sdl-error result))
          (t (setf /// //
                   //  /
                   /   result
                   *** **
                   **  *
                   *   (first result))
             (display-sdl-result *)))))))

(defun display-sdl-result (result)
  (let* ((*print-circle* t)
         (*print-length* nil)
         (*print-level* nil)
         (*print-pretty* nil)
         (full-text (format nil "Result:  ~S" result))
         (text (if (> (length full-text)
                      *sdraw-display-width*)
                   (concatenate 'string
                     (subseq full-text 0 (- *sdraw-display-width* 4))
                     "...)")
                   full-text)))
    (sdraw result)
    (if (consp result)
        (format t "~%~A~%" text))
    (terpri)))

(defun display-sdl-error (error)
  (format t "~A~%~%" error))
```

```
;;;;;;;;;;;;;;;;;;;;;;;;;;;;;;;;;;;;;;;;;;;;;;;;;;;;;;;;;;;;;;;;
;;;
;;; SCRAWL and subordinate definitions.

(defparameter *scrawl-prompt-string* "SCRAWL> ")
(defvar *scrawl-object* nil)
(defvar *scrawl-current-obj*)
(defvar *extracting-sequence* nil)

(defun scrawl (obj)
  "Read-eval-print loop to travel through list"
  (format t "~&Crawl through list:  'H' for help, 'Q' to quit.~%~%")
  (setf *scrawl-object* obj)
  (setf *scrawl-current-obj* obj)
  (setf *extracting-sequence* nil)
  (sdraw obj)
  (scrawl1))

(defun scrawl1 ()
  (loop
    (format t "~&~A" *scrawl-prompt-string*)
    (let ((command (read-uppercase-char)))
      (case command
        (#\A (scrawl-car-cmd))
        (#\D (scrawl-cdr-cmd))
        (#\B (scrawl-back-up-cmd))
        (#\S (scrawl-start-cmd))
        (#\H (display-scrawl-help))
        (#\Q (return))
        (t (display-scrawl-error))))))

(defun scrawl-car-cmd ()
  (cond ((consp *scrawl-current-obj*)
         (push 'car *extracting-sequence*)
         (setf *scrawl-current-obj* (car *scrawl-current-obj*)))
        (t (format t
             "~&Can't take CAR or CDR of an atom.  Use B to back up.~%")))
  (display-scrawl-result))
```

```
(defun scrawl-cdr-cmd ()
  (cond ((consp *scrawl-current-obj*)
         (push 'cdr *extracting-sequence*)
         (setf *scrawl-current-obj* (cdr *scrawl-current-obj*)))
        (t (format t
              "~&Can't take CAR or CDR of an atom.  Use B to back up.~%")))
  (display-scrawl-result))

(defun scrawl-back-up-cmd ()
  (cond (*extracting-sequence*
         (pop *extracting-sequence*)
         (setf *scrawl-current-obj*
               (extract-obj *extracting-sequence* *scrawl-object*)))
        (t (format t "~&Already at beginning of object.")))
  (display-scrawl-result))

(defun scrawl-start-cmd ()
  (setf *scrawl-current-obj* *scrawl-object*)
  (setf *extracting-sequence* nil)
  (display-scrawl-result))

(defun extract-obj (seq obj)
  (reduce #'funcall
          seq
          :initial-value obj
          :from-end t))

(defun get-car/cdr-string ()
  (if (null *extracting-sequence*)
      (format nil "'~S" *scrawl-object*)
      (format nil "(c~Ar '~S)"
              (map 'string #'(lambda (x)
                               (ecase x
                                 (car #\a)
                                 (cdr #\d)))
                   *extracting-sequence*)
              *scrawl-object*)))
```

```lisp
(defun display-scrawl-result (&aux (*print-pretty* nil)
                                   (*print-length* nil)
                                   (*print-level* nil)
                                   (*print-circle* t))
  (let* ((extract-string (get-car/cdr-string))
         (text (if (> (length extract-string) *sdraw-display-width*)
                   (concatenate 'string
                     (subseq extract-string 0
                        (- *sdraw-display-width* 4))
                     "...)")
                   extract-string)))
    (sdraw *scrawl-current-obj*)
    (format t "~&~%~A~%~%" text)))

(defun display-scrawl-help ()
  (format t "~&Legal commands:  A)car    D)cdr  B)back up~%")
  (format t "~&                 S)start Q)quit H)help~%"))

(defun display-scrawl-error ()
  (format t "~&Illegal command.~%")
  (display-scrawl-help))

(defun read-uppercase-char ()
  (let ((response (read-line)))
    (and (plusp (length response))
         (char-upcase (char response 0)))))
```

```
;;;;;;;;;;;;;;;;;;;;;;;;;;;;;;;;;;;;;;;;;;;;;;;;;;;;;;;;;;;;;;;;
;;;
;;; The following definitions are specific to the tty implementation.

(defparameter *cons-string* "[*|*]")
(defparameter *cons-cell-flatsize* 5.)
(defparameter *cons-h-arrowshaft-char* #\-)
(defparameter *cons-h-arrowhead-char* #\>)
(defparameter *cons-v-line* "|")
(defparameter *cons-v-arrowhead* "v")

(defvar *textline-array* (make-array *sdraw-num-lines*))
(defvar *textline-lengths* (make-array *sdraw-num-lines*))

(eval-when (eval load)
  (dotimes (i *sdraw-num-lines*)
    (setf (aref *textline-array* i)
          (make-array *sdraw-display-width*
                      :element-type 'string-char))))

(defun char-blt (row start-col string)
  (let ((spos (aref *textline-lengths* row))
        (line (aref *textline-array* row)))
    (do ((i spos (1+ i)))
        ((>= i start-col))
      (setf (aref line i) #\Space))
    (replace line string :start1 start-col)
    (setf (aref *textline-lengths* row)
          (+ start-col (length string)))))

(defun draw-structure (directions)
  (fill *textline-lengths* 0.)
  (follow-directions directions)
  (dump-display))

(defun follow-directions (dirs &optional is-car)
  (ecase (car dirs)
    (cons (draw-cons dirs))
    ((atom msg) (draw-msg  (second dirs)
                           (third dirs)
                           (fourth dirs)
                           is-car))))
```

```lisp
(defun draw-cons (obj)
  (let* ((row (second obj))
         (col (third obj))
         (car-component (fourth obj))
         (cdr-component (fifth obj))
         (line (aref *textline-array* row))
         (h-arrow-start (+ col *cons-cell-flatsize*))
         (h-arrowhead-col (1- (third cdr-component)))))
    (char-blt row col *cons-string*)
    (do ((i h-arrow-start (1+ i)))
        ((>= i h-arrowhead-col))
      (setf (aref line i) *cons-h-arrowshaft-char*))
    (setf (aref line h-arrowhead-col) *cons-h-arrowhead-char*)
    (setf (aref *textline-lengths* row) (1+ h-arrowhead-col))
    (char-blt (+ row 1) (+ col 1) *cons-v-line*)
    (char-blt (+ row 2) (+ col 1) *cons-v-arrowhead*)
    (follow-directions car-component t)
    (follow-directions cdr-component)))

(defun draw-msg (row col string is-car)
  (char-blt row
            (+ col (if (and is-car
                            (<= (length string)
                                *cons-v-arrow-offset-threshold*))
                       *cons-v-arrow-offset-value*
                       0))
            string))

(defun dump-display ()
  (terpri)
  (dotimes (i *sdraw-num-lines*)
    (let ((len (aref *textline-lengths* i)))
      (if (plusp len)
          (format t "~&~A"
                  (subseq (aref *textline-array* i) 0 len))
          (return nil))))
  (terpri))
```

Appendix B
The DTRACE Tool

DTRACE provides a more detailed trace display than most manufacturer-supplied implementations of TRACE. The program exports two functions, DTRACE and DUNTRACE, whose syntax is the same as TRACE and UNTRACE, respectively. See page 217 and all of Chapter 8 for examples.

The generic version of DTRACE shown here contains only one implementation-dependent function: FETCH-ARGLIST. FETCH-ARGLIST takes a symbol as input and returns the argument list of the function named by that symbol. Versions of DTRACE for various Lisp implementations, with appropriate FETCH-ARGLIST functions, are available on diskette from the publisher. Some of these versions also produce nicer output than the generic version, for example, by using the IBM PC graphic character set to draw arrows.

To produce a version of DTRACE for a new Lisp implementation, you will have to find out how to extract argument list information from function cells and/or property lists. See the examples at the end of the program. You can also define FETCH-ARGLIST to simply return NIL, in which case arguments will be displayed as Arg-1, Arg-2, and so on.

One other note about the DTRACE software: it lives in package DTRACE, and uses SHADOWING-IMPORT to inject the symbols DTRACE and DUNTRACE into the USER package. This can be disabled by deleting the first four forms in the file.

```
;;; -*- Mode: Lisp; Package: DTRACE -*-

;;; DTRACE is a portable alternative to the Common Lisp TRACE and UNTRACE
;;; macros.  It offers a more detailed display than most tracing tools.
;;;
;;; From the book "Common Lisp:  A Gentle Introduction to
;;;      Symbolic Computation" by David S. Touretzky.
;;; The Benjamin/Cummings Publishing Co., 1989.
;;;
;;; User-level routines:
;;;    DTRACE  - same syntax as TRACE
;;;    DUNTRACE - same syntax as UNTRACE

(in-package "DTRACE" :use '("LISP"))

(export '(dtrace::dtrace dtrace::duntrace
          *dtrace-print-length* *dtrace-print-level*
          *dtrace-print-circle* *dtrace-print-pretty*
          *dtrace-print-array*))

(shadowing-import '(dtrace::dtrace dtrace::duntrace) (find-package "USER"))

(use-package "DTRACE" "USER")

;;;;;;;;;;;;;;;;;;;;;;;;;;;;;;;;;;;;;;;;;;;;;;;;;;;;;;;;;;;;;;;;;;;;;;;
;;;
;;; DTRACE and subordinate routines.

(defparameter *dtrace-print-length* 7)
(defparameter *dtrace-print-level*  4)
(defparameter *dtrace-print-circle* t)
(defparameter *dtrace-print-pretty* nil)
(defparameter *dtrace-print-array* *print-array*)

(defvar *traced-functions* nil)
(defvar *trace-level* 0)

(defmacro dtrace (&rest function-names)
  "Turns on detailed tracing for specified functions.  Undo with DUNTRACE."
  (if (null function-names)
      (list 'quote *traced-functions*)
      (list 'quote (mapcan #'dtrace1 function-names)))))
```

```
(defun dtrace1 (name)
  (unless (symbolp name)
    (format *error-output* "~&~S is an invalid function name." name)
    (return-from dtrace1 nil))
  (unless (fboundp name)
    (format *error-output* "~&~S undefined function." name)
    (return-from dtrace1 nil))
  (eval '(untrace ,name))       ;; if they're tracing it, undo their trace
  (duntrace1 name)              ;; if we're tracing it, undo our trace
  (when (special-form-p name)
    (format *error-output*
            "~&Can't trace ~S because it's a special form." name)
    (return-from dtrace1 nil))
  (if (macro-function name)
      (trace-macro name)
      (trace-function name))
  (setf *traced-functions* (nconc *traced-functions* (list name)))
  (list name))

;;; The functions below reference DISPLAY-xxx routines that can be made
;;; implementation specific for fancy graphics.  Generic versions of
;;; these routines are defined later in this file.

(defun trace-function (name)
  (let* ((formal-arglist (fetch-arglist name))
         (old-defn (symbol-function name))
         (new-defn
          #'(lambda (&rest argument-list)
              (let ((result nil))
                (display-function-entry name)
                (let ((*trace-level* (1+ *trace-level*)))
                  (with-dtrace-printer-settings
                   (show-function-args argument-list formal-arglist))
                  (setf result (multiple-value-list
                                 (apply old-defn argument-list))))
                (display-function-return name result)
                (values-list result)))))
    (setf (get name 'original-definition) old-defn)
    (setf (get name 'traced-definition) new-defn)
    (setf (get name 'traced-type) 'defun)
    (setf (symbol-function name) new-defn)))
```

```
(defun trace-macro (name)
  (let* ((formal-arglist (fetch-arglist name))
         (old-defn (macro-function name))
         (new-defn
          #'(lambda (macro-args env)
              (let ((result nil))
                (display-function-entry name 'macro)
                (let ((*trace-level* (1+ *trace-level*)))
                  (with-dtrace-printer-settings
                    (show-function-args macro-args formal-arglist))
                  (setf result (funcall old-defn macro-args env)))
                (display-function-return name (list result) 'macro)
                (values result)))))
    (setf (get name 'original-definition) old-defn)
    (setf (get name 'traced-definition) new-defn)
    (setf (get name 'traced-type) 'defmacro)
    (setf (macro-function name) new-defn)))

(defun show-function-args (actuals formals &optional (argcount 0))
  (cond ((null actuals) nil)
        ((null formals) (handle-args-numerically actuals argcount))
        (t (case (first formals)
             (&optional (show-function-args
                          actuals (rest formals) argcount))
             (&rest (show-function-args
                      (list actuals) (rest formals) argcount))
             (&key (handle-keyword-args actuals))
             (&aux (show-function-args actuals nil argcount))
             (t (handle-one-arg (first actuals) (first formals))
                (show-function-args (rest actuals)
                                    (rest formals)
                                    (1+ argcount)))))))

(defun handle-args-numerically (actuals argcount)
  (dolist (x actuals)
    (incf argcount)
    (display-arg-numeric x argcount)))

(defun handle-one-arg (val varspec)
  (cond ((atom varspec) (display-one-arg val varspec))
        (t (display-one-arg val (first varspec))
           (if (third varspec)
               (display-one-arg t (third varspec))))))
```

```
(defun handle-keyword-args (actuals)
  (cond ((null actuals))
        ((keywordp (first actuals))
         (display-one-arg (second actuals) (first actuals))
         (handle-keyword-args (rest (rest actuals))))
        (t (display-one-arg actuals "Extra args:")))))


;;;;;;;;;;;;;;;;;;;;;;;;;;;;;;;;;;;;;;;;;;;;;;;;;;;;;;;;;;;;;;;;
;;;
;;; DUNTRACE and subordinate routines.

(defmacro duntrace (&rest function-names)
  "Turns off tracing for specified functions.
   With no args, turns off all tracing."
  (setf *trace-level* 0)   ;; safety precaution
  (list 'quote
        (mapcan #'duntrace1 (or function-names *traced-functions*))))

(defun duntrace1 (name)
  (unless (symbolp name)
    (format *error-output* "~&~S is an invalid function name." name)
    (return-from duntrace1 nil))
  (setf *traced-functions* (delete name *traced-functions*))
  (let ((orig-defn (get name 'original-definition 'none))
        (traced-defn (get name 'traced-definition))
        (traced-type (get name 'traced-type 'none)))
    (unless (or (eq orig-defn 'none)
                (not (fboundp name))
                (not (equal traced-defn   ;; did it get redefined?
                       (ecase traced-type
                         (defun (symbol-function name))
                         (defmacro (macro-function name))))))
      (ecase traced-type
        (defun (setf (symbol-function name) orig-defn))
        (defmacro (setf (macro-function name) orig-defn)))))
  (remprop name 'traced-definition)
  (remprop name 'traced-type)
  (remprop name 'original-definition)
  (list name))
```

```
;;;;;;;;;;;;;;;;;;;;;;;;;;;;;;;;;;;;;;;;;;;;;;;;;;;;;;;;;;;;;;;;
;;;
;;; Display routines.
;;;
;;; The code below generates vanilla character output for ordinary
;;; displays.  It can be replaced with special graphics code if the
;;; implementation permits, e.g., on a PC you can use the IBM graphic
;;; character set to draw nicer-looking arrows.  On a color PC you
;;; can use different colors for arrows, for function names, for
;;; argument values, and so on.

(defmacro with-dtrace-printer-settings (&body body)
  `(let ((*print-length* *dtrace-print-length*)
         (*print-level* *dtrace-print-level*)
         (*print-circle* *dtrace-print-circle*)
         (*print-pretty* *dtrace-print-pretty*)
         (*print-array* *dtrace-print-array*))
     ,@body))

(defparameter *entry-arrow-string* "----")
(defparameter *vertical-string*    "|   ")
(defparameter *exit-arrow-string*  " \\--")

(defparameter *trace-wraparound* 15)

(defun display-function-entry (name &optional ftype)
  (space-over)
  (draw-entry-arrow)
  (format *trace-output* "Enter ~S" name)
  (if (eq ftype 'macro)
      (format *trace-output* " macro")))

(defun display-one-arg (val name)
  (space-over)
  (format *trace-output*
          (typecase name
            (keyword "  ~S ~S")
            (string  "  ~A ~S")
            (t "  ~S = ~S"))
          name val))

(defun display-arg-numeric (val num)
  (space-over)
  (format *trace-output* "  Arg-~D = ~S" num val))
```

```lisp
(defun display-function-return (name results &optional ftype)
  (with-dtrace-printer-settings
    (space-over)
    (draw-exit-arrow)
    (format *trace-output* "~S ~A"
            name
            (if (eq ftype 'macro) "expanded to" "returned"))
    (cond ((null results))
          ((null (rest results))
           (format *trace-output* " ~S" (first results)))
          (t (format *trace-output* " values ~{~S, ~}~s"
                     (butlast results)
                     (car (last results)))))))

(defun space-over ()
  (format *trace-output* "~&")
  (dotimes (i (mod *trace-level* *trace-wraparound*))
    (format *trace-output* "~A" *vertical-string*)))

(defun draw-entry-arrow ()
  (format *trace-output* "~A" *entry-arrow-string*))

(defun draw-exit-arrow ()
  (format *trace-output* "~A" *exit-arrow-string*))



;;;;;;;;;;;;;;;;;;;;;;;;;;;;;;;;;;;;;;;;;;;;;;;;;;;;;;;;;;;;;;;;;;;;;;;
;;;
;;; The function FETCH-ARGLIST is implementation dependent.  It
;;; returns the formal argument list of a function as it would
;;; appear in a DEFUN or lambda expression, including any lambda
;;; list keywords.  Here are versions of FETCH-ARGLIST for three
;;; Lisp implementations.

;;; Lucid version
#+LUCID
  (defun fetch-arglist (fn)
    (system::arglist fn))
```

```
;;; GCLisp 3.1 version
#+GCLISP
(defun fetch-arglist (name)
  (let* ((s (sys:lambda-list name))
         (a (read-from-string s)))
    (if s
        (if (eql (elt s 0) #\Newline)
            (edit-arglist (rest a))
            a))))


#+GCLISP
(defun edit-arglist (arglist)
  (let ((result nil)
        (skip-non-keywords nil))
    (dolist (arg arglist (nreverse result))
      (unless (and skip-non-keywords
                   (symbolp arg)
                   (not (keywordp arg)))
        (push arg result))
      (if (eq arg '&key) (setf skip-non-keywords t)))))


;;; CMU Common Lisp version.  This version looks in a symbol's
;;; function cell and knows how to take apart lexical closures
;;; and compiled code objects found there.
#+CMU
  (defun fetch-arglist (x &optional original-x)
    (cond ((symbolp x) (fetch-arglist (symbol-function x) x))
          ((compiled-function-p x)
           (read-from-string
            (lisp::%primitive header-ref x
                              lisp::%function-arg-names-slot)))
          ((listp x) (case (first x)
                       (lambda (second x))
                       (lisp::%lexical-closure% (fetch-arglist (second x)))
                       (system:macro '(&rest "Form ="))
                       (t '(&rest "Arglist:"))))
          (t (cerror (format nil
                       "Use a reasonable default argument list for ~S"
                       original-x)
                  "Unkown object in function cell of ~S:  ~S" original-x x)
             '())))
```

Appendix C
Answers to Exercises

CHAPTER 1 ANSWERS

1.1.

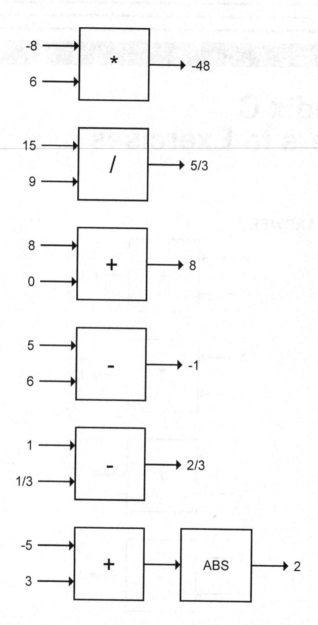

1.2.   Symbols: AARDVARK, PLUMBING, 1-2-3-GO, ZEROP, ZERO, SEVENTEEN. Numbers: 87, 1492, 3.14159265358979, 22/7, 0, −12.

1.3.

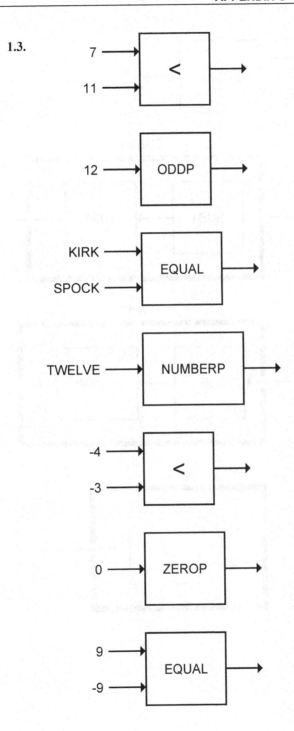

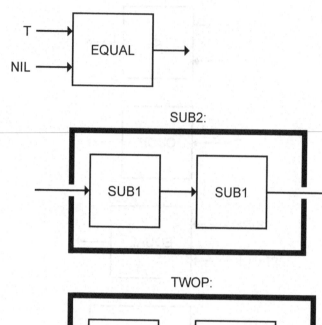

1.4.

SUB2:

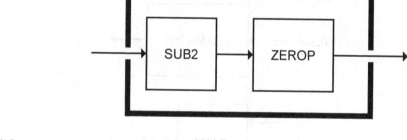

1.5.

TWOP:

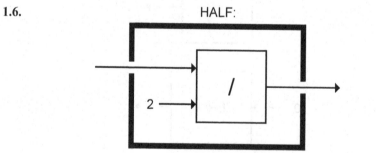

1.6.

HALF:

HALF:

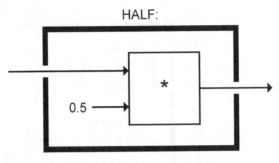

1.7.   MULTI-DIGIT-P:

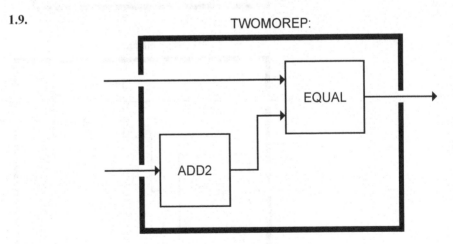

1.8.   The function computes the negation of a number, in other words, it switches the sign from positive to negative and vice versa.

1.9.   TWOMOREP:

1.10.

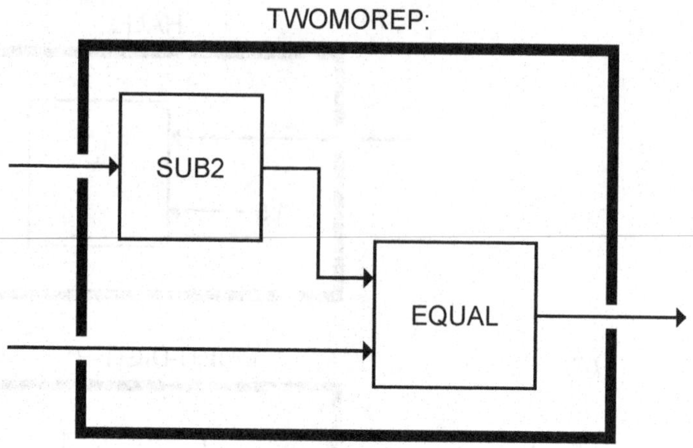

TWOMOREP:

1.11.

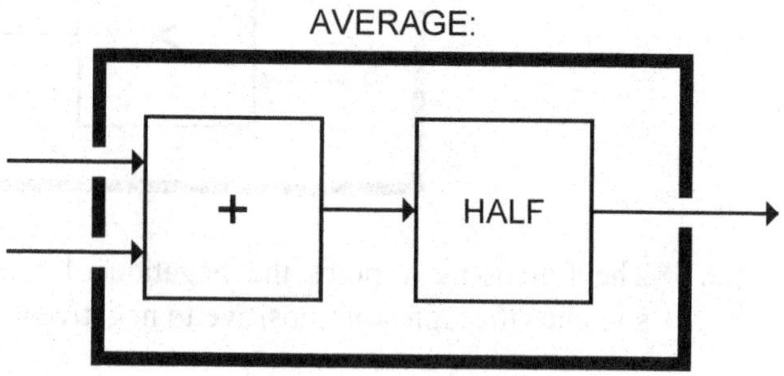

AVERAGE:

1.12.

MORE-THAN-HALF-P:

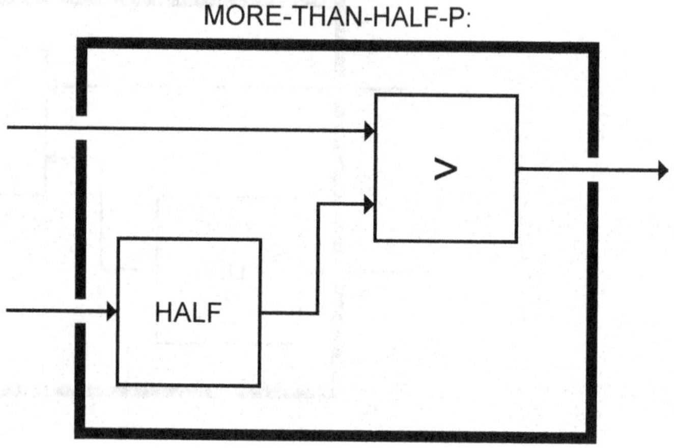

1.13. The function always returns T, since the output of NUMBERP (either T or NIL) is always a symbol.

1.14.

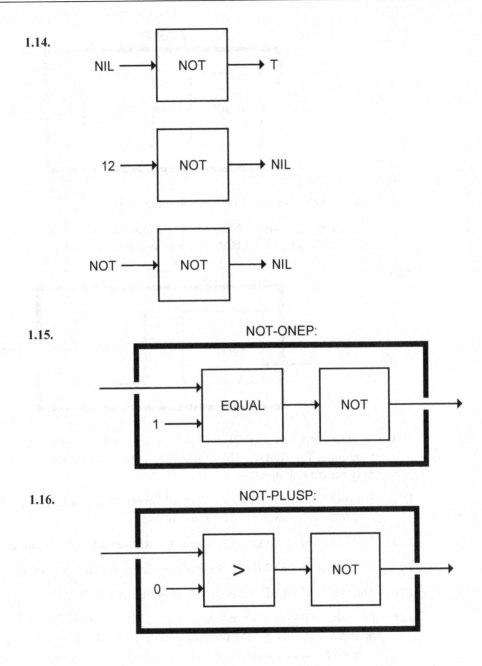

1.15.   NOT-ONEP:

1.16.   NOT-PLUSP:

1.17.

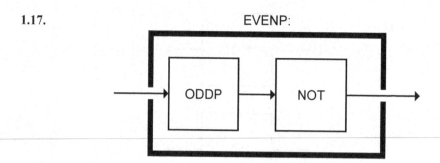

EVENP:

1.18. The predicate returns T only when its input is −2.

1.19. The function outputs NIL when its input is NIL. All other inputs, including T and RUTABAGA, result in an output of T.

1.20.

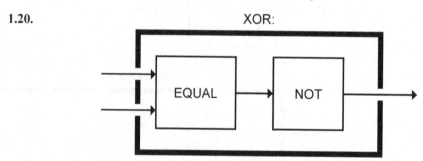

XOR:

1.21. (a) The output of ZEROP will be either T or NIL, which is the wrong type input for ADD1. (b) EQUAL requires two inputs. (c) NOT can only accept one input.

1.22. All predicates are functions. Not all functions are predicates, since not all functions answer yes or no questions.

1.23. EQUAL, NOT, < and > are predicates whose names don't end in "P".

1.24. NUMBER and SYMBOL are both symbols. Neither is a number.

1.25. The symbol FALSE is true in Lisp because it is non-NIL.

1.26. (a) False: ZEROP does not accept T or NIL as input. (b) True: all the predicates studied so far produce either T or NIL as output. Lisp has only a few exceptions to this rule.

1.27.

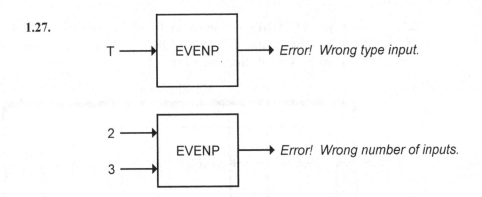

CHAPTER 2 ANSWERS

2.1.

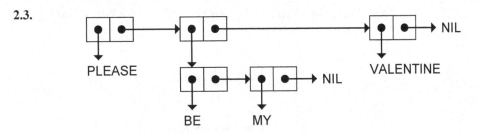

2.2.   Well-formed: the second list, ((A) (B)), the fifth, (A (B (C))), and the sixth, (((A) (B)) (C)).

2.3.

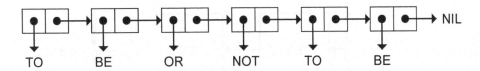

2.4.   ((BOWS ARROWS) (FLOWERS CHOCOLATES)).

2.5.   Six, three, four, four, five, six.

2.6.
Parenthesis Form	Corresponding NIL Form
()	NIL
(())	(NIL)
((()))	((NIL))
(() ())	(NIL NIL)
(() (()))	(NIL (NIL))

2.7.   Inside MY-SECOND, the input to REST is (HONK IF YOU LIKE GEESE). The output, (IF YOU LIKE GEESE), forms the input to FIRST, which outputs the symbol IF.

2.8.

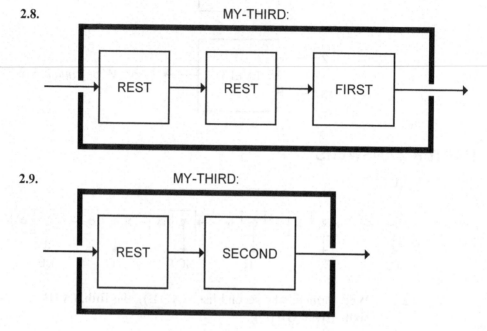

MY-THIRD:

2.9.   MY-THIRD:

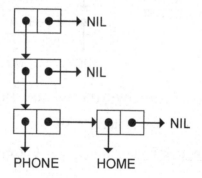

2.10.   The CAR of (((PHONE HOME))) is ((PHONE HOME)), and the CDR is NIL.

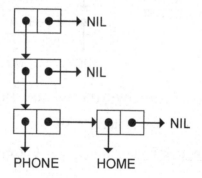

2.11.

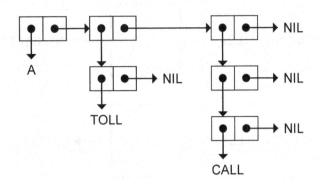

2.12. CADDDR returns the fourth element of a list.   It is pronounced *"ka-dih-dih-der"*.

2.13. FUN is the CAAAR; IN is the CAADR; THE is the CADADR; SUN is the CAADDR.

2.14. CAADR of ((BLUE CUBE) (RED PYRAMID)) is RED.   But if we read the As and Ds in the wrong direction (from left to right), we would take the CAR of the list, then take the CAR of that, and then take the CDR of that.   The first CAR would return (BLUE CUBE), the CAR of that would be BLUE, and the CDR of that would cause an error.

2.15.

Function	Result
CAR	(A B)
CDDR	((E F))
CADR	(C D)
CDAR	(B)
CADAR	B
CDDAR	NIL
CAAR	A
CDADDR	(F)
CADADDR	F

2.16. CAAR takes the CAR of the CAR.   The CAR of (FRED NIL) is FRED, and the CAR of that causes an error.

2.17.

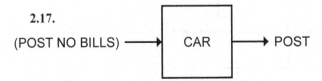

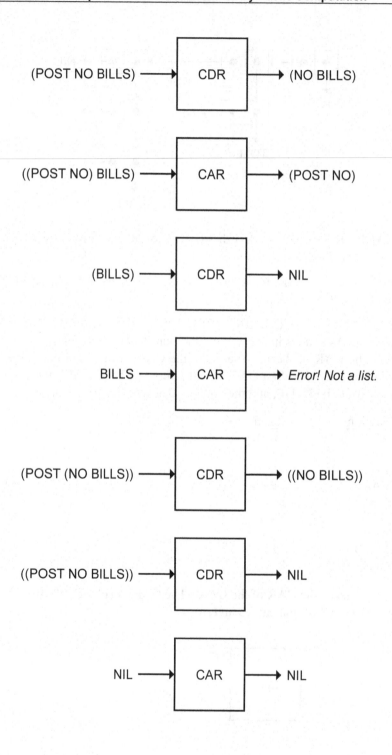

2.18.

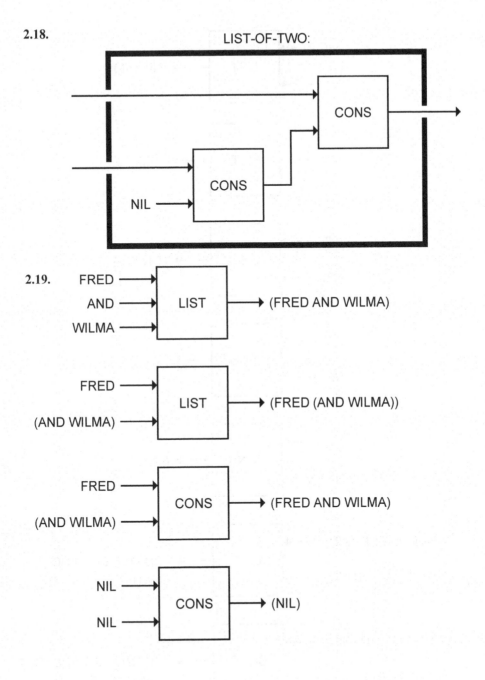

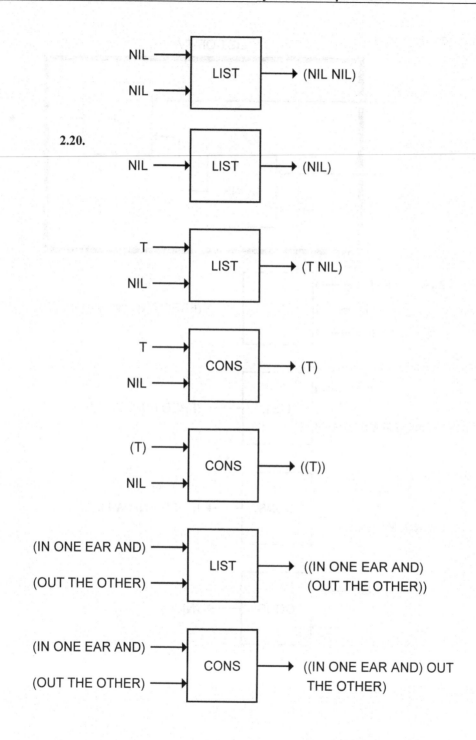

2.20.

2.21.

PAIR-OF-PAIRS:

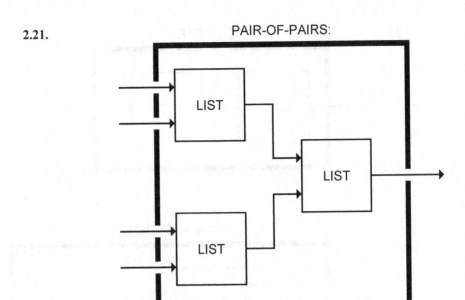

2.22.

DUO-CONS:

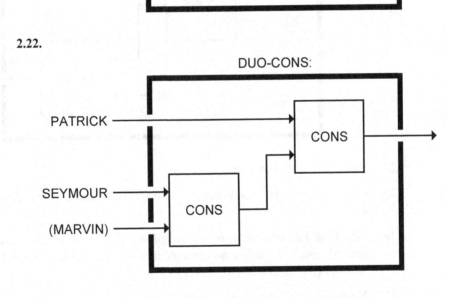

2.23.

TWO-DEEPER:

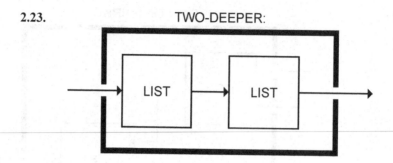

TWO-DEEPER:

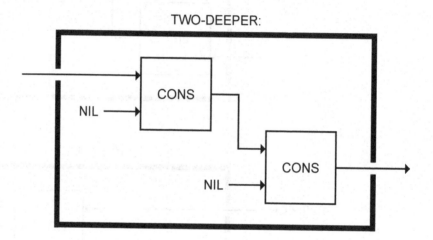

2.24. The CAAADR function.

2.25. CONS stands for ''construct.'' It constructs and returns a new cons cell.

2.26. The first function returns the length of the CDR of its input. The second function causes an error because it tries to take the CDR of a number (the output of LENGTH).

2.27. Nested lists require more cons cells than the list has top-level elements. Flat lists always have exactly as many cons cells as elements.

2.28. It's not possible to write a function to extract the last element of a list of unknown length using just CAR and CDR, because we don't know how many CDRs to use. The function needs to keep taking successive CDRs until it reaches a cell whose CDR is NIL; then it should return the CAR of that cell. We'll learn how to do this in Chapter 8.

2.29.

UNARY-ADD1:

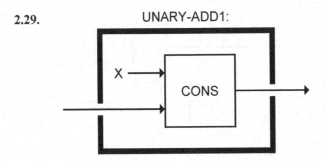

2.30. CDDR subtracts two from a unary number.

2.31. NULL is the unary ZEROP predicate.

2.32.

UNARY-GREATERP:

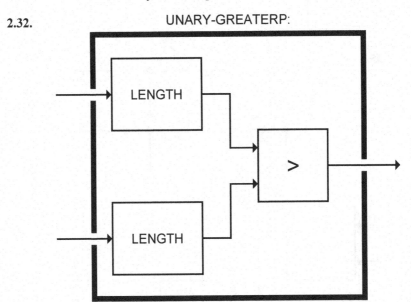

2.33. CAR returns a true value for any unary number greater than zero, so it is the unary equivalent of PLUSP.

2.34.

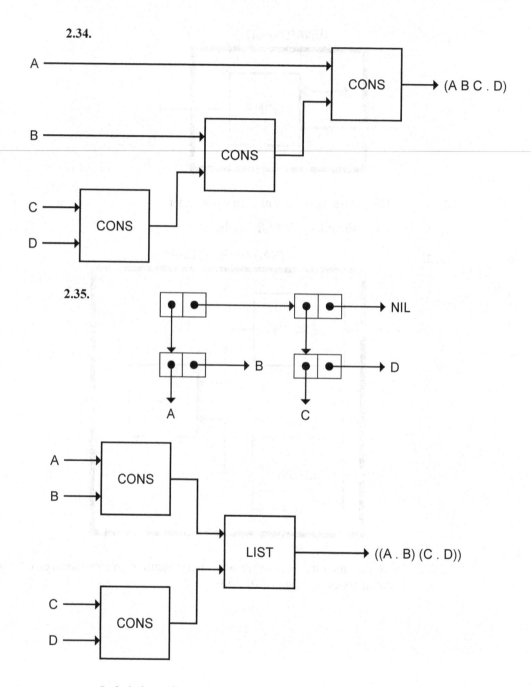

2.35.

2.36. Label the cells *a*, *b*, and *c*. Since cell *a* points to cell *b*, it must have been consed after cell *b*, because *b* would have had to be one of the inputs to CONS when cell *a* was created. By similar reasoning, cell *b* must have been consed after cell *c*. Therefore, cell *a* must have been consed after cell *c*. But cell *c* points to cell *a*, so *a* would have to have been consed before *c*, not after it. This contradiction proves that the list could not have been constructed using just CONS.

CHAPTER 3 ANSWERS

3.1.   `(not (equal 3 (abs −3)))` ⟹ `nil`

3.2.   `(/ (+ 8 12) 2)`

3.3.   `(+ (* 3 3) (* 4 4))`

3.4.

> (- 8 2)
>   8 evaluates to 8
>   2 evaluates to 2
> Enter - with inputs 8 and 2
> Result of - is 6

> (not (oddp 4))
>   > (oddp 4)
>   4 evaluates to 4
>   Enter ODDP with input 4
>   ODDP returns NIL
> Enter NOT with input NIL
> Result of NOT is T

> (> (* 2 5) 9)
>   > (* 2 5)
>   Enter * with inputs 2 and 5
>   Result of * is 10
> Enter > with inputs 10 and 9
> Result of > is T

```
→ (not (equal 5 (+ 1 4)))
   → (equal 5 (+ 1 4))
      → (+ 1 4)
      ▸ Enter + with inputs 1 and 4
      ▸ Result of + is 5
   ▸ Enter EQUAL with inputs 5 and 5
   ▸ Result of EQUAL is T
▸ Enter NOT with input T
▸ Result of NOT is NIL
```

3.5.   (defun half (n) (/ n 2))

(defun cube (n) (* n n n))

(defun onemorep (x y)
  (equal x (+ y 1)))

3.6.   (defun pythag (x y)
  (sqrt (+ (* x x) (* y y))))

3.7.   (defun miles-per-gallon
         (initial-odometer-reading
          final-odometer-reading
          gallons consumed)
  (/ (- final-odometer-reading
        initial-odometer-reading)
     gallons-consumed))

3.8.   SQUARE:

3.9.   (cons 5 (list 6 7))   ⇒   (5 6 7)

(cons 5 '(list 6 7))   ⇒   (5 list 6 7)

```
(list 3 'from 9 'gives (- 9 3))
  ⇒  (3 from 9 gives 6)

(+ (length '(1 foo 2 moo))
   (third '(1 foo 2 moo)))  ⇒  6

(rest '(cons is short for construct))
  ⇒  (is short for construct)
```

3.10. THIRD's argument must be quoted.
```
(third '(the quick brown fox))
```

Symbols used as data must be quoted.
```
(list 2 'and 2 'is 4)
```

No quote before LENGTH.
```
(+ 1 (length (list t t t t)))
```

Lists used as data must be quoted.
```
(cons 'patrick '(seymour marvin))
```

Symbols used as data must be quoted.
```
(cons 'patrick (list 'seymour 'marvin))
```

3.11.
```
(defun longer-than (x y)
  (> (length x) (length y)))
```

3.12.
```
(defun addlength (x)
  (cons (length x) x))

(addlength (addlength '(a b c)))  ⇒  (4 3 a b c)
```

3.13. The CALL-UP function requires two arguments, named CALLER and CALLEE. (CALL-UP 'FRED 'WANDA) returns (HELLO WANDA THIS IS FRED CALLING).

3.14. The CRANK-CALL function makes no use of its inputs, because its entire body is quoted. It returns (HELLO CALLEE THIS IS CALLER CALLING).

3.15. The symbol WORD is used both as a piece of data, when quoted, and as a variable name, when not quoted. (SCRABBLE 'AARDVARK) returns (AARDVARK IS A WORD). (SCRABBLE 'WORD) returns (WORD IS A WORD).

3.16. The result is (MOE (MOE LARRY) LARRY LARRY).

3.17. T and NIL are the names of constants that always evaluate to themselves. Therefore they can't be used to name variables that hold the inputs to a function.

3.18. EVAL notation is concise, and it is easy to type on a computer keyboard. It allows us to use the same notation for both functions and data. Some ideas that can be expressed in EVAL notation have no equivalent in box notation.

3.19. `(cons 'grapes '(of wrath))`
  ⇒ `(grapes of wrath)`

`(list t 'is 'not nil)` ⇒ `(t is not nil)`

`(first '(list moose goose))` ⇒ `list`

`(first (list 'moose 'goose))` ⇒ `moose`

`(cons 'home ('sweet 'home))`
  ⇒ *Error! 'SWEET undefined function.*

3.20. `(mystery '(dancing bear))` ⇒ `(bear dancing)`

`(mystery 'dancing 'bear)` ⇒ *Error! Too many inputs.*

`(mystery '(zowie))` ⇒ `(nil zowie)`

`(mystery (list 'first 'second))`
  ⇒ `(second first)`

3.21. Variables shouldn't be quoted:
  `(defun speak (x y) (list 'all 'x 'is 'y))`

A function can't have two argument lists:
  `(defun speak (x) (y) (list 'all x 'is y))`

Don't parenthesize variables in the argument list; don't quote variables; do quote ALL and IS:
  `(defun speak ((x) (y)) (list all 'x is 'y))`

3.22. Part a. On my computer, I type `lisp` and hit return.

Part b.
`(+ 3 5)` ⇒ `8`

```
(3 + 5)   ⇒   Error! 3 undefined function.

(+ 3 (5 6))   ⇒   Error! 5 undefined function.

(+ 3 (* 5 6))   ⇒   33

'(morning noon night)
   ⇒   (morning noon night)

('morning 'noon 'night)
   ⇒   Error! 'MORNING undefined function.

(list 'morning 'noon 'night)
   ⇒   (morning noon night)

(car nil)   ⇒   nil

(+ 3 foo)   ⇒   Error! FOO unassigned variable.

(+ 3 'foo)   ⇒   Error! Wrong type input to +.
```

Part c.
```
(defun myfun (x y)
   (list (list x) y))
```

Part d.
```
(defun firstp (x y)
   (equal x (first y)))
```

Part e.
```
(defun mid-add1 (x)
   (list (first x)
         (+ 1 (second x))
         (third x)))
```

Part f.
```
(defun f-to-c (temperature)
   (* (- temperature 32.0) 5/9))
```

Part g. The function tries to add one to the truth value output by ZEROP (either T or NIL). This causes a wrong type input error.

3.23. DOUBLE: $\lambda n. \; n \times 2$
SQUARE: $\lambda n. \; n \times n$
ONEMOREP: $\lambda (x,y). \; x = (y+1)$

3.24.

```
→ (alpha 3)
    Enter ALPHA with input 3
      create variable X, with value 3
  → (bravo (+ x 2) (charlie x 1))
      → (+ x 2)
      → 5
      → (charlie x 1)
      → Enter CHARLIE with inputs 3 and 1
          create variable Y, with value 3
          create variable X, with value 1
        → (- y x)
        → 2
      → Result of CHARLIE is 2
    → Enter BRAVO with inputs 5 and 2
        create variable Y, with value 5
        create variable Z, with value 2
      → (* y z)
      → 10
    → Result of BRAVO is 10
→ Result of ALPHA is 10
```

3.25. (list 'cons t nil) $\Rightarrow$ (cons t nil)

(eval (list 'cons t nil)) $\Rightarrow$ (t)

(eval (eval (list 'cons t nil)))
   $\Rightarrow$ *Error! T undefined function.*

(apply #'cons '(t nil)) $\Rightarrow$ (t)

(eval nil) $\Rightarrow$ nil

(list 'eval nil) $\Rightarrow$ (eval nil)

(eval (list 'eval nil)) $\Rightarrow$ nil

CHAPTER 4 ANSWERS

4.1.
```
(defun make-even (n)
   (if (evenp n) n
       (+ n 1)))
```

4.2.
```
(defun further (n)
   (if (< n 0)
       (- n 1)
       (+ n 1)))
```

4.3.
```
(defun my-not (x)
   (if x nil t))
```

4.4.
```
(defun ordered (x y)
   (if (< x y)
       (list x y)
       (list y x)))
```

4.5. The third clause; the second clause; the first clause.

4.6.
```
(defun my-abs (n)
   (cond ((< n 0) (- n))
         (t n)))
```

4.7. The second COND expression is correct. The rest have a variety of parenthesis errors.

4.8.
```
(defun emphasize3 (x)
   (cond ((equal (first x) 'good)
          (cons 'great (rest x)))
         ((equal (first x) 'bad)
          (cons 'awful (rest x)))
         (t (cons 'very x))))

(emphasize3 '(very long day))
  ⇒   (very very long day)
```

4.9. The function always returns its input unchanged, because the first COND clause is always true. The second clause is never tried. To fix the problem, swap the two clauses.

4.10.
```
(defun constrain (x max min)
   (cond ((< x min) min)
         ((> x max) max)
         (t x)))
```

```
(defun constrain (x max min)
  (if (< x min) min
      (if (> x max) max
          x)))
```

4.11.
```
(defun firstzero (x)
  (cond ((zerop (first x)) 'first)
        ((zerop (second x)) 'second)
        ((zerop (third x)) 'third)
        (t 'none)))
```

Calling FIRSTZERO with three separate numbers as input, instead of one list of three numbers, would cause a wrong number of inputs error.

4.12.
```
(defun cycle (n)
  (cond ((equal n 99) 1)
        (t (+ n 1))))
```

4.13.
```
(defun howcompute (x y z)
  (cond ((equal (+ x y) z) 'sum-of)
        ((equal (* x y) z) 'product-of)
        (t '(beats me))))
```

4.14.
```
(and 'fee 'fie 'foe)  ⇒  foe

(or 'fee 'fie 'foe)  ⇒  fee

(or nil 'foe nil)  ⇒  foe

(and 'fee 'fie nil)  ⇒  nil

(and (equal 'abc 'abc) 'yes)  ⇒  yes

(or (equal 'abc 'abc) 'yes)  ⇒  t
```

4.15. There is a built-in predicate called >= that has this behavior, but we can build our own predicate using OR.

```
(defun geq (x y)
  (or (> x y) (equal x y)))
```

4.16.
```
(defun crunch (n)
  (cond ((and (oddp n) (> n 0)) (* n n))
        ((and (oddp n) (< n 0)) (* n 2))
        (t (/ n 2))))
```

4.17.
```
(defun age (x y)
   (or (and (or (equal x 'boy)
                (equal x 'girl))
            (equal y 'child))
       (and (or (equal x 'man)
                (equal x 'woman))
            (equal y 'adult))))
```

4.18.
```
(defun play (x y)
   (cond ((equal x y) 'tie)
         ((or (and (equal x 'rock)
                   (equal y 'scissors))
              (and (equal x 'scissors)
                   (equal y 'paper))
              (and (equal x 'paper)
                   (eqyal y 'rock)))
          'first-wins)
         (t 'second-wins)))
```

4.19.
```
(cond ((not x) nil)
      ((not y) nil)
      ((not z) nil)
      (t w))
```
```
(if x
   (if y
      (if z w)))
```

4.20.
```
(defun compare (x y)
   (if (equal x y)
       'numbers-are-the-same
       (if (< x y)
           'first-is-smaller
           'first-is-bigger)))
```
```
(defun compare (x y)
   (or (and (equal x y) 'numbers-are-the-same)
       (and (< x y) 'first-is-smaller)
       'first-is-bigger))
```

4.21.
```
(defun gtest (x y)
   (if (> x y) t
       (if (zerop x) t
           (zerop y))))
```

```
(defun gtest (x y)
  (cond ((> x y) t)
        ((zerop x) t)
        (t (zerop y))))
```

4.22.
```
(defun boilingp (temp scale)
  (or (and (> temp 212)
           (equal scale 'fahrenheit))
      (and (> temp 100)
           (equal scale 'celsius))))
```

4.23.   If WHERE-IS had eight COND clauses, WHERE-IS-2 would need 7 IFs. WHERE-IS-3 would need one OR and seven ANDs.

4.24.   Conditionals are important because they allow a function to vary its behavior in response to different input conditions.

4.25.   When IF is given only two inputs, it uses NIL for its third input.

4.26.   A COND with any number of clauses can be rewritten to use IF because we can write nested IFs.

4.27.   (COND) evaluates to NIL, since it has no true clauses.

4.28.   Once IF determines that *test* is true, it always returns the value of *true-part*, even if that value is NIL. The translation using AND and OR does not have this property: since (EVENP 7) is NIL, the AND returns NIL, so the OR goes on to the next clause, which is 'FOO. We can correct this by writing:
```
(or (and test true-part)
    (and (not test) false-part))
```

4.29.
```
(defun logical-and (x y)
  (if x (if y t)))
```
```
(defun logical-and (x y)
  (cond (x (cond (y t)))))
```

4.30.
```
(defun logical-or (x y)
  (cond (x t)
        (y t)
        (t nil)))
```

4.31.   NOT is not a conditional: It always evaluates its input. NOT *is* a boolean function because it returns T or NIL, so we do not need to write a LOGICAL-NOT function.

4.32.

x	y	(LOGICAL-OR x y)
T	T	T
T	NIL	T
NIL	T	T
NIL	NIL	NIL

4.33.   There would be 2^3 or eight lines in the truth table.

4.34.

x	y	z	(LOGICAL-IF x y z)
T	T	T	T
T	T	NIL	T
T	NIL	T	NIL
T	NIL	NIL	NIL
NIL	T	T	T
NIL	T	NIL	NIL
NIL	NIL	T	T
NIL	NIL	NIL	NIL

4.35.   `(and x y z) = (not (or (not x) (not y) (not z)))`

`(or x y z) = (not (and (not x) (not y) (not z)))`

4.36.

x	y	(NAND x y)
T	T	NIL
T	NIL	T
NIL	T	T
NIL	NIL	T

4.37.
```
(defun logical-and (x y)
   (nand (nand x y) (nand x y)))

(defun logical-or (x y)
   (nand (nand x x) (nand y y)))
```

4.38.
```
(defun not2 (x)
   (nor x x))

(defun logical-and (x y)
   (nor (nor x x) (nor y y)))
```

```
(defun logical-or (x y)
  (nor (nor x y) (nor x y)))

(defun nand (x y)
  (nor (nor (nor x x) (nor y y))
       (nor (nor x x) (nor y y))))
```

4.39. LOGICAL-AND is not logically complete: there is no way to construct the NOT function from combinations of LOGICAL-ANDs. Therefore we also can't construct OR, NAND, and NOR.

CHAPTER 5 ANSWERS

5.1.
```
(defun good-style (p)
  (let ((q (+ p 5)))
    (list 'result 'is q)))
```

5.2. A side effect is something a function does besides returning a value. Assignment is an example of a side effect.

5.3. A local variable is only accessible within the body of the form that defines it, such as DEFUN, LET, or LET*. A global variable is defined at top-level, not inside one of these forms, so it is accessible everywhere.

5.4. SETF cannot be an ordinary function because it does not evaluate its first argument.

5.5. Yes. The difference between LET and LET* is only apparent when they are used to create more than one local variable.

5.6.
```
(defun throw-die ()
  (+ 1 (random 6)))

(defun throw-dice ()
  (list (throw-die) (throw-die)))

(defun snake-eyes-p (throw)
  (equal throw '(1 1)))

(defun boxcars-p (throw)
  (equal throw '(6 6)))
```

```
(defun throw-value (throw)
   (+ (first throw) (second throw)))
```
THROW-VALUE is a helping function used by several of the functions
that follow.

```
(defun instant-win-p (throw)
  (member (throw-value throw) '(7 11)))

(defun instant-loss-p (throw)
  (member (throw-value throw) '(2 3 12)))

(defun say-throw (throw)
   (cond ((snake-eyes-p throw) 'snake-eyes)
         ((boxcars-p throw) 'boxcars)
         (t (throw-value throw))))

(defun craps ()
   (let ((throw (throw-dice)))
     (append
       (list 'throw (first throw)
             'and (second throw)
             '--
             (say-throw throw)
             '--)
       (cond ((instant-win-p throw) '(you win))
             ((instant-loss-p throw) '(you lose))
             (t (list 'your 'point 'is
                      (throw-value throw)))))))

(defun try-for-point (point)
   (let* ((throw (throw-dice))
          (val (throw-value throw)))
     (append
       (list 'throw (first throw)
             'and (second throw)
             '--
             (say-throw throw)
             '--)
       (cond ((equal val point) '(you win))
             ((equal val 7) '(you lose))
             (t '(throw again))))))
```

CHAPTER 6 ANSWERS

6.1.   (NTH 4 '(A B C)) involves four successive CDRs of a three-element list. The fourth CDR produces a NIL result, and the CAR of that is NIL.

6.2.   (NTH 3 '(A B C . D)) produces an error. It takes three successive CDRs of its input, which yields the symbol D. Taking the CAR of D then causes an error because D is not a list.

6.3.   (LAST '(ROSEBUD)) returns (ROSEBUD).

6.4.   (LAST '((A B C))) returns ((A B C)). This is a list of one element, so the last cell in the top-level chain is the first cell.

6.5.
```
(setf line '(roses are red))

(reverse line)   ⇒   (red are roses)

(first (last line))   ⇒   red

(nth 1 line)   ⇒   are

(reverse (reverse line))   ⇒   (roses are red)

(append line (list (first line)))
   ⇒   (roses are red roses)

(append (last line) line)
   ⇒   (red roses are red)

(list (first line) (last line))
   ⇒   (roses (red))

(cons (last line) line)
   ⇒   ((red) roses are red)

(remove 'are line)   ⇒   (roses red)

(append line '(violets are blue))
   ⇒   (roses are red violets are blue)
```

6.6.
```
(defun last-element (x) (first (last x)))

(defun last-element (x) (first (reverse x)))
```

```
(defun last-element (x)
  (and x   ; to handle NIL correctly
       (nth (- (length x) 1) x)))
```

6.7.
```
(defun next-to-last (x) (second (reverse x)))
```

```
(defun next-to-last (x)
  (and (rest x) ; to handle short lists
       (nth (- (length x) 2) x)))
```

6.8.
```
(defun my-butlast (x)
  (reverse (rest (reverse x))))
```

6.9. MYSTERY is the same as FIRST.

6.10.
```
(defun palindromep (x)
  (equal x (reverse x)))
```

6.11.
```
(defun make-palindromep (x)
  (append x (reverse x)))
```

6.12. MEMBER never has to copy its input. It simply returns a pointer to one of the cons cells that make up its input, or NIL.

6.13. The result of intersecting a set with NIL is NIL.

6.14. The result of intersecting a set with itself is the set.

6.15.
```
(defun contains-article-p (sent)
  (intersection sent '(the a an)))
```

```
(defun contains-article-p (sent)
  (or (member 'the sent)
      (member 'a sent)
      (member 'an sent)))
```

```
(defun contains-article-p (sent)
  (not (and (not (member 'the sent))
            (not (member 'a sent))
            (not (member 'an sent)))))
```

6.16. The union of a set with NIL is the set.

6.17. The union of a set with itself is the set.

6.18.
```
(defun add-vowels (x)
  (union x '(a e i o u)))
```

6.19. If NIL is the first input to SET-DIFFERENCE, the result is NIL. If NIL is the second input, the result is the first input.

6.20. SET-DIFFERENCE copies (parts of) its first input. It never has to copy its second input because none of the elements of the second input appear in its result.

6.21.
```
(defun my-subsetp (x y)
  (null (set-difference x y)))
```

6.22.
```
(union a '(no soap radio))
  ⇒ (soap water no radio)

(intersection a (reverse a))   ⇒   (soap water)

(set-difference a '(stop for water))   ⇒   (soap)

(set-difference a a)   ⇒   nil

(member 'soap a)   ⇒   (soap water)

(member 'water a)   ⇒   (water)

(member 'washcloth a)   ⇒   nil
```

6.23. LENGTH returns the cardinality of a set.

6.24.
```
(defun set-equal (x y)
  (and (subsetp x y)
       (subsetp y x)))
```

6.25.
```
(defun proper-subsetp (x y)
  (and (subsetp x y)
       (not (subsetp y x))))
```

6.26.
```
(defun right-side (x)
  (rest (member '-vs- x)))

(defun left-side (x)
  (right-side (reverse x)))
```

```
(defun count-common (x)
   (length (intersection (left-side x)
                         (right-side x))))

(defun compare (x)
   (list (count-common x) 'common 'features))
```

6.27. ASSOC may be considered a predicate on the same grounds as MEMBER. ASSOC returns a true value if a given input appears in a table.

6.28.
```
(assoc 'banana produce)   ⇒   (banana . fruit)

(rassoc 'fruit produce)   ⇒   (apple . fruit)

(assoc 'lettuce produce)  ⇒   (lettuce . veggie)

(rassoc 'veggie produce)  ⇒   (celery . veggie)
```

6.29. LENGTH returns the number of entries in a table.

6.30.
```
(setf books
   '((war-and-peace leo-tolstoy)
     (oliver-twist  charles-dickens)
     (tom-sawyer    mark-twain)
     (kidnapped     robert-louis-stevenson)
     (candide       voltaire)))
```

6.31.
```
(defun who-wrote (title)
   (second (assoc title books)))
```

6.32. The WHO-WROTE function will behave exactly the same, because the order of entries in a table is unimportant when the keys (in this case the book titles) are unique.

6.33. We can't create WHAT-WROTE using the current table. However, if we rewrote the table to use dotted pairs, we could create WHAT-WROTE by using RASSOC.

6.34.
```
(setf redesigned-atlas
   '((pennsylvania (pittsburgh johnstown))
     (new-jersey (newark princeton trenton))
     (ohio (columbus))))
```

6.35. This problem can be solved using either a flat list (with MEMBER) or a table (with ASSOC).

```
(setf nerd-states
  '((sleeping    . eating)
    (eating      . waiting)
    (waiting     . programming)
    (programming . debugging)
    (debugging   . sleeping)))

(defun nerdus (x)
  (cdr (assoc x nerd-states)))

(nerdus 'playing-guitar)  ⇒  nil

(defun sleepless-nerd (x)
  (let ((y (nerdus x)))
    (if (equal y 'sleeping)
        (nerdus y)
        y)))

(defun nerd-on-caffeine (x)
  (nerdus (nerdus x)))
```

Starting in state PROGRAMMING, the nerd would go to state SLEEPING, then to WAITING, and then to DEBUGGING.

6.36.
```
(defun swap-first-last (x)
  (let* ((a (reverse (rest x)))
         (b (reverse (rest a))))
    (cons (first a)
          (append b (list (first x))))))
```

6.37.
```
(defun rotate-left (x)
  (append (rest x) (list (first x))))

(defun rotate-right (x)
  (let ((r (reverse x)))
    (cons (first r)
          (reverse (rest r)))))
```

6.38. Equal results: X and Y can be any two identical sets, including NIL. The order of elements need not be the same in the two sets. Unequal results: X and Y must not be equal sets, for example, X could be (A) and Y could be (A B).

6.39.   APPEND performs unary addition.

6.40.
```
((a b c d)
 (b c d)
 (c d)
 (d))
```

6.41.
```
(defun choices (room)
  (rest (assoc room rooms)))

(defun look (dir room)
  (second (assoc dir (choices room))))

(setf loc 'pantry)

(defun how-many-choices ()
  (length (choices loc)))

(defun upstairsp (x)
  (or (equal x 'library)
      (equal x 'upstairs-bedroom)))

(defun onstairsp (x)
  (or (equal x 'back-stairs)
      (equal x 'front-stairs)))

(defun where ()
  (if (onstairsp loc)
      (list 'robbie 'is 'on 'the loc)
      (list 'robbie 'is
            (if (upstairsp loc)
                'upstairs
                'downstairs)
            'in 'the loc)))

(defun move (dir)
  (let ((new-loc (look dir loc)))
    (cond ((null new-loc)
           '(ouch! robbie hit a wall))
          (t (set-robbie-location new-loc)
             (where)))))
```

6.42.   `(defun royal-we (sent)`
`  (subst 'we 'i sent))`

CHAPTER 7 ANSWERS

7.1.   `(defun add1 (n) (+ 1 n))`

`(mapcar #'add1 '(1 3 5 7 9))  ⇒  (2 4 6 8 10)`

7.2.   `(mapcar #'third daily-planet)  ⇒`
`  (123-76-4535 089-52-6787 951-26-1438`
`  355-16-7439)`

7.3.   `(mapcar #'zerop '(2 0 3 4 0 -5 -6))`
`  ⇒  (nil t nil nil t nil nil)`

7.4.   `(defun greater-than-five-p (n) (> n 5))`

`(mapcar #'greater-than-five-p`
`        '(2 0 3 4 0 -5 -6))`

7.5.   `(lambda (n) (- n 7))`

7.6.   `(lambda (x)`
`  (or (null x) (equal x t)))`

7.7.   `(defun flip-element (list)`
`  (mapcar #'(lambda (e)`
`             (if (equal e 'up) 'down 'up))`
`          list))`

7.8.   `(defun roughly-equal (e k)`
`  (and (not (< e (- k 10)))`
`       (not (> e (+ k 10)))))`

`(defun find-first-roughly-equal (x k)`
`  (find-if #'(lambda (e) (roughly-equal e k))`
`           x))`

7.9.   `(defun find-nested (x)`
`  (find-if #'consp x))`

7.10.
```
(setf note-table
  '((c         . 1)   (f-sharp .  7)
    (c-sharp . 2)     (g       .  8)
    (d         . 3)   (g-sharp .  9)
    (d-sharp . 4)     (a       . 10)
    (e         . 5)   (a-sharp . 11)
    (f         . 6)   (b       . 12)))

(defun numbers (x)
  (mapcar #'(lambda (e)
              (cdr (assoc e note-table)))
          x))

(defun notes (x)
  (mapcar #'(lambda (e)
              (car (rassoc e note-table)))
          x))
```

(NOTES (NOTES X)) and (NUMBERS (NUMBERS X)) both return a list of NILs the same length as the input list.

```
(defun raise (n x)
  (mapcar #'(lambda (e) (+ e n))
          x))

(defun normalize (x)
  (mapcar #'(lambda (e)
              (cond ((< e 1) (+ e 12))
                    ((> e 12) (- e 12))
                    (t e)))
          x))

(defun transpose (n x)
  (notes (normalize (raise n (numbers x)))))
```

7.11.
```
(defun pick (x)
  (remove-if-not #'(lambda (x) (< 1 x 5))
                 x))
```

7.12.
```
(defun count-the (sent)
  (length (remove-if-not
            #'(lambda (x) (equal x 'the))
            sent)))
```
A shorter solution is possible using COUNT, which is not covered in this book.

7.13.
```
(defun pick-pairs (x)
   (remove-if
      #'(lambda (x)
           (not (equal (length x) 2)))
      x))
```

7.14.
```
(defun my-intersection (x y)
   (remove-if-not
      #'(lambda (e)
           (member e y))
      x))
```

```
(defun my-union (x y)
   (append x
           (remove-if
              #'(lambda (e)
                   (member e x))
              y)))
```

7.15.
```
(defun rank (card) (first card))
```

```
(defun suit (card) (second card))
```

```
(defun count-suit (s hand)
   (length (remove-if-not
              #'(lambda (card)
                   (equal (suit card) s))
              hand)))
```
A shorter solution is possible using COUNT-IF, which is not covered in this book.

```
(defun color-of (cards)
   (second (assoc (suit cards) colors)))
```

```
(defun first-red (hand)
   (find-if #'(lambda (card)
                 (equal (color-of card) 'red))
            hand))
```

```
(defun black-cards (hand)
   (remove-if-not
      #'(lambda (card)
           (equal (color-of card) 'black))
      hand))
```

```
(defun what-ranks (s hand)
  (mapcar #'rank
    (remove-if-not
      #'(lambda (card)
          (equal (suit card) s))
      hand)))

(defun higher-rank-p (card1 card2)
  (beforep (rank card2)
           (rank card1)
           all-ranks))

(defun high-card (hand)   ;FIND-IF version
  (assoc (find-if
           #'(lambda (r)
               (assoc r hand))
           (reverse all-ranks))
         hand))

(defun high-card (hand)   ;REDUCE version
  (reduce
    #'(lambda (card1 card2)
        (if (higher-rank-p card1 card2)
            card1
            card2))
    hand))
```

7.16. UNION.

7.17.
```
(defun total-length (x)   ;conses a lot
  (length (reduce #'append x)))

(defun total-length (x)   ;conses less
  (reduce #'+ (mapcar #'length x)))
```

7.18. Suppose x and y are lists. (REDUCE #'+ (APPEND x y)) should produce the same value as the sum of (REDUCE #' + x) and (REDUCE #' + y). If y is NIL, then (APPEND x y) equals x, so (REDUCE #' + y) has to return zero. Zero is the **identity value** for addition. That's why calling + with no arguments returns zero. Similarly, calling * with no arguments returns one because one is the multiplicative identity.

7.19.
```
(defun all-odd (x)
   (every #'oddp x))
```

7.20.
```
(defun none-odd (x)
   (every #'evenp x))
```

7.21.
```
(defun not-all-odd (x)
   (find-if #'evenp x))
```

7.22.
```
(defun not-none-odd (x)
   (find-if #'oddp x))
```

7.23. All four functions are distinct. NOT-ALL-ODD should be called FIND-EVEN, and NOT-NONE-ODD should be called FIND-ODD.

7.24. An applicative operator is a function that takes another function as input, and applies it to some data.

7.25. Lambda expressions allow us to define nameless functions that can be passed to applicative operators. We can also define functions separately with DEFUN, but in that case they would not be able to refer to any of the local variables of the parent function, the way the lambda expression in MY-ASSOC refers to the local variable KEY.

7.26.
```
(defun my-find-if (pred x)
   (first (remove-if-not pred x)))
```

7.27.
```
(defun my-every (pred x)
   (null (remove-if pred x)))
```

7.28. The triangle shape below indicates a truth value (T or NIL).

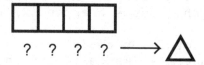

7.29.
```
(defun match-element (e q)
   (or (equal e q)
       (equal q '?)))

(defun match-triple (x pat)
   (every #'match-element
          x
          pat))
```

```
(defun fetch (pat)
  (remove-if-not
    #'(lambda (x) (match-triple x pat))
    database))

(fetch '(b4 shape ?))
(fetch '(? shape brick))
(fetch '(b2 ? b3))
(fetch '(? color ?))
(fetch '(b4 ? ?))

(defun color-pattern (block)
  (list block 'color '?))

(defun supporters (block)
  (mapcar #'first
          (fetch (list '? 'supports block))))

(defun supp-cube (block)
  (member 'cube
    (mapcar
      #'(lambda (b) (third (first (fetch
                          (list b 'shape '?)))))
      (supporters block))))

(defun desc1 (block)
  (fetch (list block '? '?)))

(defun desc2 (block)
  (mapcar #'rest (desc1 block)))

(defun description (block)
  (reduce #'append (desc2 block)))

(description 'b1)  ⇒
  (shape brick color green size small
   supported-by b2 supported-by b3)

(description 'b4)  ⇒
  (shape pyramid color blue size large
   supported-by b5)
```

Add to the database the lists (B1 COMPOSITION WOOD) and (B2 COMPOSITION PLASTIC).

7.30.
```
(mapcar #'(lambda (x y) (append x (list y)))
         words
         '(uno dos tres quatro cinco))
```

CHAPTER 8 ANSWERS

8.1. The second COND clause is never true, since none of the numbers is odd.

8.2.
```
(defun anyoddp (x)
  (if x
      (if (oddp (first x))
          t
          (anyoddp (rest x)))))
```

8.3. In (FACT 20.0) the result is computed using floating point arithmetic, which has limited precision. (FACT 20) uses integers called **bignums** instead. Bignums have unlimited precision. (FACT 0) and (FACT 0.0) both satisfy the first COND clause, which always returns the integer 0.

8.4.
```
(defun laugh (n)
  (cond ((zerop n) nil)
        (t (cons 'ha (laugh (- n 1))))))
```

8.5.
```
(defun add-up (x)
  (cond ((null x) 0)
        (t (+ (first x) (add-up (rest x))))))
```

8.6.
```
(defun all-oddp (x)
  (cond ((null x) t)
        ((evenp (first x)) nil)
        (t (all-oddp (rest x)))))
```

8.7.
```
(defun rec-member (e x)
  (cond ((null x) nil)
        ((equal e (first x)) x)
        (t (rec-member e (rest x)))))
```

8.8.
```
(defun rec-assoc (key table)
  (cond ((null table) nil)
        ((equal key (car (first table)))
         (first table))
        (t (rec-assoc key (rest table)))))
```

8.9.
```
(defun rec-nth (n x)
   (cond ((zerop n) (first x))
         (t (rec-nth (- n 1) (rest x)))))
```

8.10.
```
(defun add1 (n) (+ n 1))

(defun sub1 (n) (- n 1))

(defun rec-plus (x y)
   (cond ((zerop y) x)
         (t (rec-plus (add1 x) (sub1 y)))))
```

8.11.
```
(defun fib (n)
   (cond ((equal n 0) 1)
         ((equal n 1) 1)
         (t (+ (fib (- n 1))
               (fib (- n 2))))))
```

8.12. ANY-7-P doesn't know to stop when its input is NIL. It will work correctly as long as its input contains at least one seven; in that case it stops and returns T. Otherwise it will recurse infinitely.

8.13. Calling FACT with a negative number causes an infinite recursion.

8.14.
```
(defun infinite-recursion ()
   (infinite-recursion))
```

8.15. The car of the list is the symbol X, and the cdr is the list itself. COUNT-SLICES will recurse infinitely when given this list as input, since it can never reach the ''end'' of the cons cell chain.

8.16. Switching the first and second COND clauses would cause an error: ODDP would signal ''wrong type input'' when X is NIL.

8.17.
```
(defun find-first-odd (x)
   (cond ((null x) nil)
         ((oddp (first x)) (first x))
         (t (find-first-odd (rest x)))))
```

8.18.
```
(defun last-element (x)
   (cond ((atom (cdr x)) (car x))
         (t (last-element (cdr x)))))
```

8.19. ANYODDP will work correctly as long as there is at least one odd number in the list. If there are no odd numbers, it will get an error when it tries to compute ODDP of NIL.

8.20. FACT uses single-test augmenting recursion. The template values are:

End-test:	(ZEROP N)
End-value:	1
Aug-fun:	*
Aug-val:	N
Reduced-x:	(- N 1)

8.21.
```
(defun add-nums (n)
  (cond ((zerop n) 0)
        (t (+ n (add-nums (- n 1))))))
```

8.22.
```
(defun all-equal (x)
  (cond ((null (rest x)) t)
        ((not (equal (first x) (second x))) nil)
        (t (all-equal (rest x)))))
```
This problem does not require augmentation, since the return value is always just T or NIL. It is solved with double-test tail recursion.

8.23. The table has six entries, one for each invocation of LAUGH in (LAUGH 5).

N	First Input to CONS	Second Input to CONS	Result
5	5	(LAUGH 4)	(HA HA HA HA HA)
4	4	(LAUGH 3)	(HA HA HA HA)
3	3	(LAUGH 2)	(HA HA HA)
2	2	(LAUGH 1)	(HA HA)
1	1	(LAUGH 0)	(HA)
0			NIL

8.24.
```
(defun count-down (n)
  (cond ((zerop n) nil)
        (t (cons n (count-down (- n 1))))))
```

8.25.
```
(defun applic-fact (n)
  (reduce #'* (count-down n)))
```

8.26.
```
(defun count-down (n)
   (cond ((equal n -1) nil)
         (t (cons n (count-down (- n 1))))))

(defun count-down (n)
   (cond ((zerop n) (list 0))
         (t (cons n (count-down (- n 1))))))
```

8.27.
```
(defun square-list (x)
   (cond ((null x) nil)
         (t (cons (* (first x) (first x))
                  (square-list (rest x))))))
```

8.28.
```
(defun my-nth (n x)
   (cond ((null x) nil)   ;stop if list is empty
         ((zerop n) (first x))
         (t (my-nth (- n 1) (rest x)))))
```

8.29.
```
(defun my-member (e x)
   (cond ((null x) nil)
         ((equal e (first x)) x)
         (t (my-member e (rest x)))))
```

8.30.
```
(defun my-assoc (key table)
   (cond ((null table) nil)
         ((equal key (car (first table)))
          (first table))
         (t (my-assoc key (rest table)))))
```

8.31.
```
(defun compare-lengths (x y)
   (cond ((and (null x) (null y)) 'same-length)
         ((null x) 'second-is-longer)
         ((null y) 'first-is-longer)
         (t (compare-lengths (rest x)
                             (rest y)))))
```

8.32.
```
(defun sum-numeric-elements (x)
   (cond ((null x) 0)
         ((numberp (first x))
          (+ (first x)
             (sum-numeric-elements (rest x))))
         (t (sum-numeric-elements (rest x)))))
```

8.33.
```
(defun my-remove (e x)
   (cond ((null x) nil)
         ((equal e (first x))
          (my-remove e (rest x)))
         (t (cons e (my-remove e (rest x)))))))
```

8.34.
```
(defun my-intersection (x y)
   (cond ((null x) nil)
         ((member (first x) y)
          (cons (first x)
                (my-intersection (rest x) y)))
         (t (my-intersection (rest x) y))))
```

8.35.
```
(defun my-set-difference (x y)
   (cond ((null x) nil)
         ((not (member (first x) y))
          (cons (first x)
                (my-set-difference (rest x) y)))
         (t (my-set-difference (rest x) y))))
```

8.36.
```
(defun count-odd (x)
   ;; conditional augmentation version
   (cond ((null x) nil)
         ((oddp (first x))
          (+ 1 (count-odd (rest x))))
         (t (count-odd (rest x)))))

(defun count-odd (x)
   ;; regular augmenting version
   (cond ((null x) nil)
         (t (+ (if (oddp (first x))
                   1
                   0)
               (count-odd (rest x))))))
```

8.37.
```
(defun combine (x y) (+ x y))

(defun fib (n)
   (cond ((equal n 0) 1)
         ((equal n 1) 1)
         (t (combine (fib (- n 1))
                     (fib (- n 2))))))
```

Every nonterminal call to FIB makes one call to COMBINE, and every call to COMBINE combines the results of two more calls to FIB. Since terminal calls to FIB always return one, we can prove that the total number of calls to COMBINE is equal to Fib(N) − 1. The proof is based on the realization that every binary tree with k terminal nodes has exactly $k − 1$ nonterminal nodes.

8.38. If the first COND clause is omitted, the NILs at the end of cons cell chains will also be converted to Qs. So (ATOMS-TO-Q '(A (B) C)) will return (A (B . Q) C . Q).

8.39.
```
(defun count-atoms (tree)
   (cond ((atom tree) 1)
         (t (+ (count-atoms (car tree))
               (count-atoms (cdr tree))))))
```

8.40.
```
(defun count-cons (tree)
   (cond ((atom tree) 0)
         (t (+ 1
               (count-cons (car tree))
               (count-cons (cdr tree))))))
```

8.41.
```
(defun sum-tree (tree)
   (cond ((numberp tree) tree)
         ((atom tree) 0)
         (t (+ (sum-tree (car tree))
               (sum-tree (cdr tree))))))
```

8.42.
```
(defun my-subst (new old tree)
   (cond ((equal tree old) new)
         ((atom tree) tree)
         (t (cons (my-subst
                     new old (car tree))
                  (my-subst
                     new old (cdr tree))))))
```

8.43.
```
(defun flatten (x)
   (cond ((atom x) (list x))
         (t (append (flatten (car x))
                    (and (cdr x) (flatten (cdr
x)))))))
```

8.44.
```
(defun tree-depth (tree)
   (cond ((atom tree) 0)
         (t (+ 1 (max (tree-depth
                         (car tree))
                      (tree-depth
```

```
                                       (cdr tree)))))))
```

8.45. (defun paren-depth (list)
       (cond ((atom list) 0)
             (t (max (+ 1 (paren-depth (first list)))
                     (paren-depth (rest list))))))

8.46. (defun count-up (n)
       (cond ((zerop n) nil)
             (t (append (count-up (- n 1))
                        (list n)))))

8.47. (defun make-loaf (n)
       (if (zerop n) nil
           (cons 'x (make-loaf (- n 1)))))

8.48. (defun bury (x n)
       (cond ((zerop n) x)
             (t (list (bury x (- n 1))))))
This solution uses single-test augmenting recursion, with no
augmentation value. The augmentation function is LIST.

8.49. (defun pairings (x y)
       (cond ((null x) nil)
             (t (cons (list (first x) (first y))
                      (pairings (rest x)
                                (rest y))))))

8.50. (defun sublists (x)
       (cond ((null x) nil)
             (t (cons x (sublists (rest x))))))

8.51. (defun my-reverse (x)
       (reverse-recursively x nil))

       (defun reverse-recursively (x y)
         (cond ((null x) y)
               (t (reverse-recursively
                    (rest x)
                    (cons (first x) y)))))

8.52. (defun my-union (x y)
       (append x (union-recursively x y)))
```
```

```
(defun union-recursively (x y)
  (cond ((null y) nil)
        ((member (first y) x)
         (union-recursively x (rest y)))
        (t (cons (first y)
                 (union-recursively
                   x
                   (rest y))))))
```

This solution returns all the elements of X in their original order, followed by those elements of Y (in original order) that do not appear in X.

8.53.
```
(defun largest-even (x)
  (cond ((null x) 0)
        ((oddp (first x))
         (largest-even (rest x)))
        (t (max (first x)
                (largest-even (rest x))))))
```

8.54.
```
(defun huge (x)
  (huge-helper x x))

(defun huge-helper (x n)
  (cond ((equal n 0) 1)
        (t (* x (huge-helper x (- n 1))))))
```

8.55. A recursive function calls itself, or calls another function which in turn calls it.

8.56.
```
(defun every-other (x)
  (cond ((null x) nil)
        (t (cons (first x)
                 (every-other    (rest    (rest
x)))))))
```

8.57.
```
(defun left-half (x)
  (left-half-helper
    x (/ (length x) 2)))

(defun left-half-helper (x n)
  (cond ((not (plusp n)) nil)
        (t (cons (first x)
                 (left-half-helper
                   (rest x)
```

```
                              (- n 1))))))
8.58.  (defun merge-lists (x y)
          (cond ((null x) y)
                ((null y) x)
                ((< (first x) (first y))
                 (cons (first x)
                       (merge-lists (rest x) y)))
                (t (cons (first y)
                         (merge-lists x (rest y))))))))
```

```
8.59.  (defun faulty-fact (n)
          (cond ((zerop n) 1)
                (t (/ (fact (+ n 1))
                      (+ n 1)))))
```

The equations are correct, and the function will return the correct value
for an input of zero. For inputs greater than zero it recurses infinitely,
because each recursive call generates a larger value of N. It thus violates
the third rule of recursion: The journey gets bigger with each step
instead of smaller.

```
8.60.  (defun father (x) (second (assoc x family)))

       (defun mother (x) (third (assoc x family)))

       (defun parents (x)
         (union (and (father x) (list (father x)))
                (and (mother x) (list (mother x)))))

       (defun children (parent)
         (and parent
              (mapcar #'first
                   (remove-if-not
                     #'(lambda (entry)
                         (member parent (rest entry)))
                     family))))

       (defun siblings (x)
         (set-difference (union (children (father x))
                                (children (mother x)))
                 (list x)))

       (defun mapunion (fn x)
         (and x (reduce #'union (mapcar fn x))))
```

```
(defun grandparents (x)
  (mapunion #'parents (parents x)))

(defun cousins (x)
  (mapunion #'children
    (mapunion #'siblings (parents x))))

(defun descended-from (p1 p2)
  (cond ((null p1) nil)
        ((member p2 (parents p1)) t)
        (t (or (descended-from
                 (father p1) p2)
               (descended-from
                 (mother p1) p2)))))

(defun ancestors (x)
  (cond ((null x) nil)
        (t (union
             (parents x)
             (union (ancestors (father x))
                    (ancestors (mother x)))))))

(defun generation-gap (x y)
 (g-gap-helper x y 0))

(defun g-gap-helper (x y n)
  (cond ((null x) nil)
        ((equal x y) n)
        (t (or (g-gap-helper
                 (father x) y (1+ n))
               (g-gap-helper
                 (mother x) y (1+ n))))))

(descended-from 'robert 'deirdre)  ⇒  nil

(ancestors 'yvette)  ⇒
  (quentin julie george ellen arthur kate linda
   frank bruce suzanne colin deirdre wanda
   vincent zelda robert)

(generation-gap 'olivia 'frank)  ⇒  3

(cousins 'peter)  ⇒  (joshua robert)
```

```
(grandparents 'olivia)
  ⇒  (andre hillary ellen george)
```

8.61.
```
(defun tr-count-up (n)
  (tr-count-up1 n nil))

(defun tr-count-up1 (n result)
  (cond ((zerop n) result)
        (t (tr-count-up1
             (- n 1)
             (cons n result)))))
```

8.62.
```
(defun tr-fact (n)
  (tr-fact1 n 1))

(defun tr-fact1 (n result)
  (cond ((zerop n) result)
        (t (tr-fact1 (- n 1) (* n result)))))
```

8.63.
```
(defun tr-union (x y)
  (cond ((null x) y)
        ((member (first x) y)
         (tr-union (rest x) y))
        (t (tr-union
             (rest x)
             (cons (first x) y)))))

(defun tr-intersection (x y)
  (tr-intersect1 x y nil))

(defun tr-intersect1 (x y result)
  (cond ((null x) result)
        ((member (first x) y)
         (tr-intersect1
           (rest x)
           y
           (cons (first x) result)))
        (t (tr-intersect1
             (rest x) y result))))

(defun tr-set-difference (x y)
  (tr-setdiff1 x y nil))
```

```
     (defun tr-setdiff1 (x y result)
       (cond ((null x) result)
             ((not (member (first x) y))
              (tr-setdiff1
                (rest x)
                y
                (cons (first x) result)))
             (t (tr-setdiff1
                  (rest x) y result)))))
```

8.64.
```
     (defun tree-find-if (pred tree)
       (cond ((and tree
                   (atom tree)
                   (funcall pred tree))
              tree)
             ((atom tree) nil)
             (t (or (tree-find-if
                      pred (car tree))
                    (tree-find-if
                      pred (cdr tree))))))
```

8.65.
```
     (defun tr-count-slices (x)
       (labels ((trc1 (x n)
                  (if x
                      (trc1 (rest x)
                            (+ n 1))
                      n)))
         (trc1 x 0)))
```

```
     (defun tr-reverse (x)
       (labels ((trrev1 (x r)
                  (if x
                      (trrev1
                        (rest x)
                        (cons (first x) r))
                      r)))
         (trrev1 x nil)))
```

8.66.
```
     (defun arith-eval (exp)
       (cond ((numberp exp) exp)
             (t (funcall (second exp)
                         (arith-eval (first exp))
                         (arith-eval (third exp))))))
```

8.67.
```
(defun legalp (exp)
   (cond ((numberp exp) t)
         ((atom exp) nil)
         (t (and (equal (length exp) 3)
                 (legalp (first exp))
                 (member (second exp)
                         '(+ - * /))
                 (legalp (third exp))))))
```

8.68. NIL is a proper list, and so is any cons cell whose *cdr is a proper list.*

8.69. A positive integer greater than one is either a prime, or the product of a prime and a positive integer greater than one.

8.70.
```
(defun factor-tree (n)
   (fact-tree-help n 2))

(defun fact-tree-help (n p)
   (cond ((equal n p) n)
         ((zerop (rem n p))
          (list n p (fact-tree-help (/ n p) p)))
         (t (fact-tree-help n (+ p 1)))))
```

8.71. To view this diagram as a binary tree instead of a list, turn the page 45 degrees clockwise. The terminal nodes of the tree are the atoms A, B, C, D, E, and NIL. The nonterminal nodes are the cons cells.

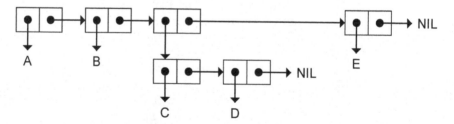

8.72. A book can be described as a tree whose nodes have varying numbers of branches. The nonterminal nodes are chapters, sections, subsections, paragraphs, sentences, and words. The terminal nodes are characters.

CHAPTER 9 ANSWERS

9.1.
```
(defun saying ()
  (format t "~&There are old pilots,~%")
  (format t "and there are bold pilots,~%")
  (format t
    "but there are no old bold pilots.~%"))
```

9.2.
```
(defun draw-line (n)
  (cond ((zerop n) (format t "~%"))
        (t (format t "*")
           (draw-line (- n 1)))))
```

9.3.
```
(defun draw-box (width height)
  (cond ((zerop height) nil)
        (t (draw-line width)
           (draw-box width (- height 1)))))
```

9.4.
```
(defun ninety-nine-bottles (n)
  (cond ((zerop n)
         (format t "~&Awww, no more beer!"))
        (t (do-verse n)
           (ninety-nine-bottles (- n 1)))))

(defun do-verse (n)
  (format t
    "~&~S bottles of beer on the wall,~%" n)
  (format t "~S bottles of beer!~%" n)
  (format t "Take one down,~%Pass it around,~%")
  (format t
    "~S bottles of beer on the wall.~%~%"
    (- n 1)))
```

9.5.
```
(defun print-board (b)
  (let ((b2 (sublis '((x   . "X")
                      (o   . "O")
                      (nil . " "))
                    b)))
    (format t "~&")
    (print-line b2)
    (format t "-----------~%")
    (print-line (nthcdr 3 b2))
    (format t "-----------~%")
    (print-line (nthcdr 6 b2))))
```

```
(defun print-line (line)
    (format t " ~A | ~A | ~A~%"
      (first line)
      (second line)
      (third line)))
```

9.6.   ```
(defun compute-pay ()
    (format t "~&What is the hourly wage? ")
    (let ((wage (read)))
      (format t "~&How many hours worked? ")
      (let ((hours (read)))
        (format t
          "~&The worker earned ~S dollars."
          (* wage hours)))))
```

9.7.   ```
(defun cookie-monster ()
    (format t "~&Give me cookie!!!!~%")
    (format t "Cookie? ")
    (let ((response (read)))
      (cond ((equal response 'cookie)
             (format t "~&Thank you!")
             (format t "...Munch munch munch")
             (format t "... BURP"))
            (t (format t "~&No want ~S...~%~%"
                  response)
               (cookie-monster)))))
```

9.8.   A symbol is a block of five pointers. Strings are not symbols, they are
       vectors. Strings evaluate to themselves. They are written enclosed in
       double quotes, and often contain mixed upper and lower case
       characters, whereas symbol names are usually all upper case.

9.9.   ```
> (format t "a~S" 'b)
aB
NIL

> (format t "always~%broke")
always
broke
NIL

> (format t "~S~S" 'alpha 'bet)
ALPHABET
NIL
```
```

9.10.
```
(defun space-over (n)
  (cond ((plusp n)
          (format t " ")
          (space-over (- n 1)))
        ((zerop n) nil)
        (t (format t "Error!"))))

(defun plot-one-point (plotting-string y-val)
  (space-over y-val)
  (format t "~A~%" plotting-string))

(defun plot-points (plotting-string y-vals)
  (mapcar
    #'(lambda (y)
        (plot-one-point plotting-string y))
    y-vals))

(defun generate (m n)
  (cond ((equal m n) (list n))
        (t (cons m (generate (+ m 1) n)))))

(defun make-graph ()
  (let* ((func
            (prompt-for "Function to graph? "))
         (start
            (prompt-for "Starting x value? "))
         (end (prompt-for "Ending x value? "))
         (plotting-string
            (prompt-for "Plotting string? ")))
    (plot-points plotting-string
      (mapcar func (generate start end)))
    t))

(defun prompt-for (prompt-string)
  (format t "~A" prompt-string)
  (read))
```

9.11.
```
(defun dot-prin1 (x)
  (cond ((atom x) (format t "~S" x))
        (t (format t "(")
           (dot-prin1 (car x))
           (format t " . ")
           (dot-prin1 (cdr x))
           (format t ")"))))
```

9.12. (DOT-PRIN1 '(A . (B . C))) should print (A . (B . C)).

9.13. Lisp prefers to write (A . NIL) in list notation, as (A). But (A . B) must be written in dot notation, because the cdr of the cons cell doesn't point to NIL or another cons cell. Lisp prints (A . B).

9.14. Both these structures cause infinite loops. For the first one, DOT-PRIN1 prints "(FOO . (FOO . (FOO . ..." until you stop it or some sort of stack overflow error occurs. For the second one, DOT-PRIN1 tries to print an infinite series of left parentheses.

9.15.
```
(defun hybrid-prin1 (x)
  (cond ((atom x) (format t "~S" x))
        (t (hybrid-print-car (car x))
           (hybrid-print-cdr (cdr x)))))

(defun hybrid-print-car (x)
  (format t "(")
  (hybrid-prin1 x))

(defun hybrid-print-cdr (x)
  (cond ((null x) (format t ")"))
        ((atom x) (format t " . ~S)" x))
        (t (format t " ")
           (hybrid-prin1 (car x))
           (hybrid-print-cdr (cdr x)))))
```

CHAPTER 10 ANSWERS

10.1. If *TOTAL-GLASSES* was not initialized, we would get an unassigned variable error when we called SELL. If it was initialized to the symbol FOO instead of to zero, we would get a wrong type input error when SELL tried to increment the total.

10.2.
```
(defun sell (n)
  (incf *total-glasses* n)
  (format t
    "~&That makes ~S glasses so far today."
    *total-glasses*))
```

10.3. (setf *met-before* 0)

```
  (defun meet (person)
    (cond ((equal person (first *friends*))
           (incf *met-before*)
           'we-just-met)
          ((member person *friends*)
           (incf *met-before*)
           'we-know-each-other)
          (t (push person *friends*)
             'pleased-to-meet-you)))
```

10.4.
```
  (defun forget (person)
    (cond ((member person *friends*)
           (setf *friends*
                 (remove person *friends*))
           'forgotten)
          (t (list 'dont 'know person))))
```

10.5.
```
  (defun pretty (x y)
    (let* ((biggest (max x y))
           (smallest (min x y))
           (avg (/ (+ x y) 2.0))
           (pct (* 100 (/ avg biggest))))
      (list 'average avg 'is pct 'percent
            'of 'max biggest)))
```

10.6.
```
  (setf x nil)

  (push x x)   ⇒   (nil)

  (push x x)   ⇒   ((nil) nil)

  (push x x)   ⇒   (((nil) nil) (nil) nil)
```

10.7. (LENGTH X) is not a valid place description: It does not name a place where a pointer is stored. SETF will complain that it has no SETF method for LENGTH.

10.8.
```
  (setf *corners* '(1 3 7 9))

  (setf *sides* '(2 4 6 8))

  (defun block-squeeze-play (board)
    (sq-and-2 board *computer* *sides* 12
      "block squeeze play"))
```

```
(defun block-two-on-one (board)
  (sq-and-2 board *opponent* *corners* 12
    "block two-on-one"))

(defun try-squeeze-play (board)
  (sq-and-2 board *opponent* nil 11
    "set up a squeeze play"))

(defun try-two-on-one (board)
  (sq-and-2 board *computer* nil 11
    "set up a two-on-one"))

(defun sq-and-2 (board player pool v strategy)
  (when (equal (nth 5 board) player)
    (or (sq-helper board 1 9 v strategy pool)
        (sq-helper board 3 7 v strategy pool))))

(defun sq-helper (board c1 c2 val strategy pool)
  (when (equal val (sum-triplet
                     board
                     (list c1 5 c2)))
    (let ((pos (find-empty-position
                 board
                 (or pool (list c1 c2)))))
      (and pos (list pos strategy)))))

(defun exploit-two-on-one (board)
  (when (equal (nth 5 board) *computer*)
    (or (exploit-two board 1 2 4 3 7)
        (exploit-two board 3 2 6 1 9)
        (exploit-two board 7 4 8 1 9)
        (exploit-two board 9 6 8 3 7))))

(defun exploit-two (board pos d1 d2 c1 c2)
  (and (equal (sum-triplet
                board
                (list c1 5 c2)) 21)
       (zerop (nth pos board))
       (zerop (nth d1 board))
       (zerop (nth d2 board))
       (list pos "exploit two-on-one")))
```

```
(defun choose-best-move (board)
  (or (make-three-in-a-row board)
      (block-opponent-win board)
      (block-squeeze-play board)
      (block-two-on-one board)
      (exploit-two-on-one board)
      (try-squeeze-play board)
      (try-two-on-one board)
      (random-move-strategy board)))
```

10.9.
```
(defun chop (x)
  (if (consp x) (setf (cdr x) nil))
  x)
```

10.10.
```
(defun ntack (x e)
  (nconc x (list e)))
```

10.11. The SETF operation constructs a circular list by making the cdr of the last cons cell point back to the first cell.

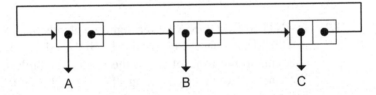

10.12. (APPEND H H) returns the list (HI HO HI HO). It copies its first input, and the result shares structure with H. (NCONC H H) turns the list H into a circular list by destructively setting the cdr of the last cell of its first argument to point to its second argument.

CHAPTER 11 ANSWERS

11.1.
```
(defun it-member (item x)
  (dolist (e x)
    (when (equal item e) (return t))))
```

11.2.
```
(defun it-assoc (key table)
  (dolist (entry table)
    (when (equal key (first entry))
      (return entry))))
```

11.3.
```
(defun check-all-odd (x)
  (cond ((null x) t)
        (t (format t "~&Checking ~S..."
             (first x))
           (unless (evenp (first x))
             (check-all-odd (rest x)))))))
```

11.4.
```
(defun it-length (x)
  (let ((n 0))
    (dolist (e x n)
      (incf n))))
```

11.5.
```
(defun it-nth (n x)
  (dotimes (i n (first x))
    (pop x)))
```

11.6.
```
(defun it-union (x y)
  (dolist (e x y)
    (unless (member e y)
      (push e y))))
```

11.7. IT-INTERSECTION went throught the elements of X from left to right, and PUSHed selected ones onto RESULT-SET. Because PUSH adds elements to the front of a list, the result was built up in reverse order. We can correct this by making IT-INTERSECTION reverse the result before returning it.

11.8.
```
(defun it-reverse (x)
  (let ((result nil))
    (dolist (e x result)
      (push e result))))
```

11.9.
```
(defun check-all-odd (x)
  (do ((z x (rest z)))
      ((null z) t)
    (format t "~&Checking ~S..." (first z))
    (if (evenp (first z)) (return nil))))
```

11.10.
```
(defun launch (n)
  (dotimes (i n)
    (format t "~S..." (- n i)))
  (format t "Blast off!"))
```

11.11.
```
(defun find-largest (list-of-numbers)
    (do* ((largest (first list-of-numbers))
          (z (rest list-of-numbers) (rest z))
          (element (first z) (first z)))
        ((null z) largest)
      (when (> element largest)
        (setf largest element))))
```

11.12.
```
(defun power-of-2 (n)
    (do ((result 1 (+ result result))
         (i 0 (+ i 1)))
        ((equal i n) result)))
```

11.13.
```
(defun first-non-integer (x)
    "Returns the first non-integer element of X."
    (dolist (e x 'none)
      (when (not (integerp e))
        (return e))))
```

11.14. The function would not work if we changed the DO* to a DO. When evaluating the expression (FIRST X) to get the initial value for E, Lisp would try to reference the global variable X, because the expression is not within the lexical scope of any local variable named X. This will probably result in an unassigned variable error.

11.15. If only the last number in the list is odd, this version of FFO-WITH-DO will return NIL instead of the number. Due to the use of parallel assignment, E is assigned the last number in the list at the same time that Z becomes NIL. When Z is NIL the DO's termination test is true, so the body is never evaluated and the last element of the list is never tested for being odd.

11.16. The variable list of a LET contains pairs of form *(variable value)*. The variable list of a DO contains triples of form *(variable initial-value update-expression)*. If the third element is omitted, the variable is not updated each time through the loop. In this case DO treats the variable just as LET would.

11.17. The value of the expression is 5.

11.18.
```
(do ((i 0 (+ i 1)))
    ((equal i 5) i)
  (format t "~%I = ~S" i))
```

The DO goes through its body five times, with the index variable I equal to zero through four. The loop terminates when I reaches five. Since the expression to be returned is I, the DO returns five. Many Lisp implementations automatically translate DOTIMES expressions into a DO expression such as this one.

11.19. The entries in a DO's variable list may appear in any order. They are completely independent due to the use of parallel assignment. With DO*, though, the order of entries is important, because sequential assignment permits dependencies to exist among the variables.

11.20. If a loop uses only one index variable, DO and DO* are equivalent.

11.21.
```
(defun fib (n)   ; version with DO*
  (do* ((cnt 0 (+ cnt 1))
        (i 1 j)
        (j 1 k)
        (k 2 (+ i j)))
       ((equal cnt n) i)))

(defun fib (n)   ; version with DO
  (do ((cnt 0 (+ cnt 1))
       (i 1 j)
       (j 1 (+ i j)))
      ((equal cnt n) i)))
```

11.22.
```
(defun complement-base (base)
  (second
    (assoc base '((a t) (t a) (g c) (c g)))))

(defun complement-strand (strand)
  (do ((s strand (rest s))
       (result nil
          (cons (complement-base (first s))
                result)))
      ((null s) (reverse result))))
```

```
        (defun make-double (strand)
          (do ((s strand (rest s))
               (result nil
                 (cons (list (first s)
                             (complement-base
                               (first s)))
                       result)))
              ((null s) (reverse result)))))

(defun count-bases (dna)
  (let ((acnt 0) (tcnt 0) (gcnt 0) (ccnt 0))
    (labels ((count-one-base (base)
               (cond ((equal base 'a) (incf acnt))
                     ((equal base 't) (incf tcnt))
                     ((equal base 'g) (incf gcnt))
                     ((equal base 'c) (incf ccnt)))))
      (dolist (element dna)
        (cond ((atom element) (count-one-base element))
              (t (count-one-base (first element))
                 (count-one-base (second element)))))
      (list (list 'a acnt)
            (list 't tcnt)
            (list 'g gcnt)
            (list 'c ccnt)))))
```

The LABELS special function used in COUNT-BASES was described
in Advanced Topics section 8.18 on page 282. This problem can be
solved—somewhat less elegantly—without LABELS, for example, by
making COUNT-ONE-BASE a separate function and keeping the
counts in global variables.

```
(defun prefixp (strand1 strand2)
  (do ((s1 strand1 (rest s1))
       (s2 strand2 (rest s2)))
      ((null s1) t)
    (unless (equal (first s1) (first s2))
      (return nil))))

(defun appearsp (strand1 strand2)
  (do ((s2 strand2 (rest s2)))
      ((null s2) nil)
    (if (prefixp strand1 s2)
        (return t))))
```

```lisp
(defun coverp (strand1 strand2)
  (do* ((len1 (length strand1))
        (s2 strand2 (nthcdr len1 s2)))
       ((null s2) t)
    (unless (prefixp strand1 s2)
      (return nil))))

(defun prefix (n strand)
  (do ((i 0 (+ i 1))
       (res nil (cons (nth i strand) res)))
      ((equal i n) (reverse res))))

(defun kernel (strand)
  (do ((i 1 (+ i 1)))
      ((coverp (prefix i strand) strand)
       (prefix i strand))))

(defun draw-dna (strand)
  (let ((n (length strand)))
    (draw-string n "-----")
    (draw-string n "  !  ")
    (draw-bases strand)
    (draw-string n "  .  ")
    (draw-string n "  .  ")
    (draw-bases (complement-strand strand))
    (draw-string n "  !  ")
    (draw-string n "-----")))

(defun draw-string (cnt string)
  (format t "~&")
  (dotimes (i cnt)
    (format t "~A" string)))

(defun draw-bases (strand)
  (format t "~&")
  (dolist (base strand)
    (format t "  ~A  " base)))
```

CHAPTER 12 ANSWERS

12.1. The symbol CAPTAIN is used in the DEFSTRUCT expression to name one of the fields of a starship. The keyword :CAPTAIN is used as an argument to MAKE-STARSHIP to specify a value for this field. The symbol STARSHIP-CAPTAIN names the accessor function that extracts this field from a starship object.

12.2. (STARSHIP-P 'STARSHIP) returns NIL, because STARSHIP is just a symbol, not a structure of type STARSHIP.

12.3.
```
(type-of 'make-starship)  ⇒  symbol
```

(TYPE-OF #'MAKE-STARSHIP) returns a value that depends on the representation of functions in your particular Lisp implementation. Some possible values are LEXICAL-CLOSURE, COMPILED-FUNCTION, or even CONS.

```
(type-of (make-starship))  ⇒  starship
```

12.4.
```
(defstruct node
  name
  question
  yes-case
  no-case)

(setf *node-list* nil)

(defun init ()
  (setf *node-list* nil)
  'initialized)

(defun add-node (name question yes-case no-case)
  (push (make-node :name name
                   :question question
                   :yes-case yes-case
                   :no-case no-case)
        *node-list*)
  name)

(defun find-node (x)
  (find-if #'(lambda (node)
               (equal (node-name node) x))
           *node-list*))
```

```
(defun process-node (name)
  (let ((nd (find-node name)))
    (if nd
        (if (y-or-n-p "~&~A "
              (node-question nd))
            (node-yes-case nd)
            (node-no-case nd))
        (format t
          "~&Node ~S not yet defined." name))))

(defun run ()
  (do ((current-node 'start
        (process-node current-node)))
      ((null current-node) nil)
    (cond ((stringp current-node)
           (format t "~&~A" current-node)
           (return nil)))))

(defun interactive-add ()
  (let* ((name (prompt-for "Node name? "))
         (quest (prompt-for "Question? "))
         (yes-action (prompt-for "If yes? "))
         (no-action (prompt-for "If no? ")))
    (add-node name quest yes-action no-action)))

> (interactive-add)
Node name? engine-will-run-briefly
Question? "Does the engine stall when cold
but not when warm? "
If yes? stalls-only-when-cold
If no? stalls-even-when-warm
ENGINE-WILL-RUN-BRIEFLY

> (interactive-add)
Node name? stalls-only-when-cold
Question? "Is the cold idle speed at
least 700 rpm? "
If yes? cold-idle-speed-normal
If no? "Adjust the cold idle speed."
STALLS-ONLY-WHEN-COLD
```

12.5. `(setf s1 (make-starship :name "Enterprise"))`

```
(defstruct (captain
              (:print-function print-captain))
   (name nil)
   (age nil)
   (ship nil))


(defun print-captain (x stream depth)
   (format stream "#<CAPTAIN ~S>"
       (captain-name x)))

(setf jim (make-captain
              :name "James T. Kirk"
              :age 35
              :ship s1))

(setf (starship-captain s1) jim)
```

CHAPTER 13 ANSWERS

13.1.
```
(defun subprop (symbol item property)
   (setf (get symbol property)
         (remove item (get symbol property))))
```

13.2.
```
(defun forget-meeting (person1 person2)
   (subprop person1 person2 'has-met)
   (subprop person2 person1 'has-met)
   'forgotten)
```

13.3.
```
(defun my-get (symbol property)
   (do ((p (symbol-plist symbol) (cddr p)))
       ((null p) nil)
     (if (equal property (first p))
         (return (second p)))))
```

13.4.
```
(defun hasprop (symbol property)
   (do ((p (symbol-plist symbol) (cddr p)))
       ((null p) nil)
     (if (equal property (first p))
         (return t))))
```

13.5. You can access any element of an array in constant time. For lists, the amount of time it takes to access an element is proportional to its distance from the beginning of the list. Another advantage of arrays is that they generally use only about half as much storage as lists.

13.6. Lists are easy to create one element at a time, using CONS. And we can splice new items into a list at any position, or snip them out, using destructive operations. Arrays can not be constructed or manipulated this easily. Also, lists can share structure in ways that are not possible for arrays.

13.7. Both structures require the same number of cons cells. However, if the association list were not dotted, for example, ((CAT MEOW) (DOG WOOF)) instead of ((CAT . MEOW) (DOG . WOOF)), then it would require one more cons cell per entry than the corresponding property list representation.

13.8.
```
(setf *hist-array* nil)
(setf *total-points* 0)

(defun new-histogram (bins)
  (setf *total-points* 0)
  (setf *hist-array*
    (make-array bins :initial-element 0))
  t)

(defun record-value (v)
  (incf *total-points*)
  (if (and (>= v 0)
           (< v (length *hist-array*)))
      (incf (aref *hist-array* v))
      (error "Value ~S out of bounds." v)))

(defun print-hist-line (i)
  (let ((val (aref *hist-array* i)))
    (format t "~&~2D [~3D] " i val)
    (dotimes (j val)
      (format t "*"))))

(defun print-histogram ()
  (dotimes (i (length *hist-array*))
    (print-hist-line i))
  (format t "~&   ~3D total" *total-points*))
```

13.9. Text of the cryptogram:

```
(setf crypto-text
 '("zj ze kljjls jf slapzi ezvlij pib kl jufwxuj p hffv jupi jf"
   "enlpo pib slafml pvv bfwkj"))


      (setf *encipher-table* (make-hash-table))
      (setf *decipher-table* (make-hash-table))

      (defun make-substitution (code clear)
        (setf (gethash clear *encipher-table*) code)
        (setf (gethash code *decipher-table*) clear))

      (defun undo-substitution (code clear)
        (setf (gethash clear *encipher-table*) nil)
        (setf (gethash code *decipher-table*) nil))

      (defun clear ()
        (clrhash *encipher-table*)
        (clrhash *decipher-table*))

      (defun decipher-string (string)
        (do* ((len (length string))
              (new-string (make-string len
                             :initial-element #\Space))
              (i 0 (1+ i)))
             ((equal i len) new-string)
          (let* ((char (aref string i))
                 (new-char
                   (gethash char *decipher-table*)))
            (when new-char
              (setf (aref new-string i) new-char)))))

      (defun show-line (line)
        (format t "~%~A~%~A~%"
          line
          (decipher-string line)))

      (defun show-text ()
        (format t "~&---------------")
        (dolist (line crypto-text)
          (show-line line))
        (format t "~&---------------"))
```

```lisp
(defun get-first-char (x)
  (char-downcase
    (char (format nil "~A" x) 0)))

(defun read-letter ()
  (let ((obj (read)))
    (if (member obj '(end undo))
        obj
        (get-first-char obj))))

(defun sub-letter (code)
  (when (gethash code *decipher-table*)
    (format t "~&'~A' has already been" code)
    (format t " deciphered as '~A'!"
      (gethash code *decipher-table*))
    (return-from sub-letter nil))
  (format t "What does '~A' decipher to? " code)
  (let ((clear (read-letter)))
    (cond ((not (characterp clear))
           (format t "~&Invalid response."))
          ((gethash clear *encipher-table*)
           (format t "But '~A' already"
             (gethash clear *encipher-table*))
           (format t " deciphers as '~A'!"
             clear))
          (t (make-substitution code clear)))))

(defun undo-letter ()
  (format t "~&Undo which letter? ")
  (let* ((code (read-letter))
         (clear (gethash code
                  *decipher-table*)))
    (cond ((not (characterp code))
           (format t "~&Invalid input."))
          (clear (undo-substitution code clear))
          (t (format t
               "~&But '~A' wasn't deciphered!"
               code)))))
```

```
(defun solve ()
  (do ((resp nil))
      ((equal resp 'end))
    (show)
    (format t "~&Substitute which letter? ")
    (setf resp (read-letter))
    (cond ((characterp resp) (sub-letter resp))
          ((equal resp 'undo) (undo-letter))
          ((equal resp 'end) nil)
          (t (format t "~&Invalid input.")))))
```

Solution to the cryptogram: *It is better to remain silent and be thought a fool than to speak and remove all doubt.*

CHAPTER 14 ANSWERS

14.1. (POP X) typically expands to something like (PROG1 (CAR X) (SETQ X (CDR X))), but the exact expansion varies from one Lisp implementation to the next.

14.2. The expansion of a DEFSTRUCT is long and complicated, since there are so many details to be handled. You will see definitions for accessor functions, SETF methods, a constructor function, a type predicate, and other things.

14.3.
```
(defmacro set-nil (var)
  (list 'setf var nil))
```

14.4.
```
(defmacro simple-rotatef (var1 var2)
  `(let ((temp1 ,var1)
         (temp2 ,var2))
     (setf ,var1 temp2)
     (setf ,var2 temp1)))
```

14.5.
```
(defmacro set-mutual (var1 var2)
  `(progn
     (setf ,var1 ',var2)
     (setf ,var2 ',var1)))
```

14.6.
```
(defmacro variable-chain (&rest vars)
  '(progn
     ,@(do ((v vars (rest v))
            (res nil))
           ((null (rest v)) (reverse res))
         (push '(setf ,(first v)
                      ',(second v))
               res))))
```

14.7. To solve this problem we must create a new state, HAVE-25, and then define all the legal transitions into and out of this state. Besides putting in quarters, we can also reach this state with an appropriate combination of nickels and dimes.

```
(defnode have-25)

(defarc start    quarter have-25 "Ker-chunk!")
(defarc have-15 dime    have-25 "Clink!")
(defarc have-20 nickel  have-25 "Clunk!")

(defarc have-5  quarter have-25
  "Nickel returned.")
(defarc have-10 quarter have-25
  "Returned ten cents.")
(defarc have-15 quarter have-25
  "Returned fifteen cents.")
(defarc have-20 quarter have-25
  "Returned twenty cents.")
(defarc have-25 quarter have-25
  "Quarter returned.")

(defarc have-25 chocolate-bar-button end
  "Deliver chocolate bar.")
(defarc have-25 coint-return start
  "Returned twenty-five cents.")
```

14.8. It's unwise to write macros that have side effects because you don't necessarily know when or how often the macro will be expanded. Some implementations expand macros once and save the result for reuse; others reexpand the macro at each macro call. Some implementations even attempt to expand macro calls in function bodies at the time the function is DEFUNed.

14.9. As of mid-1989, the 24 built-in Common Lisp special functions are: BLOCK, CATCH, COMPILER-LET, DECLARE, EVAL-WHEN, FLET, FUNCTION, GO, IF, LABELS, LET, LET*, MACROLET, MULTIPLE-VALUE-CALL, MULTIPLE-VALUE-PROG1, PROGN, PROGV, QUOTE, RETURN-FROM, SETQ, TAGBODY, THE, THROW, and UNWIND-PROTECT. This list may change with future revisions of the Common Lisp standard.

14.10. Lisp programs typically run 10 to 100 times faster after compilation.

14.11.
```
(defun compile-arc (arc)
  (let ((a (arc-action arc)))
    `((equal this-input ',(arc-label arc))
      (format t "~&~A" ,a)
      (,(node-name (arc-to arc))
       (rest input-syms)))))


(defun compile-node (node)
  (let ((name (node-name node))
        (arc-clauses
          (mapcar #'compile-arc
                  (node-outputs node))))
    `(defun ,name (input-syms
                   &aux (this-input
                          (first input-syms)))
       (cond ((null input-syms) ',name)
             ,@arc-clauses
             (t (format t
    "~&There is no arc from ~A with label ~S"
                 ',name this-input))))))


(defmacro compile-machine ()
  `(progn ,@(mapcar #'compile-node *nodes*)))

> (compile-machine)
END

> (start '(dime dime dime gum-button))
Clink!
Clink!
Dime returned.
Deliver gum, nickel change.
END
```

Glossary

a-list

See **association list.**

accessor function

A function such as STARSHIP-SPEED, defined automatically by DEFSTRUCT, that allows you to access a particular field of a structure.

address

A number describing the location of an object in memory.

applicative operator

A function that takes another function as input, and applies it to some data. Examples include MAPCAR and FIND-IF.

applicative programming

A style of programming in which functions are frequently passed as data to other functions, and explicit assignment is avoided. Repetitive operations are performed by passing functions to **applicative operators.**

APPLY

The Lisp primitive that applies a function to a set of arguments. EVAL and APPLY are the two basic functions from which Lisp interpreters are constructed. Applicative operators are all constructed from APPLY (or from FUNCALL.)

argument

A piece of data serving as input to a function, or an expression which, when evaluated, will produce that piece of data. The term is also used to refer to the names a function uses for its inputs, as in ''AVERAGE is a function of two arguments: X and Y.''

argument list

A list that specifies the names a function gives to each of its inputs, and how many inputs it requires. When defining a new function, the second input to DEFUN is the new function's argument list.

array

A contiguous block of storage whose elements are accessed by numeric subscripts. One-dimensional arrays are called **vectors**, and are a type of **sequence**.

array header

A small amount of storage at the beginning of an array where Lisp keeps information about the organization of the array, such as its length, and the number of dimensions it has.

assignment-free style

A style of programming that avoids explicit assignment to variables. Once a variable is given a value, such as by a function call or by LET, that value never changes. Assignment-free programs are considered very elegant and easy to read.

association list

A list of pairs, or, more generally, of lists, called **entries**. The car of each entry is the **key** searched for by ASSOC.

atom

Any Lisp object that is not a cons cell. All non-lists are atoms. So is the empty list, NIL.

augmentation

The process of adding something on to a result to derive a new result.

augmenting recursion

A type of recursion in which the final result is built up bit by bit, by adding something to the result of each recursive call.

backtrace

A display of the execution stack showing the function currently being evaluated, the function that called it, the function that called that function, and so on. Backtraces are displayed by the debugger upon command.

bignum

An **integer** with an arbitrary number of digits. Internally, bignums are usually represented as a special type of sequence. Compare **fixnum**.

binary tree

A tree in which each nonterminal node has exactly two children. Lists may be viewed as binary trees whose nonterminal nodes are cons cells and whose terminal nodes are atoms.

binding

An archaic term with conflicting uses. Essentially, binding means creating a variable and assigning it a value. See also **rebinding**.

block

A named sequence of Lisp expressions, forming the body of a BLOCK expression. Blocks may be exited using RETURN-FROM.

block name

A symbol serving as the name of a block. DO, DO*, DOTIMES, and DOLIST create implicit blocks named NIL. Functions defined by DEFUN or LABELS surround their bodies with implicit blocks whose name is the same as the function.

body

The body of a form, such as a function definition or a LET, LABELS, or DO expression, contains expressions to be evaluated sequentially within the lexical context of the form. Normally, the value of the last expression in the body is returned by the form.

Boolean function

A function whose inputs and outputs are truth values. Boolean functions are **truth functions**.

bucket

One of the slots of a hash table, in which a chain of items is stored. The more buckets a hash table has, the fewer items each bucket must hold, and the faster the access time for an item will be.

call

To call or invoke a function means to pass it some inputs and ask it to produce an output or side effect.

CAR

The left half of a cons cell. Also, a function that returns the contents of the left half of a cons cell. The name stands for Contents of Address portion of Register.

cardinality

The cardinality of a set is the number of elements it contains.

CDR

The right half of a **cons cell**. Also, a function that returns the contents of the right half of a cons cell. The name stands for Contents of Decrement portion of Register. See also **CAR**, **cons**.

character object

A Lisp object such as #\A that denotes a character.

character string

See **string**.

circular list

A cons cell structure in which there is a path from some cons cell back to itself, possibly via intervening cons cells. If a list is circular, it is not a tree.

clause

An element of a COND, AND, or OR conditional expression. A conditional can decide which of its clauses will be evaluated.

comment

A remark included in a program to make it more understandable to humans. Lisp comments are preceded by at least one semicolon.

composite number

An integer that is the product of a prime and some other integer. The opposite of a **prime number**.

conditional

A special function or macro function that makes decisions about which parts of its input to evaluate. Examples include IF, COND, AND, and OR.

conditional augmentation

A style of recursion in which the results of each recursive call are sometimes augmented and sometimes not, under the control of a conditional.

cons cell

The unit of computer memory from which lists are composed. Each cons cell holds two pointers, one in the car half, and one in the cdr half.

constructor function

A function such as MAKE-STARSHIP that constructs new instances of a structure type.

cryptogram

A puzzle in which a piece of text is encoded by a **substitution cipher**. Cryptograms can be solved by applying knowledge of letter frequencies, and the limited number of words with three or fewer letters, to arrive at a partial decoding.

data

Data means information. Lisp data comes in several forms, including numbers, symbols, and lists.

database

A data structure that holds a collection of facts. For example, a "blocks world" database would contain facts about the properties of individual blocks and their relationships to each other.

debugger

A tool for examining the state of Lisp programs after an error occurs. It is used to find and eliminate the "bug" responsible for the error.

DeMorgan's Theorem

A theorem showing the interchangeability of AND and OR when combined with NOT.

destructive operation

An operation that replaces one pointer with another, thereby changing the value of a variable or altering the contents of a cons cell.

destructuring

A macro function's breaking up one of its unevaluated inputs (a list) into its component elements. Destructuring may be requested by writing a list in place of a symbol in the macro's argument list.

discrimination net

A network of nodes, each containing a question, that may be used to solve diagnostic problems. The user's response to the current node's question determines which of the node's descendants will become the new current node.

documentation string

A character string serving as the online documentation for a function or variable. Documentation strings may be established using DEFUN or DEFVAR.

dot notation

A notation for writing lists in which cons cells are written as dotted pairs, that is, each cons cell is displayed as a car and cdr separated by a dot, enclosed in parentheses. The list (A (B) C) is written (A . ((B . NIL) . (C . NIL))) in dot notation. See also **hybrid notation.**

dotted list

A cons cell chain ending in an atom other than NIL. For example, (A B C . D) is a chain of three cons cells ending in the symbol D. This list *must* be written with a dot to show that the D is the cdr of the third cell, not the car of a fourth cell.

dotted pair

A single cons cell written in do notation. Usually the cdr is a non-NIL atom. A typical dotted pair is (A . B).

double-test recursion

A style of recursion in which there are two end tests. Often, one test is for success, such as finding a particular element when searching a list, and the other is for failure, such as running off the end of the list.

dynamic scoping

A scoping discipline in which a symbol is dynamically associated with the most recently-created variable with that name still in existence, independent of the lexical context in which the variable was created. **Special variables** are dynamically scoped. Compare **lexical scoping**.

element

The elements of a list are the cars of its top-level cons cells, that is, the things that appear within only one level of parentheses.

empty list

The list with no elements. It is written () or NIL.

end-of-file error

When READ tries to read an object beyond the last object in the file, an end-of-file error is signalled. This error can be disabled by supplying an optional argument to READ.

entry

An element of an association list (such as a dotted pair), or of a hash table.

escape character

Characters such as the double quote used to enclose strings. Escape characters are necessary to print objects containing special characters; otherwise it will not be possible to read the objects back in again using READ.

EVAL

The heart of Lisp: EVAL is a function that evaluates a Lisp expression according to a set of evaluation rules, and returns the result.

EVAL notation

A way to write Lisp expressions as lists. The first element of a list specifies a function, and the remaining elements specify arguments to be evaluated before the function is called.

evaltrace diagram

A graphical notation unique to this book. Evaltrace diagrams illustrate the evaluation of expressions, and are particularly useful for explaining lexical and dynamic scoping.

evaluation

The process of deriving a result from an expression.

expression

Expressions are the Lisp equivalent of sentences in English. Every Lisp object (number, symbol, list) is an expression. Lisp interpreters read expressions from the keyboard, evaluate them, and print the results.

finite state machine

A theoretical machine consisting of a finite number of nodes, representing states, connected by labeled arcs. The machine moves from one state to the next depending on which arc label matches its input. Finite state machines are useful as abstract descriptions of the mechanisms governing devices such as traffic lights, vending machines, and bits of computer circuitry.

fixnum

An integer small enough to be represented more efficiently than a **bignum.** In most implementations, fixnums are represented as binary integers 24 to 32 bits long. Larger integers must be represented as bignums.

flat list

A list of atoms. Since it contains no lists as elements, it is called flat rather than **nested**.

floating point number
A number containing a decimal point, such as 5.0 or 3.14159.

form
An expression. Forms are evaluated to yield results. The term is also used to refer to macros and special functions themselves, as in "LET* is a form for sequentially binding variables."

format control string
A string given as a second argument to format containing text to be printed, interspersed with format directives such as ~S. Several other functions also accept format control string arguments, such as YES-OR-NO-P, BREAK, and ERROR.

function
Functions transform inputs to outputs. Lisp functions are defined with DEFUN. Lisp programs are organized as collections of functions.

function cell
One of the five components of a symbol. The function cell holds a pointer to the function object representing the global function named by that symbol. (Local functions created by LABELS do not reside in the function cell.)

function object
A piece of Lisp data that is a function, and can be applied to arguments. The representation of function objects is implementation dependent.

garbage collection
The process of reclaiming storage that is no longer in use, so that it may be reused.

generalized variable
Any place a pointer may reside, such as an ordinary variable, the car or cdr half of a cons cell, an element of an array, a slot in a structure, or one of the five cells making up a symbol.

gensym
A symbol created automatically, with a name such as #:G0037, that is not registered in any package. Gensyms are often found in the expansions of complex macros such as SETF.

global variable
A variable that exists in the global lexical context rather than being local to some particular function or LET expression.

hash table
A Lisp data structure that efficiently associates keys with entries. Hash tables serve the same purpose as association lists, but they provide faster lookups of items when the number of entries is large.

hashing algorithm

The method by which a hash table assigns an entry to a **bucket**. When looking up a key in the table, the hashing algorithm determines which of the table's buckets to look in.

hybrid notation

A notation for writing lists in which dots are used only when necessary, that is, only when a cons cell chain ends in an atom other than NIL. The dot notation list ((A . NIL) . (C . (D . E))) is written in hybrid notation as ((A) C D . E).

i/o

Input/output. The process of transferring information between the computer and an external device, such as the keyboard, the display, or a disk file.

indicator

An atom (normally a symbol) that serves as the name of a property on a **property list.**

input

The inputs to a function are the pieces of data it receives. The term **input** also refers to the act of reading an object or character string from the keyboard, or from a file.

integer

A whole number, such as two. Integers are divided into **fixnums** and **bignums**. See also **floating point number** and **ratio**.

intersection

The intersection of two sets contains only those elements that appear in both sets. See **union**.

invoke

To invoke a function means to call it, in other words, to give it some inputs and ask it to produce an output.

iteration

To iterate means to repeat. Iteration in Lisp is accomplished by macros such as DO, DO*, DOTIMES, and DOLIST, which ultimately rely on the LOOP special function.

key

The item that names an entry in an association list or hash table. Entries can be retrieved (by ASSOC or GETHASH) given the associated key.

keyword

A special kind of symbol that is written with a preceding colon, such as :TEST. Keywords evaluate to themselves.

keyword argument

An optional argument named by a keyword. For example, the MEMBER function takes an optional :TEST argument.

lambda

A marker indicating that a list is a lambda expression and is to be interpreted as a description of a function.

lambda-list keyword

A special symbol such as &OPTIONAL or &REST that has a special meaning when it appears in the argument list of a function.

lambda calculus

A logical formalism defined by the mathematician Alonzo Church. John McCarthy, the creator of Lisp (and a former student of Church), borrowed lambda notation from the lambda calculus and used it for describing functions in Lisp.

lambda expression

A list that describes a function. Its first element must be the symbol LAMBDA, its second element must be an argument list, and its remaining elements constitute the body of the function. Lambda expressions must be quoted with #'. For example, #'(LAMBDA (N) (* N 2)).

lexical closure

A type of function. Lexical closures are created automatically by Lisp when functions passed as arguments to other functions need to remember their lexical context.

lexical scoping

A scoping discipline in which the only variables a function can see are those it defined itself, plus those defined by forms that contain the function, as when a function defined with DEFUN contains a lambda expression inside it. Compare **dynamic scoping.**

list

A chain of cons cells. One of the fundamental data structures of Lisp.

list surgery

Destructive modification of a list by changing the pointers stored in its cons cells. Used to efficiently insert or delete elements because it avoids copying the list.

local variable

A lexically scoped variable whose scope is limited to the body of a function or LET expression. Compare **global variable.**

logically complete

A truth function is logically complete if all other truth functions can be constructed from combinations of it. NAND is logically complete; AND is not because you cannot construct the NOT function by putting ANDs together.

macro function

A special kind of function whose arguments are not evaluated. Macro functions must return Lisp expressions, which are then evaluated.

macro expansion

The act of invoking a macro on some inputs to obtain a Lisp expression. For example, the macro call (INCF A) may expand to the expression (SETQ A (+ A 1)).

member

An item is a member of a set if it appears in (is an element of) the set.

multiple recursion

A style of recursion in which each call to a function results in several recursive calls, for example, to examine both the car and cdr halves of a tree.

nested IF

An IF appearing as the true-part or false-part of an enclosing IF. Nested IFs may be used to duplicate the multiple-clause capabilities of COND.

nested list

A list that contains other lists as elements.

NIL

The only way to say *false* in Lisp. NIL is also the empty list. It is both a symbol and a list, and it evaluates to itself.

nondestructive function

A function that does not change the value of any variable or modify pointers stored in any existing Lisp object, such as cons cells. APPEND is nondestructive; NCONC is the destructive version.

nonterminal node

A node of a tree with at least one descendant.

output

The output of a function is the result it returns. The term may also refer to a program's outputting information to the display, or to a file.

package

Packages are the name spaces in which symbols are registered. The default package is called USER. Lisp functions and variables are named by symbols in package LISP.

package name

A character string giving the name of a package, such as "USER". APROPOS takes a package name as an optional second argument.

pattern matcher

A function that matches an input list against a pattern that may contain wildcards. For example, the input (B1 COLOR GREEN) matches the pattern (B1 COLOR ?).

pointer

A pointer to an object gives the address of that object in memory. Pointers are drawn as arrows in cons cell diagrams.

predicate

A function that answers a question by returning T (or some other non-NIL value) for *true*, or NIL for *false*.

predicate expression

An expression whose value is interpreted as *true* or *false*. Used with conditionals.

prime number

An integer that is not divisible by any other integers except one and itself. Every **composite number** is a product of two or more primes.

primitive

An elementary function that is built in to Lisp, not defined by the user. CONS and + are primitives.

proper list

A cons cell chain ending in NIL. NIL itself is also a proper list (a chain of zero cons cells.)

proper subset

A proper subset is a subset that is not equal to the whole set. Set x is a proper subset of set y if x is a subset of y but y is not a subset of x.

property list

A list composed of alternating property **indicators** and values, such as (SIBLINGS (GEORGE WANDA) AGE 23 SEX MALE). Every symbol contain a plist cell that points to its associated property list. Properties can be retrieved using the GET function.

pushdown stack

A data structure where new elements are pushed on or popped off only at one end, usually called the ''top'' of the stack. Named after spring loaded stacks of dishes in cafeterias. Also called a LIFO (Last In, First Out) stack. Stacks are implemented as lists or vectors in Lisp.

ratio

A fractional number composed of a numerator and denominator, both of which are integers. For example: 3/4. The denominator cannot be zero. In Common Lisp, the denominator also cannot be one, or the number would be an **integer**, not a ratio.

rational

A number expressible as the ratio of two integers. In Common Lisp, **rationals** are either **integers** or **ratios**.

read-eval-print loop

The part of a Lisp interpreter that reads expressions from the keyboard, evaluates them, and prints the result.

rebinding

Rebinding a **special variable** means creating a new dynamic variable with the same name, such as with LET. The name is then dynamically associated with the new variable when it appears anywhere in the program, and the old variable is inaccessible until the form that bound the new variable returns.

reciprocal

The reciprocal of a number is one divided by that number. For example, the reciprocal of two is one-half.

recursion

A thing is recursive if it contains a reference to itself in its definition. Recursive functions call themselves.

recursion template

A fill-in-the-blanks description of a class of recursive functions. For example, CAR/CDR recursion describes a class of functions for searching binary trees.

result

The output of (or value returned by) a function or expression.

return

When a function "returns a value," it is outputting a piece of data.

root node

The topmost node of a tree. The only node with no parent.

S-expression

Lisp objects in printed form used to be called S-expressions, meaning *symbolic* expressions.

scope

The scope of an object is the region of the program in which the object can be referenced. For example, if a variable names the input to some function, the scope of the variable is limited to the body of that function. See also **lexical scoping** and **dynamic scoping**.

sequence

A linear collection of elements. Sequences in Lisp include lists and vectors, and hence strings, which are a type of vector. Many functions that worked only lists in previous Lisp dialects work on all types of sequences in Common Lisp. Examples include LENGTH and REVERSE.

set

An unordered collection of elements, each of which appears only once in the set. In Lisp, sets are implemented as lists.

set difference

The set difference (or set subtraction) of sets x and y is the set of elements that appear in x and do not appear in y.

set exclusive or

The exclusive or of two sets is the set of elements that appear in one set but not the other.

side effect

Any action a function takes other than returning a value. Assignment to variables, and input/output operations are examples of side effects.

single-test recursion

A style of recursive function in which there is only one end test. Single-test recursion is used when the function is guaranteed to eventually find what it's looking for, so there is no need to check for failure. An example would be the recursive definition of FACT, where the end test is ZEROP.

special form

See **special function**.

special function

A built-in function that does not evaluate its arguments. Special functions provide the primitive constructs, such as assignment, block structure, looping, and variable binding, from which the rest of Lisp is built. They do not return Lisp expressions to be evaluated, as macros do. Lisp programmers can create new macros, but they cannot create new special functions.

special variable

A dynamically scoped variable. When a name is declared special, all variables with that name will be dynamically scoped.

stream object

A Lisp object describing a connection to a file. Lisp programs read and write files by supplying an appropriate stream object as optional input to READ or FORMAT.

string

A sequence of characters enclosed in double quotes, such as the string "Foo Bar". Strings are vectors of character objects.

structure

A user-defined datatype composed of named slots. An example is the STARSHIP structure, whose slots are NAME, CAPTAIN, SPEED, SHIELDS, and CONDITION.

subset

A set x is a subset of a set y if every element of x is an element of y. See also **proper subset**.

substitution cipher

A method of secret writing in which one letter is substituted for another throughout the message. Substitution ciphers are easy to crack using letter frequency information. For example, E is the most frequently occurring letter in English text, so if a coded message contains more Qs than any other letter, Q probably deciphers to E.

symbol

One of the fundamental Lisp datatypes. Internally, symbols are composed of five cells: the name, value, function, plist, and package cells. Besides serving as data, symbols also serve as *names* for things, such as functions, variables, types, and blocks.

symbol name

Symbols are named by character strings. Each symbol contains a name cell that holds a pointer to the character string that is the symbol's name.

T

The standard way to say *true* in Lisp. T is a symbol. It evaluates to itself.

tail recursive

A function is tail recursive if it does all its work before making the recursive call. Tail recursive functions return the result of the recursive call without augmenting (modifying) it, or doing any other additional work. Clever Lisp compilers turn tail recursive calls into jump instructions, eliminating the need for a call stack.

terminal call

A call to a function that results in no further recursive calls. The function simply returns a value.

terminal node

A node of a tree with no descendants, in other words, a bottommost node. Terminal nodes are also called *leaves*.

top-level prompt

A prompt character such as ``>'' or ``*'' that indicates to the user that he or she is typing to the top-level read-eval-print loop.

TRACE

A tool for displaying function entries and exits.

tree

A structure composed of nodes and links, where each node has zero or more children and exactly one parent. An exception is the topmost node, or **root**, which has no parent. Trees may be represented as lists in Lisp.

truth function

A function whose inputs and output are truth values, that is, *true* or *false*.

type predicate
A predicate such as NUMBERP or CONSP that returns true if its input is a particular type of data.

type system
The set of datatypes a language offers, and their organization. The Lisp type system includes type predicates, a TYPE-OF function for generating type descriptions, and a facility for creating new datatypes with DEFSTRUCT.

unassigned variable
A variable that has no value.

unbound variable
''Unbound'' is an archaic term for ''unassigned,'' and is avoided in this book. See **unassigned variable.**

union
The union of two sets contains all the elements of each set. Each element appears only once in the union. See also **intersection**.

value cell
A cell in the internal representation of a symbol where Lisp keeps the value of the global lexical variable (or the currently accessible dynamic variable) named by that symbol.

variable
A place where a value is stored. Ordinary variables are named by symbols. Generalized variables are named by place descriptions, which may be Lisp expressions.

vector
A one-dimensional **array**.

Further Reading

Reference Works

Franz Inc., *Common Lisp: The Reference*. Addison-Wesley, Reading, MA, 1988.

Steele, Guy L. Jr., *Common Lisp: The Language*, 2nd edition. Digital Press, Woburn, MA, 1990.

Historical Material

Barstow, David R., Shrobe, Howard, E., and Sandewall, Erik (eds.), *Interactive Programming Environments,* McGraw-Hill, New York, 1984.

Gabriel, Richard P., "Lisp," in Stuart C. Shapiro (ed.), *Encyclopedia of Artificial Intelligence*, volume 1, pp. 508–528, John Wiley & Sons, New York, 1987.

McCarthy, John, "Recursive functions of symbolic expressions and their computation by machine," *Communications of the ACM* **3**(4), 184–195 (1960).

McCarthy, John, "History of Lisp," in D. Wexelblat (ed.), *History of Programming Languages,* Academic Press, New York, 1978.

McCarthy, John, Abrahams, Paul W., Edwards, Daniel J., Hart, Timothy P., and Levin, Michael I., *Lisp 1.5 Programmer's Guide*, 2nd ed., MIT Press, Cambridge, MA, 1965.

Advanced Material

Charniak, Eugene, Riesbeck, Christopher K., McDermott, Drew, and Meehan, James R., *Artificial Intelligence Programming,* 2nd ed., Lawrence Erlbaum Associates, Hillsdale, NJ, 1987.

Charniak, Eugene, and McDermott, Drew, *Artificial Intelligence*, Addison-Wesley, Reading, MA, 1985.

Gabriel, Richard P., *Performance and Evaluation of Lisp Systems,* MIT Press, Cambridge, MA, 1985.

Hofstadter, Douglas R., *Godel, Escher, Bach: an Eternal Golden Braid,* Basic Books, New York, 1979.

Keene, Sonya E., *Object-Oriented Programming in Common Lisp,* Addison-Wesley, Reading, MA, 1989.

Russell, Stuart, and Norvig, Peter, *Artificial Intelligence: A Modern Approach*, 3rd edition, Prentice Hall, Upper Saddle River, NJ, 2010.

Winston, Patrick H., *Artificial Intelligence,* 2nd ed., Addison-Wesley, Reading, MA, 1984.

Other Lisp Textbooks

Abelson, Harold, and Sussman, Gerald Jay, *Structure and Interpretation of Computer Programs,* MIT Press, Cambridge, MA, 1985.

Anderson, John R., Corbett, Albert T., and Reiser, Brian J., *Essential Lisp*, Addison-Wesley, Reading, MA, 1987.

Graham, Paul, *On Lisp: Advanced Techniques for Common Lisp*, Addison-Wesley, Reading, MA, 1993. Available free online at www.paulgraham.com.

Graham, Paul, *ANSI Common Lisp*, Prentice-Hall, Englewood Cliffs, NJ, 1995.

Wilensky, Robert, *Common LISPcraft*, W. W. Norton, New York, 1986.

Winston, Patrick H., and Horn, Berthold K. P., *Lisp*, 3rd ed., Addison-Wesley, Reading, MA, 1989.

Index

< predicate   10
= predicate   197
> predicate   10, 71
#′ notation   202, 225
( ) notation   383
#S notation   368
#\ notation   387
" character (string quotes)   288
′ character (quote)   87, 104
( )   *see* NIL
* convention   308
* function   2
+ function   2
, @ combination   414
, character (comma)   412
- function   2, 114
/ function   2
; character   150
` character (backquote)   412

A-lists (association lists)   179
ABS (absolute value)   2, 114
Accessor functions   369, 381
Actions in a COND clause   360
ADD1 example   12
Addition   2, 71
Addresses
   comparing with EQ   196
   of compiled code objects   105
   of symbols   195
   stored in cons cells   43
Algol   341
AND macro
   DeMorgan's theorem   134
   examples   122
   interchangeability with other conditionals   126
   use as a conditional   125
   vs. LOGICAL-AND   132

Apostrophe   *see* quoting
APPEND   161, 164, 169, 280, 335, 337
Applicative operators
   and eval notation   77
   creating   229, 282
   EVERY   214
   FIND-IF   207
   graphical representation   203, 207, 210, 214, 216
   lambda expressions for   205
   MAP   403
   MAPCAR   202, 224
   operating on multiple lists   224
   overview   201
   REDUCE   213
   REMOVE-IF   210
   REMOVE-IF-NOT   210
APPLY   110, 111, 225, 362
   See also   EVAL
APROPOS   151
Arcs in finite state machines   419
AREF   385
Argument list
   empty   103
   in function definition   82
Arguments
   functions without   103
   keyword   363
   names for   82
   of functions   80
   of macros   409
   of special functions   104, 411
   optional   360
   rest   361
Arithmetic
   errors   24
   expressions   283
   functions   2
   unary, with lists   70

Arrays
  accessing 385
  creating 383, 403
  headers 384
  initializing 386
  printing 385
  storing into 385
Art, recursion in 268
Artificial intelligence 27
Assertions in a database 219
Assignment
  in loops 345
  sequential, in DO* 351
  to local variables 312
  using generalized variables 315
  with INCF or DECF 309
  with PUSH and POP 310
  with SETF 138, 307
ASSOC 179, 183, 197, 200, 207
Association lists 179, 388
Asterisk convention 308
ATOM 67
Atomic datatypes 67
Augmenting recursion 252
Automobile diagnosis 374
&AUX keyword 364
Auxiliary variables 364

Backquote character 412
Backtrace 272
Binary trees 285
Binding of variables 157, 347
BLOCK special function 354
Block structure 353
Blocks world examples 178, 181, 219, 337
&BODY keyword 429
Body of a function 82, 140
Boole, George 132
boolean functions
  DeMorgan's theorem on 134
  examples 132
  logical completeness 135
  truth tables 133
Box notation 1, 77, 79
BREAK 272, 326
Buckets in hash tables 389
Built-in functions 12

CAAR 49
CADDR 46
CADR 46

CAR
  function 43
  of dotted pairs 75
  of nested lists 47
  of NIL 50
  part of cons cell 43
  storing into 315
  symmetry with CONS 57
  use in NTH 167
CAR/CDR recursion 262
Cardinality of a set 175
Cards and applicatives 212
Carriage return in FORMAT 289
Cascading of CONS 56
Case distinctions and EQUALP 380
Cat in the Hat 268
CDAR 46
CDDR 47
CDR
  function 43
  of dotted pairs 75
  of nested lists 47
  of NIL 50
  of single-element lists 44
  of unary numbers 70
  part of cons cell 43
  storing into 332
  symmetry with CONS 57
CERROR 328
Character objects 387, 402
Character strings 288
Church, Alonzo 106
Ciphers 395
Circular lists 74, 248, 304, 334
Clauses
  in AND 122
  in COND 116
  in OR 122
  use of T as the test 117
CLOS (Common Lisp Object System) 365
Closure objects 206, 225, 226, 230
CLRHASH 398
COERCE 402
Collatz's conjecture 247
Comma with backquote 412
Comments in programs 150
Common Lisp 28
COMPILE 415
COMPILE-FILE 415, 417
Compiled function objects 202
Compilers
  for finite state machines 428
  for Lisp 415

Completeness of boolean functions 135
Complex predicates 123
Composite numbers 285
Concatenation
   destructive 335
   of lists 161
COND macro
   consequent part 116, 360
   examples 116
   in complex predicates 123
   in recursive functions 242
   interchangeability with IF 126
   parenthesis errors in 119
   side effects in consequent part 308
   See also IF
Conditional augmentation 258
Conditionals
   AND 122
   COND 116
   Demorgan's theorem and 134
   IF 113
   interchangeability of 126
   making decisions 113
   nested conditionals 126
   OR 122
Cons cells
   circular structures 74, 248, 304, 334
   destructive operations on 334, 335, 336
   graphical notation for 160
   parts of 43
   picture of 43
CONS function
   and NIL 55
   asymmetry of arguments 160
   building nested lists 56
   examples 52
   relation to LIST and APPEND 164
   symmetry with CAR/CDR 57
   See also CAR, CDR, LIST
Consequent part of a COND clause 116
CONSP 66
Constants 150
Constructor functions 368, 370
Control characters 97
Control structures
   applicative operators 201
   conditionals 113
   iteration 341
   recursion 231
:COUNT keyword 199, 226
Craps (dice game) 151
Cryptogram exercise 395
CTRL key 97

Cursor position, for output 289

Data 1
Databases
   blocks world 219
   genealogical 275
   property lists used for 390
Debugger 272
DECF macro 309
Decision making in programs 113
Declarations 418, 436
DEFCONSTANT macro 439
Defining
   constants 439
   functions 12, 82
   macros 408
   parameters 439
   special variables 436
DEFMACRO macro 408
DEFPARAMETER macro 439
DEFSTRUCT macro 367, 378, 380, 406
DEFUN macro 82, 95, 205
DEFVAR macro 139, 308, 436, 439
DELETE 337
DeMorgan's Theorem 134
Deoxyribonucleic acid (DNA) 355
Depth of cons cell structure 44, 254
Depth of parenthesization 33, 266
DESCRIBE 372, 389
Destructive operations
   on lists 332, 334, 335, 336
   on sequences 386
   on strings 387
Destructuring 430
Deviant list structures 74
Diagnosis problems 374
Dice exercise 151
:DIRECTION keyword 295
Discrimination nets 374
Division
   by zero 24
   function 2
DNA exercise 355
DO macro 347, 349, 352, 427
DO with empty body 348
DO* macro 351
DOCUMENTATION 149
Documentation string 150
DOLIST macro 341, 346
Dot notation 303
DOTIMES macro 341, 427

Dotted pairs
  examples 72, 78
  hybrid notation 304
  use with SUBLIS 193
  See also Cons cell
Double-test tail recursion 250
DOVECTOR example 431
Dragon stories 232, 236, 238, 241, 244
Drawing Hands lithograph 268
DRIBBLE 298
DTRACE tool 218, 234, 408
DUNTRACE macro 218
Dynamic binding 438
Dynamic scoping 158, 435

Efficiency of list operations 194
Elegance in programming 310, 312, 344, 349
:ELEMENT-TYPE keyword 403
Elements of a list 31
Empty argument lists 103
Empty list *see* NIL
End-of-file conditions 302
Entries in tables 179
EQ 195, 302, 390
EQL 197, 336
EQUAL 10, 195, 336, 379
Equality
  implicit tests involving 197
  of addresses 196
  of lists 38
  of numbers 197
  of sets 175
  of strings 198
  of structures 379
  of symbols 10
EQUALP 198, 380
ERROR 326
Errors
  division by zero 24
  fatal 99
  misdefining functions 91
  misuse of parentheses in COND 120
  misuse of quotes 87
  non-list input 36, 40
  recovery from 98
  types of errors 24
  unassigned variable 120
  undefined function 120
  use of AND and OR to avoid 125
  wrong-number-of-inputs 24
  wrong-type input 24, 88
Escher, M. C. 268
EVAL function 78, 110

See also APPLY
EVAL notation
  defined 77
  need for quoting 87
  vs. box notation 79
Evaltrace diagrams 81, 88, 109, 127, 141, 142, 143, 144,
  145, 146, 155, 156, 227, 228, 410, 434
Evaltrace notation
  for assignment 141
  for conditionals 127
  for dynamic variables 437
  for expressions 80
  for function entry and exit 85
  for LET forms 143
  for LET* forms 144
  for lexical closures 227
  for macro expansion 407, 409
  for nested contexts 93
  for variable creation 84
  for variable references 155
  relationship to STEP 131
  scope boundaries 109, 138, 156, 228, 436
  suppression of detail 85
Evaluation rules
  for AND and OR 122
  for COND 116
  for IF 114
  for keywords 199
  for lists 80
  for NIL 80
  for numbers 80
  for quoted objects 87, 104
  for strings 288
  for symbols 86
  for T 80
  summarized 95
EVERY operator 214, 215
Exclusive-OR truth function 22
Exiting a loop 342
Expressions
  arithmetic 283
  lambda 205
  Lisp 78
  returned by macros 406
  S-expressions 283
Extracting elements from a list 39, 167

Factorial function 236, 237, 345, 349
Factorization 285
False-part (of IF conditional) 114
Falsity 7
Family trees 275
FETCH example 219

Fibonacci numbers 244, 246, 262, 281, 355
File i/o 294, 295
FIND-IF operator 207, 226, 386, 403
Finite state machines 418, 428
FIRST 39, 44
Floating point numbers 3
Format control string 289, 299
FORMAT function 289
Forms (expressions) 140, 141, 144
:FROM-END keyword 199, 226
FUNCALL 202, 225, 229, 282
Function cells 154
Function objects 202
FUNCTION special function 225
Functions
    argument list 82
    arguments 80
    as data 77
    as return values 230
    body 82, 140, 360
    boolean 132
    built-in 12
    cascading 56
    defining new 12
    defining with DEFUN 82
    inputs of 4, 80
    local 282
    logical 132
    on numbers 2
    outputs 4
    plotting 296
    recursive 231
    result 1
    side effects 140
    tail recursive 279
    what is a function 1
    without arguments 103

Games
    craps 151
    Rock-Scissors-Paper 125
    tic-tac-toe 315, 328
Garbage collection 400
Genealogy exercise 275
Generalized variables 315
Gensyms 407
GET 390
GETHASH 388
Global variables 138, 154, 155, 208, 308
GO special function 427
Graphical representation of applicatives 203, 207, 210, 214, 216
Graphing exercise 296

Hash tables 388
Helping functions 228, 266
Histogram exercise 393
Hofstadter, Douglas 268
Hybrid notation 304

I/O (input/output) 287
IF special function
    examples 113
    nested IF 126
    See also AND, COND, OR
Implicit assignment 349
Implicit blocks 353
INCF macro 309, 406
:INCLUDE keyword 380
Indicator 389
Infinite loop 352
Infinite recursion 244, 246
Inheritance and structures 380
:INITIAL-CONTENTS keyword 386
:INITIAL-ELEMENT keyword 386
Input to a function 4
Input/output 287
INSPECT 373, 389
Integers 3
Internal representation of lists 31
Internal structure of symbols 105
INTERSECTION 172, 200
Iteration 341
Iteration forms
    DO 347
    DO* 351
    DOLIST 341
    DOTIMES 341

&KEY keyword 363
Key part of table entry 179
Keyword arguments 198, 226, 363, 370
KEYWORDP 199

LABELS special functions 282
Lambda expressions 205, 225, 230
Lambda notation 106
lambda-list keywords
    &AUX 364
    &BODY 429
    &KEY 363
    &OPTIONAL 360, 409
    &REST 361
LAST 168, 169
Lemonade stand example 308

LENGTH
   and unary numbers 71
   of dotted lists 75
   of lists 35
   of NIL 38
   of sequences 386
   of strings 387
Length of arrays 384
Length of lists 35
LET special function 141, 158, 347, 360, 407
LET* special function 144, 158, 351, 364
Levels of parentheses 33
Lexical closures 206, 225, 226, 230
Lexical scoping 156, 435
Lisp
   acronym 31
   compiler 415
   Lisp 1.5 dialect 301, 339, 389
LIST 58, 164
List consing recursion 254
     See also CONS
List surgery 332
LISTP 66
Lists
   as sets 170
   as tables 179
   as trees 192, 284
   as unary numbers 70
   circular 74, 248, 304, 334
   concatenation 161
   cons cell notation 160
   constructing new 52, 58, 89
   dot notation 303
   empty list 37, 55
   hybrid notation 304
   internal representation 31
   last cons cell 168
   length 35
   LISt Processor 31
   nested 33, 56, 162
   parenthesis notation 160
   printed representation 31
   reversal 165
   shared structure 195
   substitution 192, 193
Literature, recursion in 268
Local functions 282
Local variables 137, 141, 208, 312
Logical completeness 135
Logical functions 132
Looping *see* Iteration

McCarthy, John 27, 107

Macro functions
   as shorthand 405
   computing expansion 406
   defining 408
   destructuring 430
   displaying expansion 407
   historical significance 435
   lexical scoping 433
   use during compilation 417
   vs.ordinaryfunctions' 132
MAKE-ARRAY 386, 403
MAKE-HASH-TABLE 388
MAKE-STARSHIP example 368
Map of Robbie's house 188
MAP operator 403
MAPCAR operator 202, 224, 282, 346, 354
MEMBER 170, 197, 199
Memory usage 358, 400
Multiple recursion 260
Multiplication 2
Music, transposing 208

Names
   of arguments 82
   of blocks 353
   of functions 2, 107
   of symbols 105
   of variables 84
NAND truth function 135
NCONC 335, 337
Negation
   of numbers 114
   of predicates 20
Nerd state example 184
Nested IF 126
Nested lists
   and APPEND 162
   CAR and CDR of 47
   constructing new 56
   examples 33
   internal representation 33
   recursion on 262
Newline in FORMAT 289
NIL
   as argument list 103
   as empty list 37
   as logical value 132
   as symbol 7, 37
   CAR and CDR of 50
   evaluation rule for 80
   internal representation 154
   meaning false 7
   summary of interesting properties 68

NINTERSECTION 336
Nodes
    of discrimination nets 374
    of finite state machines 419
    of trees 283
Non-list cons structures 72
Nondestructive functions 162, 168
Nonterminal nodes of a tree 283
NOT predicate 18
Notes, transposing 208
NREVERSE 336
NSET-DIFFERENCE 336
NSUBST 336
NTH 167, 169
NTHCDR 166, 169
NULL 67
Number theory 247
NUMBERP 8
Numbers
    composite 285
    definition of a number 7
    evaluation rule 80
    factorization 285
    Fibonacci 244, 246, 262, 355
    floating point 3
    integers 3
    internal representation 197
    negation 114
    prime 285
    ratios 3
    type predicate for 8
    unary 70
NUNION 336

ODDP 9
Online documentation 149
Operators, applicative 201
Opposite predicates 20
&OPTIONAL keyword 360, 409
OR macro
    DeMorgan's theorem 134
    examples 122
    interchangeability with other conditionals 126
    use as a conditional 125
Order of inputs 4
Output functions 289, 301
Output of a function 4

Palindromes 170
Parameters to format directives 299
Parenthesis notation 31, 160
Pattern matcher 219
Patterns and their interpretations 220

PI 150
Place descriptions 315
Playing cards 212
Plotting program 296
PLUSP 125
Pointers 32
POP macro 310
PPMX tool 407, 426
Predicates
    and applicatives 207, 210, 214
    complex 123
    defined 8
    negation 20
Prime numbers 285
Primitive functions 12
PRIN1 301
PRINC 301
PRINT 301
PRINT-ARRRAY 385
PRINT-BASE 441
PRINT-CIRCLE 441
:PRINT-FUNCTION keyword 378
PROG1 359, 362
PROG2 359, 362
PROGN 359, 362
Prompt string 97, 272
Proper lists 72
Proper subsets 175
Property lists 389, 401
PUSH macro 310
PUSHNEW macro 391

Query functions 293
QUOTE special function 104, 225
Quotient 2
Quoting of objects 88, 89

RANDOM 140, 147, 151, 394
RASSOC 180, 200
Ratios 3
READ 98, 292, 294
Read-eval-print loop 98, 368
Rebinding special variables 440
Reciprocal function (1/x) 23
Recording interactive sessions 298
Recursion
    art and literature examples 268
    data structures 283
    factorial example 236, 237
    Fibonacci example 244
    helping functions 266
    infinite 244, 246
    significance of 231

tail 279
three rules of 241, 243
two-part 280
vs. iteration 344, 346
Recursion templates
  augmenting tail 252
  CAR/CDR 262
  conditional augmentation 258
  double-test tail 250
  list consing 254
  multiple 260
  simultaneous 256
  single-test tail 250
REDUCE operator 213, 226, 403
REMOVE 168, 169, 199
REMOVE-DUPLICATES 183
REMOVE-IF operator 210, 226
REMOVE-IF-NOT operator 210, 226
REMPROP 391
Replacing elements in a list 63
REST 40, 44
&REST keyword 361
Result of a function 1
RETURN 342
RETURN-FROM 353
REVERSE 165, 169, 280, 386
Robbie the Robot 188
Rock-Scissors-Paper game 125
ROOM 400
Rules of recursion 241, 243

S-expressions 283
Scheme xi
Scheme dialect 435
Scope of variables 109
Scoping 226
SCRAWL 187
SDRAW tool 186
SDRAW-LOOP 187
SECOND 39, 46
Sequences 386, 402
SET 338
SET-DIFFERENCE 173, 200, 211
SETF macro
  assigning to globals 138
  assignment with 307
  macroexpansion 406
  of AREF 385
  of SYMBOL-PLIST 402
SETQ special function
  history 338
  macroexpansion of INCF 406

Sets
  cardinality 175
  equality 175
  examples 170
  functions on 170, 172, 173, 174, 183
  intersection 172
  proper subset 175
  removing duplicates 183
  set difference 173, 211
  subset predicate 174
  subsets 174, 210
  symmetric difference 181
  test for membership in 170
  union 173
Shared structure 195
SHOWVAR example 413
Side effects 140, 147, 310
SIMPLE-STRING type 366
Simultaneous recursion 256
Single-test tail recursion 250
Special forms see Special functions
Special functions
  BLOCK 354
  FUNCTION 225
  IF 113
  LET 141
  LET* 144
  QUOTE 104
  SETQ 338
  vs. macro functions 411
  vs. ordinary functions 113
Special variables 436, 440
Splicing with backquote 414
SQRT 2
Square roots 2, 416
Stacks 310
Starship example 367
State transitions 419
STEP tool 130
Stream objects 294
STRING-CHAR type 366, 403
STRINGP 288
Strings 288, 366, 387, 402
Structures
  accessing 369
  as datatypes 365
  creating 368
  defining 367
  equality tests 379
  modifying 369
  print function 378
  redefining 371
SUBLIS 193, 200

Subscripts in arrays 383
SUBSETP 174
Subsets
    of a set 174, 210
    proper subset 175
SUBST 192, 200, 336
Substitution ciphers 395
Subtraction 2
Suess, Dr. 268
SYMBOL-PLIST 391, 401
SYMBOL-VALUE 339
SYMBOLPZ 8
Symbols
    definition of a symbol 7
    examples 6
    function cell 154
    internal structure 105
    property list 389, 401
    type predicate for 8
    value cell 153, 339
Symmetric set difference 181
Symmetry of CONS and CAR/CDR 57

T
    as argument to FORMAT 289
    as logical value 132
    as output of predicates 8
    evaluation rule for 80
    in COND clauses 117
    internal representation 154
    special meaning 7
Tables
    and applicative operators 203
    ASSOC function 179
    examples 179
    extracting portions with MAPCAR 204
    functions on 193
    RASSOC function 180
    searching with FIND-IF 207
    use with SUBLIS 193
TAGBODY special function 427
Tail recursion 279
Terminal nodes of trees 283
TERMINAL-IO 294
TERPRI 301
:TEST keyword 200, 336
Test part of a COND clause 116
THIRD 39, 46
Three rules of recursion 241, 243
Tic-tac-toe 315, 328
TIME macro 358

Tools
    debugger 272
    DESCRIBE 372, 389
    DRIBBLE 298
    DTRACE 218, 234, 408
    DUNTRACE 218
    INSPECT 373, 389
    PPMX 407, 426
    ROOM 400
    SCRAWL 187
    SDRAW 186
    SDRAW-LOOP 187
    STEP 130
    TIME 358
    TRACE 216, 234
    UNTRACE 216
Top-level prompt 97
TRACE tool 216, 234
Tracing macros 408
Transposing music 208
Trees
    as nested lists 192
    binary 285
    for arithmetic expressions 283
    genealogical 275
    nonterminal nodes 283
    S-expressions 283
    terminal nodes 283
True-part (of IF conditional) 114
Truth 7
Truth functions
    LOGICAL-AND example 132
    XOR example 21
Truth tables 133
Type hierarchy 366
Type predicates
    ATOM 67
    CONSP 66
    for structures 368, 381
    KEYWORDP 199
    LISTP 66
    NULL 67
    NUMBERP 8
    STRINGP 288
    SYMBOLP 8
Type system 365
TYPE-OF 202, 366
TYPEP 366

Unary arithmetic 70
Unassigned variable error 120, 137
Unbound variable 157
Undefined function error 120

UNION   173, 200
UNLESS macro   313
UNTRACE macro   217
Updating local variables   312
Updating methods for variables   309
USER package   151

Value cells   153, 339
Variables
    assignment   307
    auxiliary   364
    binding   157
    generalized   315
    global   138, 154, 155, 208
    in DO macro   347
    in DO*   351
    in LET   141
    in LET*   144
    local   137, 141, 208, 312
    updating   312
Vectors   383, 387, 403
Vending machines   418, 428

WARN   328
WHEN macro   313
WITH-OPEN-FILE macro
    reading files   294
    writing files   295
Wrong number of inputs error   24
Wrong-type input error   24, 88

XOR truth function   22

Y-OR-N-P   293
YES-OR-NO-P   293

Zero divide error   24
Zero, factorial of   236
ZEROP   9, 71

~% directive   289
~& directive   289
~A directive   290
~D directive   300
~F directive   300
~S directive   290

A CATALOG OF SELECTED
DOVER BOOKS
IN SCIENCE AND MATHEMATICS

Engineering

FUNDAMENTALS OF ASTRODYNAMICS, Roger R. Bate, Donald D. Mueller, and Jerry E. White. Teaching text developed by U.S. Air Force Academy develops the basic two-body and n-body equations of motion; orbit determination; classical orbital elements, coordinate transformations; differential correction; more. 1971 edition. 455pp. 5 3/8 x 8 1/2.　　0-486-60061-0

INTRODUCTION TO CONTINUUM MECHANICS FOR ENGINEERS: Revised Edition, Ray M. Bowen. This self-contained text introduces classical continuum models within a modern framework. Its numerous exercises illustrate the governing principles, linearizations, and other approximations that constitute classical continuum models. 2007 edition. 320pp. 6 1/8 x 9 1/4.
0-486-47460-7

ENGINEERING MECHANICS FOR STRUCTURES, Louis L. Bucciarelli. This text explores the mechanics of solids and statics as well as the strength of materials and elasticity theory. Its many design exercises encourage creative initiative and systems thinking. 2009 edition. 320pp. 6 1/8 x 9 1/4.　　0-486-46855-0

FEEDBACK CONTROL THEORY, John C. Doyle, Bruce A. Francis and Allen R. Tannenbaum. This excellent introduction to feedback control system design offers a theoretical approach that captures the essential issues and can be applied to a wide range of practical problems. 1992 edition. 224pp. 6 1/2 x 9 1/4.　　0-486-46933-6

THE FORCES OF MATTER, Michael Faraday. These lectures by a famous inventor offer an easy-to-understand introduction to the interactions of the universe's physical forces. Six essays explore gravitation, cohesion, chemical affinity, heat, magnetism, and electricity. 1993 edition. 96pp. 5 3/8 x 8 1/2.　　0-486-47482-8

DYNAMICS, Lawrence E. Goodman and William H. Warner. Beginning engineering text introduces calculus of vectors, particle motion, dynamics of particle systems and plane rigid bodies, technical applications in plane motions, and more. Exercises and answers in every chapter. 619pp. 5 3/8 x 8 1/2.　　0-486-42006-X

ADAPTIVE FILTERING PREDICTION AND CONTROL, Graham C. Goodwin and Kwai Sang Sin. This unified survey focuses on linear discrete-time systems and explores natural extensions to nonlinear systems. It emphasizes discrete-time systems, summarizing theoretical and practical aspects of a large class of adaptive algorithms. 1984 edition. 560pp. 6 1/2 x 9 1/4.
0-486-46932-8

INDUCTANCE CALCULATIONS, Frederick W. Grover. This authoritative reference enables the design of virtually every type of inductor. It features a single simple formula for each type of inductor, together with tables containing essential numerical factors. 1946 edition. 304pp. 5 3/8 x 8 1/2.　　0-486-47440-2

THERMODYNAMICS: Foundations and Applications, Elias P. Gyftopoulos and Gian Paolo Beretta. Designed by two MIT professors, this authoritative text discusses basic concepts and applications in detail, emphasizing generality, definitions, and logical consistency. More than 300 solved problems cover realistic energy systems and processes. 800pp. 6 1/8 x 9 1/4.
0-486-43932-1

THE FINITE ELEMENT METHOD: Linear Static and Dynamic Finite Element Analysis, Thomas J. R. Hughes. Text for students without in-depth mathematical training, this text includes a comprehensive presentation and analysis of algorithms of time-dependent phenomena plus beam, plate, and shell theories. Solution guide available upon request. 672pp. 6 1/2 x 9 1/4.
0-486-41181-8

HELICOPTER THEORY, Wayne Johnson. Monumental engineering text covers vertical flight, forward flight, performance, mathematics of rotating systems, rotary wing dynamics and aerodynamics, aeroelasticity, stability and control, stall, noise, and more. 189 illustrations. 1980 edition. 1089pp. 5 5/8 x 8 1/4. 0-486-68230-7

MATHEMATICAL HANDBOOK FOR SCIENTISTS AND ENGINEERS: Definitions, Theorems, and Formulas for Reference and Review, Granino A. Korn and Theresa M. Korn. Convenient access to information from every area of mathematics: Fourier transforms, Z transforms, linear and nonlinear programming, calculus of variations, random-process theory, special functions, combinatorial analysis, game theory, much more. 1152pp. 5 3/8 x 8 1/2.
0-486-41147-8

A HEAT TRANSFER TEXTBOOK: Fourth Edition, John H. Lienhard V and John H. Lienhard IV. This introduction to heat and mass transfer for engineering students features worked examples and end-of-chapter exercises. Worked examples and end-of-chapter exercises appear throughout the book, along with well-drawn, illuminating figures. 768pp. 7 x 9 1/4.
0-486-47931-5

BASIC ELECTRICITY, U.S. Bureau of Naval Personnel. Originally a training course; best nontechnical coverage. Topics include batteries, circuits, conductors, AC and DC, inductance and capacitance, generators, motors, transformers, amplifiers, etc. Many questions with answers. 349 illustrations. 1969 edition. 448pp. 6 1/2 x 9 1/4. 0-486-20973-3

BASIC ELECTRONICS, U.S. Bureau of Naval Personnel. Clear, well-illustrated introduction to electronic equipment covers numerous essential topics: electron tubes, semiconductors, electronic power supplies, tuned circuits, amplifiers, receivers, ranging and navigation systems, computers, antennas, more. 560 illustrations. 567pp. 6 1/2 x 9 1/4. 0-486-21076-6

BASIC WING AND AIRFOIL THEORY, Alan Pope. This self-contained treatment by a pioneer in the study of wind effects covers flow functions, airfoil construction and pressure distribution, finite and monoplane wings, and many other subjects. 1951 edition. 320pp. 5 3/8 x 8 1/2.
0-486-47188-8

SYNTHETIC FUELS, Ronald F. Probstein and R. Edwin Hicks. This unified presentation examines the methods and processes for converting coal, oil, shale, tar sands, and various forms of biomass into liquid, gaseous, and clean solid fuels. 1982 edition. 512pp. 6 1/8 x 9 1/4.
0-486-44977-7

THEORY OF ELASTIC STABILITY, Stephen P. Timoshenko and James M. Gere. Written by world-renowned authorities on mechanics, this classic ranges from theoretical explanations of 2- and 3-D stress and strain to practical applications such as torsion, bending, and thermal stress. 1961 edition. 560pp. 5 3/8 x 8 1/2. 0-486-47207-8

PRINCIPLES OF DIGITAL COMMUNICATION AND CODING, Andrew J. Viterbi and Jim K. Omura. This classic by two digital communications experts is geared toward students of communications theory and to designers of channels, links, terminals, modems, or networks used to transmit and receive digital messages. 1979 edition. 576pp. 6 1/8 x 9 1/4. 0-486-46901-8

LINEAR SYSTEM THEORY: The State Space Approach, Lotfi A. Zadeh and Charles A. Desoer. Written by two pioneers in the field, this exploration of the state space approach focuses on problems of stability and control, plus connections between this approach and classical techniques. 1963 edition. 656pp. 6 1/8 x 9 1/4 0-486-46663-9

Mathematics–Bestsellers

HANDBOOK OF MATHEMATICAL FUNCTIONS: with Formulas, Graphs, and Mathematical Tables, Edited by Milton Abramowitz and Irene A. Stegun. A classic resource for working with special functions, standard trig, and exponential logarithmic definitions and extensions, it features 29 sets of tables, some to as high as 20 places. 1046pp. 8 x 10 1/2.
0-486-61272-4

ABSTRACT AND CONCRETE CATEGORIES: The Joy of Cats, Jiri Adamek, Horst Herrlich, and George E. Strecker. This up-to-date introductory treatment employs category theory to explore the theory of structures. Its unique approach stresses concrete categories and presents a systematic view of factorization structures. Numerous examples. 1990 edition, updated 2004. 528pp. 6 1/8 x 9 1/4.
0-486-46934-4

MATHEMATICS: Its Content, Methods and Meaning, A. D. Aleksandrov, A. N. Kolmogorov, and M. A. Lavrent'ev. Major survey offers comprehensive, coherent discussions of analytic geometry, algebra, differential equations, calculus of variations, functions of a complex variable, prime numbers, linear and non-Euclidean geometry, topology, functional analysis, more. 1963 edition. 1120pp. 5 3/8 x 8 1/2.
0-486-40916-3

INTRODUCTION TO VECTORS AND TENSORS: Second Edition–Two Volumes Bound as One, Ray M. Bowen and C.-C. Wang. Convenient single-volume compilation of two texts offers both introduction and in-depth survey. Geared toward engineering and science students rather than mathematicians, it focuses on physics and engineering applications. 1976 edition. 560pp. 6 1/2 x 9 1/4.
0-486-46914-X

AN INTRODUCTION TO ORTHOGONAL POLYNOMIALS, Theodore S. Chihara. Concise introduction covers general elementary theory, including the representation theorem and distribution functions, continued fractions and chain sequences, the recurrence formula, special functions, and some specific systems. 1978 edition. 272pp. 5 3/8 x 8 1/2.
0-486-47929-3

ADVANCED MATHEMATICS FOR ENGINEERS AND SCIENTISTS, Paul DuChateau. This primary text and supplemental reference focuses on linear algebra, calculus, and ordinary differential equations. Additional topics include partial differential equations and approximation methods. Includes solved problems. 1992 edition. 400pp. 7 1/2 x 9 1/4.
0-486-47930-7

PARTIAL DIFFERENTIAL EQUATIONS FOR SCIENTISTS AND ENGINEERS, Stanley J. Farlow. Practical text shows how to formulate and solve partial differential equations. Coverage of diffusion-type problems, hyperbolic-type problems, elliptic-type problems, numerical and approximate methods. Solution guide available upon request. 1982 edition. 414pp. 6 1/8 x 9 1/4.
0-486-67620-X

VARIATIONAL PRINCIPLES AND FREE-BOUNDARY PROBLEMS, Avner Friedman. Advanced graduate-level text examines variational methods in partial differential equations and illustrates their applications to free-boundary problems. Features detailed statements of standard theory of elliptic and parabolic operators. 1982 edition. 720pp. 6 1/8 x 9 1/4.   0-486-47853-X

LINEAR ANALYSIS AND REPRESENTATION THEORY, Steven A. Gaal. Unified treatment covers topics from the theory of operators and operator algebras on Hilbert spaces; integration and representation theory for topological groups; and the theory of Lie algebras, Lie groups, and transform groups. 1973 edition. 704pp. 6 1/8 x 9 1/4.   0-486-47851-3

Browse over 9,000 books at www.doverpublications.com

A SURVEY OF INDUSTRIAL MATHEMATICS, Charles R. MacCluer. Students learn how to solve problems they'll encounter in their professional lives with this concise single-volume treatment. It employs MATLAB and other strategies to explore typical industrial problems. 2000 edition. 384pp. 5 3/8 x 8 1/2.   0-486-47702-9

NUMBER SYSTEMS AND THE FOUNDATIONS OF ANALYSIS, Elliott Mendelson. Geared toward undergraduate and beginning graduate students, this study explores natural numbers, integers, rational numbers, real numbers, and complex numbers. Numerous exercises and appendixes supplement the text. 1973 edition. 368pp. 5 3/8 x 8 1/2.   0-486-45792-3

A FIRST LOOK AT NUMERICAL FUNCTIONAL ANALYSIS, W. W. Sawyer. Text by renowned educator shows how problems in numerical analysis lead to concepts of functional analysis. Topics include Banach and Hilbert spaces, contraction mappings, convergence, differentiation and integration, and Euclidean space. 1978 edition. 208pp. 5 3/8 x 8 1/2.
0-486-47882-3

FRACTALS, CHAOS, POWER LAWS: Minutes from an Infinite Paradise, Manfred Schroeder. A fascinating exploration of the connections between chaos theory, physics, biology, and mathematics, this book abounds in award-winning computer graphics, optical illusions, and games that clarify memorable insights into self-similarity. 1992 edition. 448pp. 6 1/8 x 9 1/4.
0-486-47204-3

SET THEORY AND THE CONTINUUM PROBLEM, Raymond M. Smullyan and Melvin Fitting. A lucid, elegant, and complete survey of set theory, this three-part treatment explores axiomatic set theory, the consistency of the continuum hypothesis, and forcing and independence results. 1996 edition. 336pp. 6 x 9.   0-486-47484-4

DYNAMICAL SYSTEMS, Shlomo Sternberg. A pioneer in the field of dynamical systems discusses one-dimensional dynamics, differential equations, random walks, iterated function systems, symbolic dynamics, and Markov chains. Supplementary materials include PowerPoint slides and MATLAB exercises. 2010 edition. 272pp. 6 1/8 x 9 1/4.   0-486-47705-3

ORDINARY DIFFERENTIAL EQUATIONS, Morris Tenenbaum and Harry Pollard. Skillfully organized introductory text examines origin of differential equations, then defines basic terms and outlines general solution of a differential equation. Explores integrating factors; dilution and accretion problems; Laplace Transforms; Newton's Interpolation Formulas, more. 818pp. 5 3/8 x 8 1/2.   0-486-64940-7

MATROID THEORY, D. J. A. Welsh. Text by a noted expert describes standard examples and investigation results, using elementary proofs to develop basic matroid properties before advancing to a more sophisticated treatment. Includes numerous exercises. 1976 edition. 448pp. 5 3/8 x 8 1/2.   0-486-47439-9

THE CONCEPT OF A RIEMANN SURFACE, Hermann Weyl. This classic on the general history of functions combines function theory and geometry, forming the basis of the modern approach to analysis, geometry, and topology. 1955 edition. 208pp. 5 3/8 x 8 1/2.   0-486-47004-0

THE LAPLACE TRANSFORM, David Vernon Widder. This volume focuses on the Laplace and Stieltjes transforms, offering a highly theoretical treatment. Topics include fundamental formulas, the moment problem, monotonic functions, and Tauberian theorems. 1941 edition. 416pp. 5 3/8 x 8 1/2.   0-486-47755-X

Browse over 9,000 books at www.doverpublications.com

Mathematics–Logic and Problem Solving

PERPLEXING PUZZLES AND TANTALIZING TEASERS, Martin Gardner. Ninety-three riddles, mazes, illusions, tricky questions, word and picture puzzles, and other challenges offer hours of entertainment for youngsters. Filled with rib-tickling drawings. Solutions. 224pp. 5 3/8 x 8 1/2.
0-486-25637-5

MY BEST MATHEMATICAL AND LOGIC PUZZLES, Martin Gardner. The noted expert selects 70 of his favorite "short" puzzles. Includes The Returning Explorer, The Mutilated Chessboard, Scrambled Box Tops, and dozens more. Complete solutions included. 96pp. 5 3/8 x 8 1/2.
0-486-28152-3

THE LADY OR THE TIGER?: and Other Logic Puzzles, Raymond M. Smullyan. Created by a renowned puzzle master, these whimsically themed challenges involve paradoxes about probability, time, and change; metapuzzles; and self-referentiality. Nineteen chapters advance in difficulty from relatively simple to highly complex. 1982 edition. 240pp. 5 3/8 x 8 1/2.
0-486-47027-X

SATAN, CANTOR AND INFINITY: Mind-Boggling Puzzles, Raymond M. Smullyan. A renowned mathematician tells stories of knights and knaves in an entertaining look at the logical precepts behind infinity, probability, time, and change. Requires a strong background in mathematics. Complete solutions. 288pp. 5 3/8 x 8 1/2.
0-486-47036-9

THE RED BOOK OF MATHEMATICAL PROBLEMS, Kenneth S. Williams and Kenneth Hardy. Handy compilation of 100 practice problems, hints and solutions indispensable for students preparing for the William Lowell Putnam and other mathematical competitions. Preface to the First Edition. Sources. 1988 edition. 192pp. 5 3/8 x 8 1/2.
0-486-69415-1

KING ARTHUR IN SEARCH OF HIS DOG AND OTHER CURIOUS PUZZLES, Raymond M. Smullyan. This fanciful, original collection for readers of all ages features arithmetic puzzles, logic problems related to crime detection, and logic and arithmetic puzzles involving King Arthur and his Dogs of the Round Table. 160pp. 5 3/8 x 8 1/2.
0-486-47435-6

UNDECIDABLE THEORIES: Studies in Logic and the Foundation of Mathematics, Alfred Tarski in collaboration with Andrzej Mostowski and Raphael M. Robinson. This well-known book by the famed logician consists of three treatises: "A General Method in Proofs of Undecidability," "Undecidability and Essential Undecidability in Mathematics," and "Undecidability of the Elementary Theory of Groups." 1953 edition. 112pp. 5 3/8 x 8 1/2.
0-486-47703-7

LOGIC FOR MATHEMATICIANS, J. Barkley Rosser. Examination of essential topics and theorems assumes no background in logic. "Undoubtedly a major addition to the literature of mathematical logic." – *Bulletin of the American Mathematical Society.* 1978 edition. 592pp. 6 1/8 x 9 1/4.
0-486-46898-4

INTRODUCTION TO PROOF IN ABSTRACT MATHEMATICS, Andrew Wohlgemuth. This undergraduate text teaches students what constitutes an acceptable proof, and it develops their ability to do proofs of routine problems as well as those requiring creative insights. 1990 edition. 384pp. 6 1/2 x 9 1/4.
0-486-47854-8

FIRST COURSE IN MATHEMATICAL LOGIC, Patrick Suppes and Shirley Hill. Rigorous introduction is simple enough in presentation and context for wide range of students. Symbolizing sentences; logical inference; truth and validity; truth tables; terms, predicates, universal quantifiers; universal specification and laws of identity; more. 288pp. 5 3/8 x 8 1/2. 0-486-42259-3

Mathematics–Algebra and Calculus

VECTOR CALCULUS, Peter Baxandall and Hans Liebeck. This introductory text offers a rigorous, comprehensive treatment. Classical theorems of vector calculus are amply illustrated with figures, worked examples, physical applications, and exercises with hints and answers. 1986 edition. 560pp. 5 3/8 x 8 1/2.   0-486-46620-5

ADVANCED CALCULUS: An Introduction to Classical Analysis, Louis Brand. A course in analysis that focuses on the functions of a real variable, this text introduces the basic concepts in their simplest setting and illustrates its teachings with numerous examples, theorems, and proofs. 1955 edition. 592pp. 5 3/8 x 8 1/2.   0-486-44548-8

ADVANCED CALCULUS, Avner Friedman. Intended for students who have already completed a one-year course in elementary calculus, this two-part treatment advances from functions of one variable to those of several variables. Solutions. 1971 edition. 432pp. 5 3/8 x 8 1/2.
0-486-45795-8

METHODS OF MATHEMATICS APPLIED TO CALCULUS, PROBABILITY, AND STATISTICS, Richard W. Hamming. This 4-part treatment begins with algebra and analytic geometry and proceeds to an exploration of the calculus of algebraic functions and transcendental functions and applications. 1985 edition. Includes 310 figures and 18 tables. 880pp. 6 1/2 x 9 1/4.
0-486-43945-3

BASIC ALGEBRA I: Second Edition, Nathan Jacobson. A classic text and standard reference for a generation, this volume covers all undergraduate algebra topics, including groups, rings, modules, Galois theory, polynomials, linear algebra, and associative algebra. 1985 edition. 528pp. 6 1/8 x 9 1/4.   0-486-47189-6

BASIC ALGEBRA II: Second Edition, Nathan Jacobson. This classic text and standard reference comprises all subjects of a first-year graduate-level course, including in-depth coverage of groups and polynomials and extensive use of categories and functors. 1989 edition. 704pp. 6 1/8 x 9 1/4.   0-486-47187-X

CALCULUS: An Intuitive and Physical Approach (Second Edition), Morris Kline. Application-oriented introduction relates the subject as closely as possible to science with explorations of the derivative; differentiation and integration of the powers of x; theorems on differentiation, anti-differentiation; the chain rule; trigonometric functions; more. Examples. 1967 edition. 960pp. 6 1/2 x 9 1/4.   0-486-40453-6

ABSTRACT ALGEBRA AND SOLUTION BY RADICALS, John E. Maxfield and Margaret W. Maxfield. Accessible advanced undergraduate-level text starts with groups, rings, fields, and polynomials and advances to Galois theory, radicals and roots of unity, and solution by radicals. Numerous examples, illustrations, exercises, appendixes. 1971 edition. 224pp. 6 1/8 x 9 1/4.
0-486-47723-1

AN INTRODUCTION TO THE THEORY OF LINEAR SPACES, Georgi E. Shilov. Translated by Richard A. Silverman. Introductory treatment offers a clear exposition of algebra, geometry, and analysis as parts of an integrated whole rather than separate subjects. Numerous examples illustrate many different fields, and problems include hints or answers. 1961 edition. 320pp. 5 3/8 x 8 1/2.   0-486-63070-6

LINEAR ALGEBRA, Georgi E. Shilov. Covers determinants, linear spaces, systems of linear equations, linear functions of a vector argument, coordinate transformations, the canonical form of the matrix of a linear operator, bilinear and quadratic forms, and more. 387pp. 5 3/8 x 8 1/2.
0-486-63518-X

Mathematics–Probability and Statistics

BASIC PROBABILITY THEORY, Robert B. Ash. This text emphasizes the probabilistic way of thinking, rather than measure-theoretic concepts. Geared toward advanced undergraduates and graduate students, it features solutions to some of the problems. 1970 edition. 352pp. 5 3/8 x 8 1/2.
0-486-46628-0

PRINCIPLES OF STATISTICS, M. G. Bulmer. Concise description of classical statistics, from basic dice probabilities to modern regression analysis. Equal stress on theory and applications. Moderate difficulty; only basic calculus required. Includes problems with answers. 252pp. 5 5/8 x 8 1/4.
0-486-63760-3

OUTLINE OF BASIC STATISTICS: Dictionary and Formulas, John E. Freund and Frank J. Williams. Handy guide includes a 70-page outline of essential statistical formulas covering grouped and ungrouped data, finite populations, probability, and more, plus over 1,000 clear, concise definitions of statistical terms. 1966 edition. 208pp. 5 3/8 x 8 1/2.
0-486-47769-X

GOOD THINKING: The Foundations of Probability and Its Applications, Irving J. Good. This in-depth treatment of probability theory by a famous British statistician explores Keynesian principles and surveys such topics as Bayesian rationality, corroboration, hypothesis testing, and mathematical tools for induction and simplicity. 1983 edition. 352pp. 5 3/8 x 8 1/2.
0-486-47438-0

INTRODUCTION TO PROBABILITY THEORY WITH CONTEMPORARY APPLICATIONS, Lester L. Helms. Extensive discussions and clear examples, written in plain language, expose students to the rules and methods of probability. Exercises foster problem-solving skills, and all problems feature step-by-step solutions. 1997 edition. 368pp. 6 1/2 x 9 1/4.
0-486-47418-6

CHANCE, LUCK, AND STATISTICS, Horace C. Levinson. In simple, non-technical language, this volume explores the fundamentals governing chance and applies them to sports, government, and business. "Clear and lively ... remarkably accurate." – *Scientific Monthly.* 384pp. 5 3/8 x 8 1/2.
0-486-41997-5

FIFTY CHALLENGING PROBLEMS IN PROBABILITY WITH SOLUTIONS, Frederick Mosteller. Remarkable puzzlers, graded in difficulty, illustrate elementary and advanced aspects of probability. These problems were selected for originality, general interest, or because they demonstrate valuable techniques. Also includes detailed solutions. 88pp. 5 3/8 x 8 1/2.
0-486-65355-2

EXPERIMENTAL STATISTICS, Mary Gibbons Natrella. A handbook for those seeking engineering information and quantitative data for designing, developing, constructing, and testing equipment. Covers the planning of experiments, the analyzing of extreme-value data; and more. 1966 edition. Index. Includes 52 figures and 76 tables. 560pp. 8 3/8 x 11.
0-486-43937-2

STOCHASTIC MODELING: Analysis and Simulation, Barry L. Nelson. Coherent introduction to techniques also offers a guide to the mathematical, numerical, and simulation tools of systems analysis. Includes formulation of models, analysis, and interpretation of results. 1995 edition. 336pp. 6 1/8 x 9 1/4.
0-486-47770-3

INTRODUCTION TO BIOSTATISTICS: Second Edition, Robert R. Sokal and F. James Rohlf. Suitable for undergraduates with a minimal background in mathematics, this introduction ranges from descriptive statistics to fundamental distributions and the testing of hypotheses. Includes numerous worked-out problems and examples. 1987 edition. 384pp. 6 1/8 x 9 1/4.
0-486-46961-1